AF905441

Models and Modeling in Science Education

Volume 7

Series Editor

Professor Emeritus J.K. Gilbert
Institute of Education, The University of Reading, UK

Editorial Board

Professor D.F. Treagust
Science and Mathematics Education Centre, Curtin University
Australia
Professor J.H. van Driel
ICLON, University of Leiden, The Netherlands
Dr. Rosária Justi
Department of Chemistry, University of Minas Gerais, Brazil
Dr. Janice Gobert
Worchester Polytechnic Institute

For further volumes:
http://www.springer.com/series/6931

David F. Treagust • Chi-Yan Tsui
Editors

Multiple Representations in Biological Education

Springer

Editors
David F. Treagust
Science and Mathematics
Education Centre
Curtin University
Perth, Western Australia
Australia

Chi-Yan Tsui
Science and Mathematics
Education Centre
Curtin University
Perth, Western Australia
Australia

ISSN 1871-2983 ISSN 2213-2260 (electronic)
ISBN 978-94-007-4191-1 ISBN 978-94-007-4192-8 (eBook)
DOI 10.1007/978-94-007-4192-8
Springer Dordrecht Heidelberg New York London

Library of Congress Control Number: 2012954056

© Springer Science+Business Media B.V. 2013
This work is subject to copyright. All rights are reserved by the Publisher, whether the whole or part of the material is concerned, specifically the rights of translation, reprinting, reuse of illustrations, recitation, broadcasting, reproduction on microfilms or in any other physical way, and transmission or information storage and retrieval, electronic adaptation, computer software, or by similar or dissimilar methodology now known or hereafter developed. Exempted from this legal reservation are brief excerpts in connection with reviews or scholarly analysis or material supplied specifically for the purpose of being entered and executed on a computer system, for exclusive use by the purchaser of the work. Duplication of this publication or parts thereof is permitted only under the provisions of the Copyright Law of the Publisher's location, in its current version, and permission for use must always be obtained from Springer. Permissions for use may be obtained through RightsLink at the Copyright Clearance Center. Violations are liable to prosecution under the respective Copyright Law.
The use of general descriptive names, registered names, trademarks, service marks, etc. in this publication does not imply, even in the absence of a specific statement, that such names are exempt from the relevant protective laws and regulations and therefore free for general use.
While the advice and information in this book are believed to be true and accurate at the date of publication, neither the authors nor the editors nor the publisher can accept any legal responsibility for any errors or omissions that may be made. The publisher makes no warranty, express or implied, with respect to the material contained herein.

Printed on acid-free paper

Springer is part of Springer Science+Business Media (www.springer.com)

Foreword

Biology is the study of living organisms from all perspectives, ranging from molecular and cellular to entire organisms, social groups, and populations. Some organisms are invisible to the naked eye, as are most of the processes and events that occur within living systems. Experts use language to discuss their understandings of these complex events with students and colleagues, and they typically employ pictorial representations to provide concise summaries.

This book focuses on the nature of these pictorial representations and summarizes various perspectives on the creation, use, and effectiveness of external representations in biology. The topics are wide ranging, from cell division to climate change, from mutation to evolution, from fitness to phylogenetic trees, and from simple drawings to hypermedia. The primary questions are: How well do students understand these representations? How often do they misinterpret them? And how can the representations be made more effective?

The various approaches the authors have taken are briefly summarized in the preface. The pedagogical value of such representations can be enormous, both in teaching about the microscopic and submicroscopic worlds of biology and about many of the complex but largely invisible processes that are essential for successful life on earth, including such diverse events as water cycles, gas exchanges, waste decomposition, blood flow, and nutrient transfers.

Effective conceptual representations help learners achieve successful knowledge acquisition. They can assist students in such areas as making sense of complex phenomena, constructing representations in their own minds, and correcting prior misunderstandings. They can also serve as a basis for students' discussions of the processes being learned.

There is relatively little standardization regarding the forms of external representations used in biology, as compared to those employed in chemistry and physics. But this is not a negative. It reflects the complexity and variability of the structures and processes being described. This book is very timely, providing an overview of some of the many forms of external representations in biology.

These external representations generally aim to provide succinct summaries of ideas being discussed, often conveying information more clearly and concisely than can be achieved with words alone. For biologists, biology educators, and biology teachers, there is much to learn in this publication.

Professor Emeritus of Biology
San Diego State University

Kathleen Fisher

Preface

The chapters in this volume are organized into three parts, respectively, with an emphasis on multiple representations used in *learning*, *teaching*, and *assessment,* although each chapter has, to a lesser or greater extent, aspects of all three. The introductory chapter in Part I provides a theoretical basis illustrating by means of a proposed theoretical cube model how multiple representations in biology involving three dimensions (*modes, levels of representations*, and *domain knowledge*) can serve one or more of Ainsworth's (1999) pedagogical functions of multiple external representations (MERs) in supporting learning.

The other chapters in Part I discuss the role of MERs in learning biology. Most of these chapters have a focus on various ways in which students learn biology using MERs and encompass a broad spectrum of major content areas in biology, across the symbolic, submicro, micro, and macro levels along the hierarchical organization in biology, as well as across the different modes of representations encapsulated in different platforms for learning: symbolism (Anderson, Schönborn, du Plessis, Gupthar, and Hull), pictures (Roth and Pozzer-Ardenghi), static visualizations (Eilam), hypermedia (Liu and Hmelo-Silver), and simulations (Yarden and Yarden). Some chapters also emphasize the collaboration of students and teachers in learning with MERs, which has implications for teaching and teacher education (Yarden and Yarden) and can contribute toward developing teaching materials and resolving challenges in teaching (Eilam).

The chapters in Part II examine the implications of using MERs for teaching biology and biology teacher education with each chapter having a major focus on the pedagogy of using MERs in many different instructional strategies and approaches in the major domains of biology. The importance of horizontal and vertical translations across multiple representations in domains of ecology, genetics, and evolution is highlighted by Schönborn and Bögeholz. The focuses in other chapters in Part II range from computer-based modeling for teaching 4th graders (10-year-olds) about evolution (Horwitz) to MERs of genetics in secondary school textbooks (Clément and Castéra) and complex process diagrams in premedical molecular biology (Griffard) and to phylogenetic trees (Halverson and Friedrichsen) and nested systems for teaching about photosynthesis and plant cellular respiration (Schwartz and Brown) in university classrooms. The use of phylogenetic trees in teaching about evolution explained by Halverson and Friedrichsen is vividly illustrated by the real-life example—cited by Wong, Cheng, and Yip—in which genomic sequencing of viral genome led to scientists' success in tracing the source of the Severe Acute Respiratory Syndrome (SARS) virus to bats. Wong et al.'s case study of scientists' research on SARS virus is used in biology and science teacher education for promoting teachers understanding of nature of science.

The chapters in Part III address the assessment of students' understanding of different content areas in biology using different methods and approaches in multi-representational learning environments (e.g., computer-based modeling, computer log files, interviews, conceptual mapping, two-tier tests, microgenetic methods, and others) and along a spectrum of levels. Buckley and Quellmalz illustrates—by way of three learning projects: *Science for Life* (human body systems), *BioLogica* (genetics), and *Calipers* (ecosystems)—how computer-based simulations can be harnessed for both supporting and assessing multiple representational learning of living systems. Tsui and Treagust's case studies used a two-tier diagnostic instrument and interviews to evaluate students' understanding in terms of genetics reasoning the students had learned from *BioLogica*, and their case studies also touch on the potential of bilingual representation of biological concepts in improving learning of English language learners. Encouraging more non-English native speakers to participate at all levels in science education appears to be increasingly important in the age of globalization (cf. Fensham, 2011). Niebert, Riemeier, and Gropengießer's study used interviews to explore students' metaphorical understanding of imperceptible phenomena (e.g., cell division at the *microscopic level* and climate change at the *macroscopic level*) by means of familiar representations of phenomena in the *mesocosm* (or the world of medium dimensions within human perception). Using a microgenetic method, Srivastavas and Ramadas examine how university students learned at the *symbolic or molecular level* in visualizing the double-helix structure of DNA. Using observations, Verhoeff, Boersma, and Waarlo report their critical appraisal of secondary students' systems thinking skills in two modeling studies for learning the complex living systems (cells and ecosystems).

The Conclusion chapter presents a synthesis of the themes from the chapters 2 to 18 and their analysis based on the examination of these chapters using the proposed theoretical cube model as a lens. Useful chapter examples are cited to illustrate the common themes and the ways multiple external representations (MERs) and their pedagogical functions can contribute to improving biological education across different content areas and contexts and to meet the challenges in the twenty-first century.

Our thanks go to John Gilbert, the editor of the series *Models and Modeling in Science Education,* for his valuable comments and suggestions and to Kathleen Fisher for writing the Foreword for this volume. We are also grateful to the Springer's editorial staff, particularly, Bernadette Ohmer, whose advice and support have made the volume possible. We do hope that this volume's collection of research projects on multiple representations in teaching and learning of biology can benefit biological education researchers and inform biology teachers and biology teacher educators in improving their classroom practice in one way or another.

Curtin University, Australia

David F. Treagust
Chi-Yan Tsui

References

Ainsworth, S. (1999). The functions of multiple representations. *Computers & Education, 33*(2/3), 131–152.

Fensham, P. J. (2011). Globalization of science education: Comment and a commentary. *Journal of Research in Science Teaching, 48*(6), 698–709.

Contents

Part I Role of Multiple Representations in Learning Biology

1 **Introduction to Multiple Representations: Their Importance in Biology and Biological Education** 3
Chi-Yan Tsui and David F. Treagust

2 **Identifying and Developing Students' Ability to Reason with Concepts and Representations in Biology** 19
Trevor R. Anderson, Konrad J. Schönborn, Lynn du Plessis, Abindra S. Gupthar, and Tracy L. Hull

3 **Pictures in Biology Education** 39
Wolff-Michael Roth and Lilian Pozzer-Ardenghi

4 **Possible Constraints of Visualization in Biology: Challenges in Learning with Multiple Representations** 55
Billie Eilam

5 **Promoting the Collaborative Use of Cognitive and Metacognitive Skills Through Conceptual Representations in Hypermedia** 75
Lei Liu and Cindy E. Hmelo-Silver

6 **Learning and Teaching Biotechnological Methods Using Animations** 93
Hagit Yarden and Anat Yarden

Part II Implications for Biology Teaching and Teacher Education with Multiple Representations

7 **Experts' Views on Translation Across Multiple External Representations in Acquiring Biological Knowledge About Ecology, Genetics, and Evolution** 111
Konrad J. Schönborn and Susanne Bögeholz

8 **Evolution Is a Model, Why Not Teach It That Way?** 129
Paul Horwitz

9 **Multiple Representations of Human Genetics in Biology Textbooks** 147
Pierre Clément and Jérémy Castéra

10 **Deconstructing and Decoding Complex Process Diagrams in University Biology** 165
Phyllis B. Griffard

11 **Learning Tree Thinking: Developing a New Framework of Representational Competence** 185
Kristy L. Halverson and Patricia Friedrichsen

12 **Understanding Photosynthesis and Cellular Respiration: Encouraging a View of Biological *Nested* Systems** 203
Reneé Schwartz and Mary H. Brown

13 **Scientific Models in the Severe Acute Respiratory Syndrome (SARS) Research and in the Biology Curriculum** 225
Alice Siu Ling Wong, Maurice M.W. Cheng, and Valerie W.Y. Yip

Part III Assessment of Learning and Teaching with Multiple Representations

14 **Supporting and Assessing Complex Biology Learning with Computer-Based Simulations and Representations** 247
Barbara C. Buckley and Edys S. Quellmalz

15 **Secondary Students' Understanding of Genetics Using *BioLogica*: Two Case Studies** 269
Chi-Yan Tsui and David F. Treagust

16 **The Hidden Hand that Shapes Conceptual Understanding: Choosing Effective Representations for Teaching Cell Division and Climate Change** 293
Kai Niebert, Tanja Riemeier, and Harald Gropengießer

17 **Analogy and Gesture for Mental Visualization of DNA Structure** 311
Anveshna Srivastava and Jayashree Ramadas

18 **Multiple Representations in Modeling Strategies for the Development of Systems Thinking in Biology Education** 331
Roald Pieter Verhoeff, Kerst Th Boersma, and Arend Jan Waarlo

19 Conclusion: Contributions of Multiple Representations to Biological Education 349
David F. Treagust and Chi-Yan Tsui

About the Authors 369

Index 381

Contributors

Trevor R. Anderson Division of Chemical Education, Department of Chemistry, Purdue University, West Lafayette, USA

School of Biochemistry, Genetics and Microbiology, University of KwaZulu-Natal, Berea, South Africa

Kerst Th Boersma Faculty of Science, Utrecht University, Utrecht, The Netherlands

Susanne Bögeholz Biologische Fakultät, Georg-August-Universität Göttingen, Göttingen, Germany

Mary H. Brown Science Department, Lansing Community College, Lansing, MI, USA

Barbara C. Buckley Science, Technology, Engineering and Mathematics, WestEd, Redwood City, CA, USA

Edys S. Quellmalz Science, Technology, Engineering and Mathematics, WestEd, Redwood City, CA, USA

Jérémy Castéra Science Education Centre, University of Tartu, Estonia

Maurice M.W. Cheng Faculty of Education, The University of Hong Kong, Hong Kong SAR, China

Pierre Clément The Interdisciplinary Laboratory of Didactical Research and History of Science and Techniques, University Lyon 1, Villeurbanne Cedex, France

Lynn du Plessis School of Biochemistry, Genetics and Microbiology, University of KwaZulu-Natal, Berea, South Africa

Billie Eilam Head of the Department of Learning, Instruction and Teacher Education, Faculty of Education, University of Haifa, Haifa, Israel

Patricia Friedrichsen Division of Biological Sciences, Department of Learning, Teaching and Curriculum, University of Missouri, Columbia, MO, USA

Phyllis B. Griffard Department of Natural Sciences, University of Houston – Downtown, One Main Street, Houston, USA

Harald Gropengießer Institute for Science Education, Leibniz Universität Hannover, Hannover, Germany

Abindra S. Gupthar School of Biochemistry, Genetics and Microbiology, University of KwaZulu-Natal, Berea, South Africa

Kristy L. Halverson Department of Biological Sciences, University of Southern Mississippi, Hattiesburg, MS, USA

Cindy E. Hmelo-Silver Department of Educational Psychology, Graduate School of Education, Rutgers, the State University of New Jersey, New Brunswick, NJ, USA

Paul Horwitz Modeling Center, Concord Consortium, Concord, MA, USA

Tracy L. Hull School of Biochemistry, Genetics and Microbiology, University of KwaZulu-Natal, Berea, South Africa

Lei Liu Educational Testing Service, Princeton, NJ, USA

Kai Niebert Institute for Science Education, Leibniz Universität Hannover, Hannover, Germany

Lilian Pozzer-Ardenghi Department of Curriculum, Teaching and Learning, Faculty of Education, University of Manitoba Winnipeg, Canada

Jayashree Ramadas Homi Bhabha Centre for Science Education, Tata Institute of Fundamental Research, Mumbai, India

Tanja Riemeier Institute for Science Education, Leibniz Universität Hannover, Hannover, Germany

Wolff-Michael Roth Applied Cognitive Science, University of Victoria, Victoria, BC, V8W 7G6, Canada

Konrad J. Schönborn Division of Media and Information Technology, Department of Science and Technology (ITN), Linköping University, Norrköping, Sweden

Reneé Schwartz Department of Biological Sciences, Mallinson Institute of Science Education, Western Michigan University, Kalamazoo, MI, USA

Anveshna Srivastava Homi Bhabha Centre for Science Education, Tata Institute of Fundamental Research, Mumbai, India

David F. Treagust Science and Mathematics Education Centre, Curtin University, Perth, WA, Australia

Chi-Yan Tsui Science and Mathematics Education Centre, Curtin University, Perth, WA, Australia

Roald pieter Verhoeff Department of Philosophy and Science Studies, Radboud University Nijmegen, Nijmegen, The Netherlands

Arend Jan Waarlo Centre for Science and Mathematics Education, Utrecht University, Utrecht, The Netherlands

Alice Siu Ling Wong Faculty of Education, The University of Hong Kong, the Hong Kong SAR, China

Anat Yarden Department of Science Teaching, Weizmann Institute of Science, Rehovot, Israel

Hagit Yarden Department of Education in Technology and Science, Technion Institute of Technology, Haifa, Israel

Valerie W.Y. Yip Faculty of Education, The University of Hong Kong, Hong Kong SAR, China

Part I
Role of Multiple Representations in Learning Biology

Chapters in Part I address approaches by different researchers using a range of external representations for learning biology.

Chapter 1
Introduction to Multiple Representations: Their Importance in Biology and Biological Education

Chi-Yan Tsui and David F. Treagust

Information gently but relentlessly drizzles down on us in an invisible, impalpable electric rain. . .just plug in a modem and watch a flood of information from the world's uncounted electronic memories come pouring out into your laptop. . . For better or worse the world is awash with information.

(von Baeyer, 2003, pp. 2–3)

Seeking a Unifying Theoretical Framework for Learning with Multiple Representations

A review of the extant literature shows over the past decades how science teachers and science teacher educators, as well as science education researchers, have effectively used various external representations for teaching and learning. The variety of external representations in the literature includes *analogies* (e.g., Dagher, 1994; Spiro, Feltovich, Coulson, & Anderson, 1989; Treagust, Harrison, & Venville, 1998), *metaphors* (e.g., Aubusson, Harrison, & Ritchie, 2006; Martins & Ogborn, 1997), *visualization* (see its Convention 2 definition in Gilbert, Reiner, & Nakhleh, 2008, p. 2), *discourse* (e.g., Lemke, 1990, 1998), *models and model-based learning* (e.g., Buckley, 2000; Clement & Rae-Mamirez, 2008; Gilbert & Boulter, 1998), *multilevel representations* (Johnstone, 1982, 1991), *multimodal representations* (e.g., Jaipal, 2010; Waldrip, Prain, & Carolan, 2010), and others. Given this diversity in use of external representations in various combinations

---

Unless stated otherwise, the term *multiple representations* in this chapter refers to multiple external representations (MERs) used by Ainsworth (1999).

C.-Y. Tsui • D.F. Treagust (✉)
Science and Mathematics Education Centre, Curtin University, GPO Box U1987, Perth, WA 6845, Australia
e-mail: C.Tsui@curtin.edu.au; D.Treagust@curtin.edu.au

D.F. Treagust and C.-Y. Tsui (eds.), *Multiple Representations in Biological Education*, Models and Modeling in Science Education 7, DOI 10.1007/978-94-007-4192-8_1,

© Springer Science+Business Media B.V. 2013

across different content areas and contexts in the teaching and learning of science, there is a need for a unifying theoretical framework. We have attempted to seek such a framework that could show how these seemingly disparate external representations can be harnessed for improving higher order learning, such as reasoning and problem solving, as well as for constructing learners' internal representations (mental models) of understanding (e.g., Gentner & Stevens, 1983).

On the basis of the recent literature and our previous research work on learning of genetics (Tsui & Treagust, 2003, 2007, 2010), we believe that the use of Ainsworth's (1999) pedagogical functions of multiple external representations (MERs) can be a useful framework for conceptualizing learning with different external representations in science education in general and in biological education in particular. This functional taxonomy from the research area of cognitive science and computational approaches to learning, described in the following section, forms a major theoretical framework for this volume.

MERs and Their Pedagogical Functions

Learning with MERs

Learners can benefit from learning with more than one external representation. Van Someren, Reimann, Boshuizen, and de Jong's (1998) collection of research studies in *Learning with Multiple Representations* provided many examples of learning with multiple representations in computational science, mathematics, physics, chemistry, accidentology, economics, and clinical medicine. Learners are likely to benefit when information is presented in more than one representation. This is because specific information can best be conveyed in a particular representation, several representations can be more useful in displaying a variety of information, and problem-solving expertise depends on the problem solver's repertoire of multiple representations of the same domain (de Jong et al., 1998). Furthermore, a specific sequence of learning material is beneficial for the learning process as discussed in the chapters in van Someren et al.'s (1998) book. Paivio's (1986) dual coding theory that humans have separate channels for processing visual and verbal representations has also been an important theoretical basis for using both verbal and visual representations to support learning.

Functional Taxonomy of Multiple Representations

Drawing on the research literature of using multiple representations in computer-based learning environments and her own research, Ainsworth (1999, 2006) proposed a functional taxonomy of multiple representations. Accordingly, multiple external representations (MERs) of knowledge—when co-deployed (i.e., when two or more external representations are simultaneously used) in teaching and

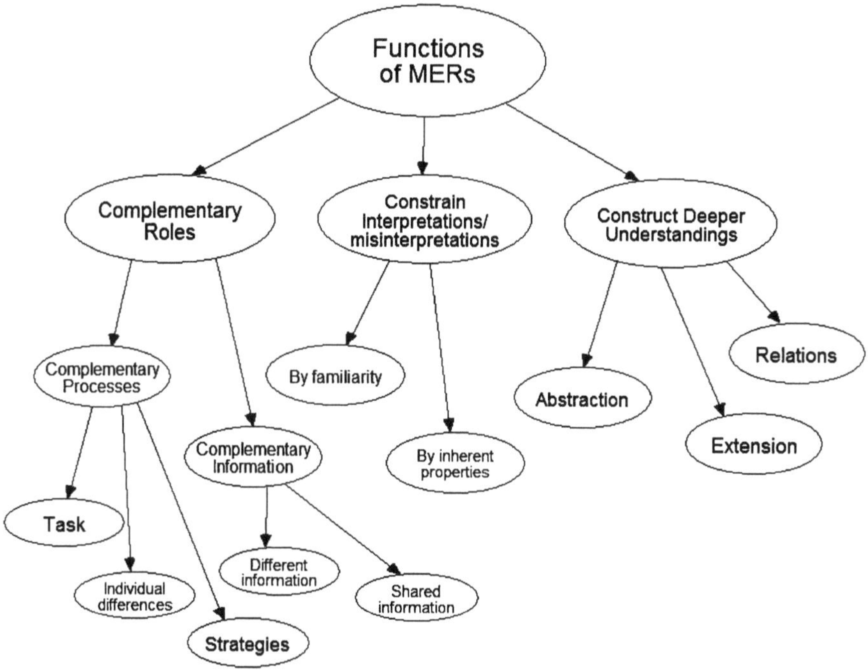

Fig. 1.1 Functional taxonomy of multiple external representations (MERs) (from Ainsworth, 1999)

learning—can serve three basic pedagogical functions (see Fig. 1.1). First, MERs support complementary processes (for multiple tasks, learners' individual differences, or strategies to improve performance) and complementary information (different or shared). For example, providing graphs, tables, equations, and pictures of biological phenomena means that each representation can de designed so that the information is presented in a way that is most appropriate for learners' needs. Different forms of representations make certain inferences easier—graphs allow perceptual patterns to be seen, tables indicate empty cells, and equations indicate precise quantitative relationships between variables (e.g., *Lotka–Volterra equations* for prey-and-predator relationship discussed in Verhoeff, Boersma, and Waarlo's Chap. 18 in this volume). Second, MERs constrain interpretation or misinterpretation of phenomena by familiarity or inherent properties. For example, when learning involves MERs, a familiar representation can support the learners' interpretation of a less familiar representation for understanding the latter (e.g., the use of metaphors and analogies), or a diagram accompanying a description, by way of its inherent properties, can visually support the learners' interpretation of an ambiguous description (e.g., about the physical location of objects). Third, MERs promote the construction of deeper *understanding* through abstraction, extension, and relations (see Fig. 1.1) as follows:

1. *Abstraction* (i.e., detecting and extracting a subset of relevant elements of information from a representation)

2. *Extension* (i.e., extending knowledge learned in one representation to new situations with other representations or making generalizations from representations)
3. *Relations* (i.e., translating between two or more unfamiliar representations)

Although the functions of MERs have been primarily used to explain learning that involves computers and multimedia, multi-representational learning environments are ubiquitous and do exist even when there are no learning technologies in the classroom. Indeed, there have been a number of recent studies that have drawn upon Ainsworth's functional taxonomy—in chemistry (e.g., Cook, Wiebe, & Carter, 2008), in mathematics (e.g., White & Pea, 2011), in physics (e.g., van der Meij & de Jong, 2011), and in biology (e.g., chapter authors in this volume).

To understand the effectiveness of using MERs to support learning, Ainsworth (2006) argued that three aspects of MERs must be considered: *design* parameters unique to learning with multiple representations, the *functions* of multiple representations that support the learning, and the *cognitive tasks* undertaken by a learner interacting with multiple representations. In the chapters in this volume, the authors have provided a number of examples to illustrate in one way or another some of these aspects concerning the learning effectiveness using MERs.

Seven chapter authors report their recent studies explicitly using or referring to Ainsworth's functional taxonomy on various content areas of biology—constraints in visualizing textbook diagrams (Eilam), tree thinking for learning evolution with phylogenetic trees (Halverson and Friedrichsen), learning genetics reasoning with *BioLogica* (Tsui and Treagust), comprehension of biotechnological tools using animations (Yarden and Yarden), learning textbook complex process diagrams (Griffard), translation processes across representations (Schönborn and Bögeholz), and the use of analogy and gesture for understanding DNA double helix (Srivastava and Ramadas). Most other chapters use similar ideas for discussing how MERs can support learning in other content areas of biology. For example, Liu and Hmelo-Silver report the use of conceptual representations through the design of function-oriented hypermedia to support student learning of human body systems. Their chapter points to, without saying, the complementary and constraining pedagogical functions of MERs. The collaborative learning environment—which allows learners to make better use of their cognitive and metacognitive skills for co-construction of deeper understanding—is also in keeping with the pedagogical functions of MERs.

Costs of Learning with MERs

When people are learning complex scientific concepts, interacting with MERs represented in different modes can bring unique benefits for the learners. "Unfortunately, there is considerable evidence to show that learners often fail to exploit these advantages, and in the worse cases inappropriate combinations of representations can completely inhibit learning" (Ainsworth, 2008a, p. 191). One of the often cited reasons is based on Sweller's (1994) cognitive load theory. Accordingly, using several modes of representations for learning a particular concept may incur extraneous cognitive overload upon the learners' short-term memory that is limited in both its capacity and duration resulting in little or no learning taking place. In this volume, several chapters address student difficulties when learning biology with MERs—for example, difficulties in understanding symbolism (Anderson, Schönborn, du Plessis, Gupthar, and Hull), interpreting static visualization (Eilam), decoding static complex process diagrams (Griffard), and identifying essential features from dynamic animations (Yarden and Yarden). Learning with multiple representations may not always be as useful as intended. In other words, as explained by Ainsworth (2008a), MERs are powerful tools but "like all powerful tools they need careful handling if learners are to use them successfully" (p. 191).

Dimensions of Multiple External Representations (MERs) for Biological Science

Our view about learning with multiple external representations (MERs) in biology involves three dimensions: *modes of representations*, *levels of representations*, and *domain knowledge of biology*.

Modes of Representations

The different external representations in the science education literature discussed earlier in this chapter involve, in different combinations, different modes of representations. Real-life objects, actions (e.g., gestures), photographs, animations, natural drawings, diagrams, graphs, charts, tables, equations, and linguistic input and output are used for externally representing biological ideas, concepts, or phenomena in their own special ways. On examining these multimodal representations, we have identified among them a continuum in terms of the degree of abstractness in ways similar to the continuum of inscriptions described in Pozzer and Roth (2003) and in their chapter. We therefore construe these modes of representations as a *continuum of increasing abstraction*, of which human language is deemed the most abstract mode of representation.

Levels of Representations

For decades, Johnstone's (1982) multilevel representations across symbolic, submicro, and macro levels has been a predominant explanatory framework in science education, particularly in chemical education (cf. Gilbert & Treagust, 2009). Johnstone (1991) also argued that these multilevel representations can be equally applied to understanding biological concepts. However, we have found that Johnstone's triple levels of representations have limitations for describing and explaining learning in biology because biological knowledge, unlike chemical knowledge, extends to multiple, hierarchically organized levels of nested but different biological entities. That is, cells are nested within tissues, which are in turn nested within organs and then within the next level—systems, organisms, populations, communities, ecosystems—and up to the top level of the biosphere. In contrast, representations at the triple levels in chemistry are different representations of the same entities. For example, as explained by Taber (2009), a chemical equation (e.g., $CH_4 + 2O_2 \rightarrow CO_2 + 2H_2O$) at the symbolic level can be used to represent either the change in substances at the macro level or the particle interactions at the submicro level (i.e., they represent the same four chemical substances).

Biology is unique in that four levels of representation need to be considered for full understanding of biological phenomena: (1) the *macroscopic level* at which biological structures are visible to the naked eye; (2) the *cellular or subcellular (microscopic) level* at which structures are only visible under a light microscope or an electron microscope; (3) the *molecular (submicroscopic) level* involving DNA, proteins, and various biochemicals (cf. Marbach-Ad & Stavy, 2000), for example, biochemicals can be identified using electrophoresis, chromatography, the centrifuge, and other analytical tools, including the latest cryogenic electron tomography (National Research Council [NRC], 2009, p. 54); and (4) the *symbolic level* that provides explanatory mechanisms of phenomena represented by symbols, formulas, chemical equations, metabolic pathways, numerical calculations, genotypes, inheritance patterns, phylogenetic trees in evolution, and so on.

Domain Knowledge of Biology

The domain knowledge of biology—a body of extensive and complex knowledge about life and living organisms—incorporates the integration of other disciplines, particularly chemistry, physics, and mathematics. Life or living systems can be conceptually represented, as suggested by the teacher's guide of the Biological Science Curriculum Study (2006), by six unifying themes:

1. Evolution: patterns and products of change in living systems
2. Homeostasis: maintaining dynamic equilibrium in living systems
3. Energy, matter, and organization: relationships in living systems
4. Continuity: reproduction and inheritance in living systems
5. Development: growth and differentiation in living systems
6. Ecology: interaction and interdependence in living systems

(pp. x–xi)

A Theoretical Model for Interpreting Learning with MERs in Biology

Drawing upon the relevant literature and the three dimensions of an MER about biology—modes of representations, levels of representations, and domain knowledge of biology—we propose a theoretical model or *the cube model* with the three dimensions and their components, respectively, on the three faces of the cube (see Fig. 1.2). We intend to use this theoretical cube model as a lens for interpreting learning with MERs in terms of how learning can take place through *translation* across the representations of biological knowledge. Translation between representations can be defined as "an information processing task, requiring understanding of the underlying concept to the extent that the individual can interpret the information provided by the initial representation and infer the details required to construct the target representation" (Geig & Rubba, 1993, p. 883).

Examining and Interpreting the Chapters with the Cube Model

Learning Through Translations Across MERs

With this proposed theoretical cube model as a lens, we have examined and interpreted the chapters in terms of how learning with MERs through three major translations between MERs can achieve one or more of the pedagogical functions. According to Ainsworth (1999), translation here refers more specifically to the learning situation where a learner must see the relation between two MERs, comprehend their relations, and act to reproduce such relations. Our position concurs with the findings reported by Schönborn and Bögeholz that experts view translation across MERs as essential for constructing knowledge in the domains of ecology, genetics, and evolution.

First, as depicted in Fig. 1.2, learning can take place through *horizontal translation across modes of representations* (*HTM*) along a continuum of representations with increasing abstraction from real-life worldly objects and actions to human language. Whereas how the biological knowledge is represented by different modes rests on the expertise of the designers—of the curriculum, educational software, or classroom instruction—whether these MERs can serve one or more of the pedagogical functions largely depends on the way they are deployed in teaching. For example, Yarden and Yarden draw on Paivio's (1986) dual coding theory to argue that information encoded in both visual and verbal representations, such as pictures, will be better remembered than information encoded in only one of the two, particularly more so than representations in words alone, thus enhancing their learning through HTM in terms of the pedagogical functions. Another example of HTM is the horizontal translation in Schönborn and Bögeholz's chapter that focuses

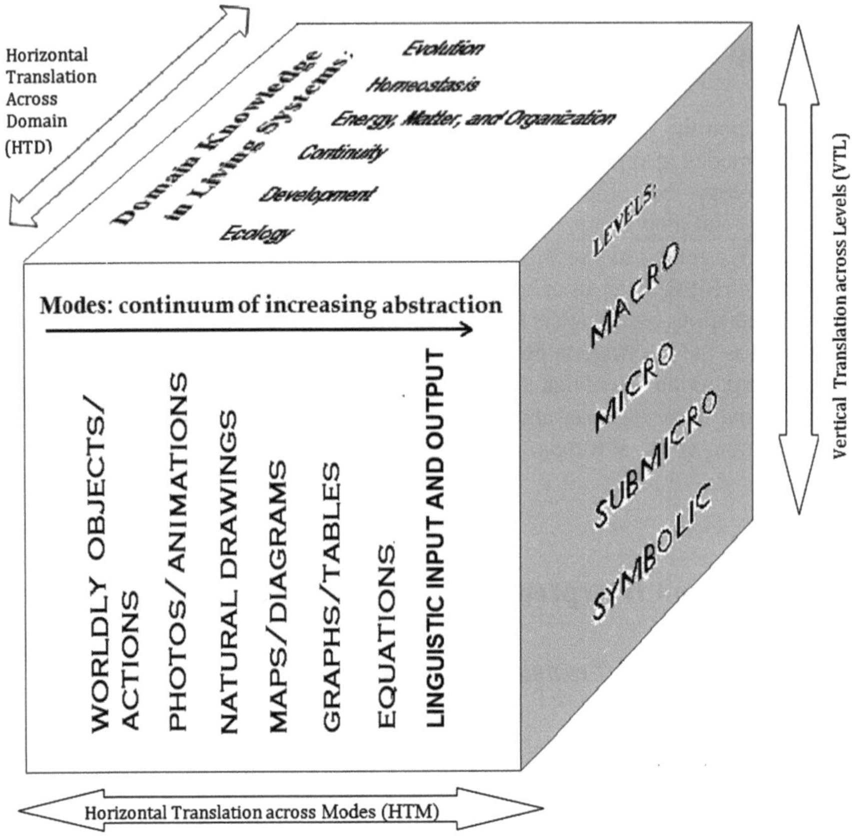

Fig. 1.2 Proposed three-dimensional theoretical model or *the cube model* for learning biology with multiple external representations (MERs) through translations

on the experts' view about the horizontal translation in constructing biological knowledge across modes of representations in the domains of ecology, genetics, and evolution. For learning ecology, HTM is from the visualizable concrete objects (e.g., preys and predators) to the underlying concept (e.g., prey–predator relationships) represented by more abstract graphs, equations, or verbal descriptions. However, as pointed out by Halverson and Friedrichsen, students who lack representational competence often fail to make the correct association between the content knowledge of biology and its abstract representations because they tend to focus on the uninformative superficial features of the representations but do not use them as a tool. Therefore, it is important that students develop their representational competence in order to learn biological knowledge through HTM.

Second, *vertical translation across levels of representations* (*VTL*) is unique to learning biology through the hierarchically organized and nested domain knowledge from the symbolic level (explanatory mechanisms), the submicro level (molecules), the micro level (organelles and cells), and the macro level (tissues, organs, systems, organisms, populations, communities, ecosystems, and biosphere). Using the example of the phylogenetic tree, Halverson and Friedrichsen use this VTL—along the hierarchical levels of biological organization from the cells to the population of the species—to explain how learners process the information of a phylogenetic tree or cladogram (a graphical mode of representation at the symbolic level) for understanding evolution by natural selection. Accordingly, at the highest level of representational competence, experts can interpret the information provided by the cladogram to construct their target understanding of evolution in terms of solving phylogenetic problems, explaining evolutionary phenomena, and making predictions.

Third, learning can also take place through *horizontal translation across the domain knowledge of biology* (*HTD*). For example, Schwartz and Brown's case studies highlight how students learned the interconnected processes of photosynthesis and plant cellular respiration to provide plants with energy. The students repeatedly used the key idea of *energy,* indicating that they used HTD to construct their understanding of energy interrelationships within and between the nested biological systems. Schönborn and Bögeholz (see examples in Appendix I of their chapter) explain this HTD in school biology curriculum for developing different knowledge components (terms, concepts, principles, and fundamentals) and their interrelationships. Such HTD can be carried out using either the same mode of representations (e.g., electromicrographs showing the fine structures of three cell organelles: chloroplast, Golgi apparatus, and mitochondrion) to illustrate the same principle of *increased surface area* or different modes of representations (e.g., different visual–graphical modes showing the interaction of enzyme–substrate, antigen–antibody, and hormone–receptor) to illustrate another same principle of *key and lock hypothesis.*

Using the cube model as a lens for examining and interpreting learning with MERs, we argue that the HTD or horizontal translation across the domain of biology can take place within and between the six unifying themes of the biological content knowledge: evolution; homeostasis; energy, matter and organization; continuity (including reproduction and genetics); development; and ecology (see Fig. 1.2). More important in biology than in other sciences is this way of moving back and forth across the structurally and functionally related content areas in order to achieve one or more the pedagogical functions of MERs. Learning with MERs in biology is often more complicated than a single translation between two representations because several translations in different combinations may be required for understanding a particular concept of the domain or for the reasoning to solve a problem because representations and problem-solving strategies interact (see examples in Ainsworth, 1999, p. 137). However, students may have difficulties in translating between representations. In computer-based interactive learning environments, dynamic links between representations are often designed to reduce the cognitive load upon learners through the computer's automatic translation between representations, enabling learners to concentrate on interacting with the representations and the consequences of such interactions (Ainsworth, 2008a). For example, the videogames of *Evolution Readiness* (Horwitz) allow the 10-year-old fourth graders to interact with the representations of plant growth under different light levels that can be easily manipulated. The students can then observe how the virtual plants grow in response to the changes they have made as the plant growth under different light levels is automatically translated to a graph showing the number of flowers of the plants in the virtual greenhouse. In addition, learners also need to map the representations in the instruction materials to the conceptual knowledge of the domain. Another example is Liu and Hmelo-Silver's study on learning with hypermedia that uses function-oriented information of complex human body systems—conceptually organized hypermedia to connect different components of the biological knowledge. The study shows that function-oriented hypermedia promotes learners' cognitive and metacognitive processing by affording and constraining how they set goals for exploring, monitoring, and evaluating their own learning as well for co-constructing a shared understanding in collaborative learning contexts.

As discussed in several other chapters, the process of learning with MERs is affected by learners' representational preferences that depend on the learner's experience and their representational competence (Griffard; Halverson and Friedrichsen) or translation competence for constructing knowledge (Schönborn and Bögeholz). Learning with MERs is also affected by more stable individual differences—"such as IQ, spatial reasoning, locus of control, field dependence, verbal ability, vocabulary, gender, and age" (Winn, as cited by Ainsworth, 1999, p. 136). Biology educators and teachers should consider how the learners' translation processes across the representations can be supported in order to maximize learning outcomes with MERs, that is, to achieve one or more of the pedagogical functions of MERs.

Limitations of the Cube Model

In the preceding sections, we propose that learners, when learning with MERs, have to explore the learning space for constructing their knowledge by way of one or more of the three translations (HTM, VTL, and HTM). However, on examining the chapters, we have found that the cube model in Fig. 1.2 has its limitations for interpreting learning with MERs in some chapters.

The cube model does not show learning that can take place via translation across levels associated with some MERs beyond the size and temporal scales familiar to humans (Eilam; Niebert, Riemeier, and Gropengießer). For example, it does not show learning by way of the vertical translations of MERs across some points in time that are beyond the temporal scales directly accessible to us within our average lifetime, such as over billions of years during the process of evolution of biological organisms (Halverson and Friedrichsen). In this regard, vertical translations of MERs of biological phenomena across the symbolic, submicro, micro, and macro levels do not show that such learning is beyond human perceptual or experiential limits without including Vollmer's (1984) mesocosmic level (Niebert et al.) which we discuss later in this chapter.

This cube model also does not allow us to interpret the translation across some complex process diagrams in molecular biology which are arranged in a temporal sequence (Griffard). It is also not useful for interpreting some other chapters, for example, Wong, Cheng, and Yip's, and Clément and Castéra's chapters, which involve philosophical, cultural, social, and political impacts on biological education, which will be discussed in length in the Conclusion chapter.

Nonetheless, with this cube model we have identified several interesting themes in some chapters that can be used to explain learning with MERs in terms of the pedagogical functions of MERs in ways that are useful for improving biological education in the twenty-first century. Some examples are discussed in the sections that follow.

Mesocosmic Representations

The novel notion of Vollmer's (1984) *mesocosm*, suggested by Niebert et al., is useful in constraining interpretation of abstract phenomena in biology. In chemistry, it is essential to use the meso level between the macro and micro level to connect student learning of chemical concepts in terms of understanding the chemistry of human activities and the related scientific and technological developments (cf. Meijer, Bulte, & Pilot, 2009). We believe that Niebert et al.'s notion that the common source domains lie in the perceptible mesocosm—for the source-to-target mapping (like a *hidden hand*) that leads to better understanding of the abstract target domains in the microcosm (e.g., cell division) or the macrocosm (e.g., climate change)—appears to be even more important for learning biological knowledge. As argued by Vollmer, this is because much of biological knowledge is within the mesocosm or a world of medium dimensions which refers to humans and their sensory abilities. Mesocosmic quantities range from seconds (e.g., heart beat) to decennia (e.g., human lifetime), from zero velocity (e.g., a bus at standstill) to 10 m/s (e.g., a sprinter in action), from grams (e.g., table sugar) to kilograms (e.g., rocks or cars), and from 0°C (e.g., ice) to 100°C (e.g., boiling water). Vollmer's mesocosm as "that section of the real world we cope with in perceiving and acting, sensually and motorially [...]" (p. 87) is unique for examining learning in biology.

The advent of information and communications technology has created many computer microworlds that allow users to visualize any biological entities as if they were in the mesocosm. The videogames of *Evolution Readiness* (Horwitz) are examples that epitomize a *virtual mesocosm* for younger students to learn about the complex and abstract concept of evolution by natural selection. So are the activities of the *BioLogica* program in Tsui and Treagust's and Buckley and Quellmalz studies on learning genetics in secondary schools through interaction with and manipulation of objects in *BioLogica*. In terms of MER functions, we believe that all these mesocosmic representations—which are less abstract and more familiar representations within learners' experiences and sensory abilities—can constrain the interpretation or misinterpretation of biological phenomena, allowing the learners to construct deeper understanding through abstraction, relation, or extensions.

Anthropocentric or Human-Centered Representations

Another common theoretical theme shared by at least four chapters is *anthropocentric or human-centered representations.* Anthropocentrism—the tendencies of humans to view themselves as central in the world—is closely related to children's learning of biology. For example, developmental cognitive science holds a predominant view that "young children possess only one markedly anthropocentric vintage point and most undergo fundamental conceptual change, overturning their initially human-centered framework before they can acquire a distinctly biological framework" (Hermann, Waxman, & Mewdin, 2010, p. 9979).

The anthropocentric representations in this volume include *self as first referent* for explaining conceptions of photosynthesis and respiration (Schwartz and Brown), the use of *human palm gesture* for understanding the three-dimensional model of the DNA double helix (Srivastava and Ramadas), *gestures and body positions* as resources for reading pictures (Roth and Pozzer-Ardenghi), and *bodily experience* on which the conceptual structure is grounded (Niebert et al.). These chapter authors argue that learners can benefit from the learning with these representations co-deployed with other MERs. Furthermore, social interactions, such as gestures or body positions, in reading pictures in biological learning, are also important in helping children and newcomers learn about biology (Roth and Pozzer-Ardenghi). We believe that the use of anthropocentric representations portrayed in these chapters is in keeping with Ainsworth's functional taxonomy because familiar everyday human examples and experiences are useful for constraining the interpretation/misinterpretation of unfamiliar or abstract phenomena in biology.

Systems Representations

A number of chapters highlight the importance of systems thinking and representations—those by Anderson et al., Schönborn and Bögeholz, Schwartz and Brown, and Buckley and Quellmalz—and particularly that by Verhoeff et al. As Palsson (2000) predicted at the beginning of the twenty-first century, "[t]he advent of high-throughput technologies, such as genomics and proteomics, is enabling biologists to study cells as systems" (p. 1147). The new millennium has seen research in biology actually taking place in this direction from traditional in vivo or in vitro biology experiments to *in silico* biology experiments where online informatics databases of genomics and proteomics are searched, and computational and modeling methods used in research, for example, in evolutionary biology (e.g., Rodrigo, Carrera, & Elena, 2010), in molecular genetics (e.g., Ligorioa, Izzotti, Pulliero, & Arrigoc, 2011), in studies on genomes and networks in cellular and multicellular systems, and in the discovery of drugs for human diseases (e.g., Werner, 2003). As Werner put it, "[b]ecause complexity of multicellular systems, *in silico* multicellular modeling and simulation will become an essential component in the drug discovery process" (p. 1121). Systems biology is highlighted by a recent US report *A New Biology for the 21st Century* (NRC, 2009) as important predictive modeling that "seeks a deep quantitative understanding of complex biological processes through dynamic interaction of components that may include multiple molecular, cellular, organismal, population, community, and ecosystem functions" (p. 61). The report called for interdisciplinary efforts to find biology-based solutions for global societal problems about food, environment, energy, and health. Should basic biological education include an introduction to systems biology? Already in the Netherlands, as reported by Verhoeff et al., the school curriculum has required students to learn some basic ideas of systems biology that appears to be a timely response to aligning school biology for meeting the challenges in the twenty-first century.

Learning New Biology with MERs in the Twenty-First Century

In conducting research, biologists have to predict the changes in a biological system in ways constrained by the incomplete information available (Palsson, 2000). Therefore, a researcher in such situations has to interpret the phenomenon to find a solution for a certain problem, depending on "building the capacity to understand, predict, and influence the responses and capabilities of complex biological systems" (NRC, 2009, p. 6). The researcher has to explore the solution space and endeavor to solve the problem through reasoning toward finding a solution. Likewise, for a learner in biology, the incomplete constraints (e.g., the various MERs of a particular biological phenomenon) often require the learner to explore within that learning space and struggle to interpret the phenomenon in order to construct a deeper understanding of this phenomenon. The chapters in Part I of this volume (on learning with MERs) provide examples of research to illustrate how the pedagogical functions of MERs can be used to improve learning of biology in various ways.

The thought-provoking metaphor of von Baeyer's (2003) *electric rain* in this chapter's epigraph brings home the message that in the information age, we are now overwhelmed with too much information that may incur cognitive overload upon our limited working memory (Sweller, 1994), and learners may not have paid enough attention to the relevant information for understanding of what is represented. As Kings et al. (2008) put it, "[this] main insight highlighted the cognitive bottleneck of the human mind: while information is no longer scarce, attention is" (p. 20). The chapters in Part II of this volume (on teaching with MERs) have addressed how instructional designs using MERs can improve the learning of the complex and abstract domain knowledge of biology in the various content areas using different strategies but not without costs. Ainsworth (2008b) pointed out that evaluation of multimedia learning environments require the right methods for the specific research questions within the particular learning contexts and that apart from the conventional experimental research designs, useful methods include computer modeling, case studies, ethnographic studies, and microgenetic studies.

Indeed, the authors of this volume, particularly those of the chapters in Part III (on assessment of learning and teaching), have discussed some of these methods for assessing learning with MERs that could bridge or narrow the gaps between learning with MERs discussed in Part I and teaching with MERs in Part II. The alignment of learning and teaching involving MERs of complex phenomena is likely to contribute to biology teacher education in terms of developing the teachers' pedagogical content knowledge.

To conclude, we hope this introduction to the theoretical perspectives can serve as a primer for you, a reader of this volume. As you read through this volume, you will know that multiple representations of information have brought with them benefits but also costs for learners in a world with too much information. We believe that this volume can contribute to a new theoretical framework that allows biology educators to capitalize on the pedagogical functions of MERs—along with the increasingly sophisticated, mobile, and portable information and communications technology available for learning and teaching—for improving biological education in the twenty-first century. In this chapter, we have provided a theoretical background of multiple representations in biology and biological education, with which we hope that readers can more fully appreciate the extent of the research reported in the following chapters.

Acknowledgments We would like to thank John Gilbert, the editor of the series *Models and Modeling in Science Education*, for his valuable and critical comments and suggestions on an earlier draft of this chapter. We are grateful to Kathleen Fisher, Anat Yarden, Kristy Halverson, and Shaaron Ainsworth, who reviewed the final draft of this chapter. Their comments and suggestions—particularly those from Shaaron Ainsworth whose theoretical framework we have adopted for this volume—have undoubtedly enabled us to improve the quality of this chapter. Any remaining inadequacies are ours.

References

Ainsworth, S. (1999). The functions of multiple representations. *Computers in Education, 33*(2/3), 131–152.

Ainsworth, S. (2006). DeFT: A conceptual framework for considering learning with multiple representations. *Learning and Instruction, 15*(3), 183–198.

Ainsworth, S. (2008a). The educational value of multiple-representations when learning complex scientific concepts. In J. K. Gilbert, M. Reiner, & M. Nakhleh (Eds.), *Visualization: Theory and practice in science education* (pp. 191–208). London: Springer.

Ainsworth, S. (2008b). How should we evaluate multimedia learning environments? In J.-F. Rouet, R. Lowe, & W. Schnotz (Eds.), *Understanding multimedia documents* (pp. 249–265). Dordrecht, The Netherlands: Springer Science & Business Media.

Aubusson, P., Harrison, A. G., & Ritchie, S. M. (Eds.). (2006). *Metaphor and analogy in science education*. Dordrecht, The Netherlands: Springer.

Biological Sciences Curriculum Study. (2006). *BSCS Biology: A human approach (teacher guide)* (3rd ed.). Dubuque, IA: Kendall/Hunt.

Buckley, C. B. (2000). Interactive multimedia and model-based learning in biology. *International Journal of Science Education, 22*(9), 895–935.

Clement, J. J., & Rae-Mamirez, M. A. (Eds.). (2008). *Model based learning and instruction in science*. Dordrecht, The Netherlands: Springer.

Cook, M., Wiebe, E., & Carter, G. (2008). The influence of prior knowledge on viewing and interpreting graphics with macroscopic and molecular representations. *Science Education, 92* (5), 848.

Dagher, Z. R. (1994). Does the use of analogies contribute to conceptual change? *Science Education, 78*(6), 601–614.

de Jong, T., Ainsworth, S. E., Dobson, M., van der Hulst, A., Levonen, J., Reimann, P., et al. (1998). Acquiring knowledge in science and mathematics: The use of multiple representations in technology-based learning environments. In M. W. van Someren, P. Reimann, H. P. A. Boshuizen, & T. de Jong (Eds.), *Learning with multiple representations* (pp. 9–40). London: Elsevier Science.

Geig, P., & Rubba, P. (1993). Translation of representations of the structure and the relationship to reasoning, gender, spatial reasoning, and specific prior knowledge. *Journal of Research in Science Teaching, 30*(8), 883–903.

Gentner, D., & Stevens, A. L. (Eds.). (1983). *Mental models*. Hillsdale, NJ: Lawrence ERlbaum Associates.

Gilbert, J. K., & Boulter, C. J. (1998). Learning science through models and modelling. In B. J. Fraser (Ed.), *International handbook of science education* (pp. 53–66). Dordrecht, The Netherlands: Kluwer.

Gilbert, J. K., Reiner, M., & Nakhleh, M. (Eds.). (2008). *Visualization: Theory and practice in science education*. New York/London: Springer.

Gilbert, J. K., & Treagust, D. (Eds.). (2009). *Multiple representations in chemical education*. Dordrecht, The Netherlands: Springer.

Hermann, P., Waxman, S. R., & Mewdin, D. L. (2010). Anthropocentrism is not the first step in children's reasoning about the natural world. *Proceedings of the National Academy of Sciences of the United States of America, 107*(22), 9979–9984.

Jaipal, K. (2010). Meaning making through multiple modalities in a biology classroom: A multimodal semiotics discourse analysis. *Science Education, 94*(1), 48–72.

Johnstone, A. H. (1982). Macro and micro chemistry. *School Science Review, 19*(3), 71–73.

Johnstone, A. H. (1991). Why is science difficult to learn? Things are seldom what they seem. *Journal of Computer Assisted Learning, 7*, 75–83.

Kings, N. J., Davies, J., Verrill, D., Aral, S., Brynjolfsson, E., & van Alstyne, M. (2008). Social networks, social computing and knowledge management. In P. Warren, J. Davies, & D. Brown (Eds.), *ICT futures: Delivering pervasive, real-time and secure services* (pp. 17–26). West Sussex, UK: Wiley.

Lemke, J. L. (1990). *Talking science: Language, learning, and values*. Norwood, NJ: Ablex Publishing Corporation.

Lemke, J. L. (1998). Multiplying meaning: Visual and verbal semiotics in scientific text. In J. R. Martin & R. Veel (Eds.), *Reading science* (pp. 87–113). London/New York: Routledge.

Ligorioa, M., Izzotti, A., Pulliero, A., & Arrigoc, P. (2011). Mutagens interfere with microRNA maturation by inhibiting DICER: An in silico biology analysis. *Mutation Research, 717*, 116–128.

Marbach-Ad, G., & Stavy, R. (2000). Students' cellular and molecular explanations of genetic phenomena. *Journal of Biological Education, 34*(4), 200–205.

Martins, I., & Ogborn, J. (1997). Metaphorical reasoning about genetics. *International Journal of Educational Research, 19*(6), 48–63.

Meijer, M. R., Bulte, A. M. W., & Pilot, A. (2009). Structure–property relations between macro and micro representations: Relevant meso-levels in authentic tasks. In J. K. Gilbert & D. Treagust (Eds.), *Multiple representations in chemical education* (pp. 195–213). Dordrecht, The Netherlands: Springer.

National Research Council [NRC]. (2009). *A new biology for the 21st century*. Washington, DC: National Academic Press.

Paivio, A. (1986). *Mental representations: A dual coding approach*. New York: Oxford University Press.

Palsson, B. (2000). The challenges of *in silico* biology. *Nature Biotechnology, 18*, 1147–1150.

Pozzer, L. L., & Roth, W.-M. (2003). Prevalence, structure, and functions of photographs in high school biology textbooks. *Journal of Research in Science Teaching, 40*(10), 1089–1114.

Rodrigo, G., Carrera, J., & Elena, S. (2010). Network design meets in silico evolutionary biology. *Biochimie, 92*, 746–752.

Spiro, R. J., Feltovich, P. J., Coulson, R. L., & Anderson, D. K. (1989). Multiple analogies for complex concepts: Antidotes for analogy-induced misconception in advanced knowledge acquisition. In S. Vosniadou & A. Ortony (Eds.), *Similarity and analogical reasoning* (pp. 498–531). Cambridge, UK: Cambridge University Press.

Sweller, J. (1994). Cognitive load theory, learning difficulty, and instructional design. *Learning and Instruction, 4*, 295–312.

Taber, K. S. (2009). Learning at the symbolic level. In J. K. Gilbert & D. Treagust (Eds.), *Multiple representations in chemical education* (pp. 75–105). Dordrecht, The Netherlands: Springer.

Treagust, D. F., Harrison, A. G., & Venville, G. J. (1998). Teaching science effectively with analogies: An approach for preservice and inservice teacher education. *Journal of Science Teacher Education, 9*(2), 85–101.

Tsui, C.-Y., & Treagust, D. F. (2003). Genetics reasoning with multiple external representations. *Research in Science Education, 33*(1), 111–135.

Tsui, C.-Y., & Treagust, D. F. (2007). Understanding genetics: Analysis of secondary students' conceptual status. *Journal of Research in Science Teaching, 44*(2), 205–235.

Tsui, C.-Y., & Treagust, D. F. (2010). Evaluating secondary students' scientific reasoning in genetics using a two-tier diagnostic instrument. *International Journal of Science Education, 32* (8), 1073–1098.

van der Meij, J., & de Jong, T. (2011). The effects of directive self-explanation prompts to support active processing of multiple representations in a simulation-based learning environment. *Journal of Computer Assisted Learning, 27*, 411–423.

van Someren, M. W., Reimann, P., Boshuizen, H. P. A., & de Jong, T. (Eds.). (1998). *Learning with multiple representations*. London: Pergamon.

Vollmer, G. (1984). Mesocosm and objective knowledge. In F. M. Wuketits (Ed.), *Concepts and approaches in evolutionary epistemology* (pp. 69–121). Dordrecht, The Netherlands: D. Reidel Publishing Company.

von Baeyer, H. C. (2003). *Information: The new language of science*. Cambridge, MA: Harvard University Press.

Waldrip, B., Prain, V., & Carolan, J. (2010). Using multi-modal representations to improve learning in junior secondary science. *Research in Science Education, 40*(1), 65–80.

Werner, E. (2003). In silico multicellular systems biology and minimal genomes. *Drug Discovery Today, 8*(24), 1121–1127.

White, T., & Pea, R. (2011). Distributed by design: On the promises and pitfalls of collaborative learning with multiple representations. *The Journal of the Learning Sciences, 20*(3), 489–547.

Chapter 2
Identifying and Developing Students' Ability to Reason with Concepts and Representations in Biology

Trevor R. Anderson, Konrad J. Schönborn, Lynn du Plessis, Abindra S. Gupthar, and Tracy L. Hull

Introduction

Life scientists are highly dependent on the use of external representations (ERs) and symbolic language to research and teach modern biology (e.g., Tsui & Treagust, 2003), particularly at the submicroscopic level in areas such as biochemistry, physiology, molecular biology, and immunochemistry. At this level of cellular organization, the abstract nature of molecules and cellular processes necessitates the use of ERs or visualization tools such as physical models, diagrams, micrographs, computer images, animations, and other symbolic language to help learners and researchers construct meaningful mental models (or internal representations within the *mind's eye*) of biological concepts and phenomena (Schönborn & Anderson, 2006).

T.R. Anderson (✉)
Divisions of Chemical Education and Biochemistry, Department of Chemistry, Purdue University, 560 Oval Drive, West Lafayette, IN 47907-2084, USA

School of Biochemistry, Genetics and Microbiology, University of KwaZulu-Natal, Private Bag X01, Scottsville 3209, South Africa
e-mail: ander333@purdue.edu

K.J. Schönborn
Division of Media and Information Technology, Department of Science and Technology (ITN), Linköping University, Campus Norrköping SE-601 74, Norrköping, Sweden
e-mail: konrad.schonborn@itn.liu.se; konsc@itn.liu.se

L. du Plessis • T.L. Hull
School of Biochemistry, Genetics and Microbiology, University of KwaZulu-Natal, Private Bag X01, Scottsville 3209, South Africa
e-mail: aubreydup@mweb.co.za

A.S. Gupthar
School of Biochemistry, Genetics and Microbiology, University of KwaZulu-Natal, Private Bag 54001, Durban 4000, South Africa
e-mail: gupthara@ukzn.ac.za

D.F. Treagust and C.-Y. Tsui (eds.), *Multiple Representations in Biological Education*, Models and Modeling in Science Education 7, DOI 10.1007/978-94-007-4192-8_2,
© Springer Science+Business Media B.V. 2013

However, the frequent use of misleading symbolism, the great variation in ER design quality, and the poor methods of teaching and learning with ERs often lead to conceptual, visual, and reasoning difficulties that can seriously affect students' understanding of biology (Schönborn & Anderson, 2010). Thus, there is an urgent need to investigate such problems so that student difficulties can be prevented or remediated and so that better quality and more standardized ERs become available to biology education practitioners and researchers.

In this chapter, we describe a conceptual-reasoning-mode (CRM) model (Schönborn & Anderson, 2009) of seven factors affecting students' ability to interpret and learn from ERs. Using the model, we classify various reasoning abilities described in the literature and illustrate how the model can guide student interpretation of an ER. We also show how the model can guide the design and validation of assessment tasks aimed at developing (formatively) and assessing (summatively) students' reasoning ability. We then describe various student difficulties and show how the model can be used as an analytical tool for identifying the nature and source of the difficulties and for designing potential remediation strategies for addressing the difficulties. We conclude by discussing the implications of our research for improving learning and teaching with ERs in biology.

Description of the CRM Model

Our research has empirically identified a predictive model of seven factors that affect students' ability to interpret, visualize, and learn from ERs in a biochemistry context (Schönborn & Anderson, 2009). We have shown that the factors are interdependent in nature and meaningfully expressed as a Venn diagram (see Fig. 2.1).

The conceptual factor (C) represents a student's conceptual knowledge of relevance to an ER, whereas the reasoning factor (R) represents all the reasoning (sense-making) abilities necessary for interpreting an ER. The representation mode factor (M) characterizes the external nature of the ER, including its constituent symbolic markings. As depicted by the Venn diagram (see Fig. 2.1), these three factors are interdependent generating four further interactive factors. This is because students cannot engage their repertoire of reasoning abilities without something to reason with, that is, with the ER (represented by factor R-M) and/or with their conceptual knowledge (factor R-C). In addition, all ERs represent some form of scientific propositional knowledge represented by factor C-M of the model. Finally, interpretation of an ER through engagement of all these factors can be represented by the C-R-M interactive factor.

In this chapter, we demonstrate how the CRM model can be used by biology instructors as a very useful guiding framework and analytical tool in a variety of important applications, particularly with respect to the identification, the development and assessment of student reasoning, and the remediation of any related difficulties.

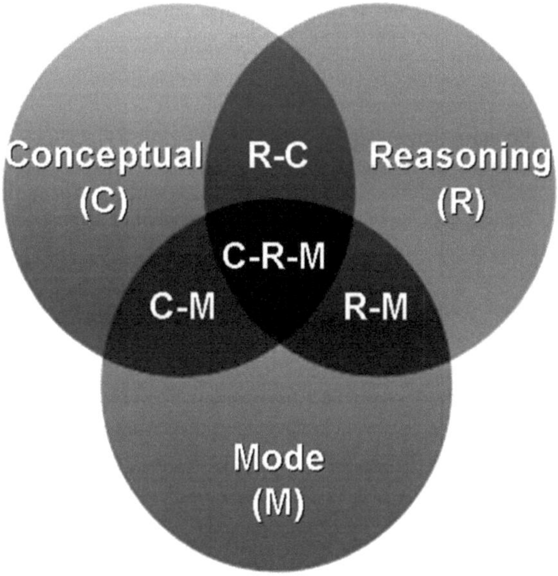

Fig. 2.1 The CRM model of seven factors affecting students' ability to interpret and visualize ERs in biology (Adapted from Schönborn & Anderson, 2009)

Using the CRM Model to Classify Expert Ways of Reasoning

In a recent synthesis of the literature (Anderson & Schönborn, 2008; Schönborn & Anderson, 2010), we identified several key ways of reasoning employed by experts in the practice of biology. In Table 2.1, we classify these cognitive skills according to the CRM model, that is, according to whether they, in our view, correspond to factors R-C or R-M.

There are several important points to note regarding the skills and their classification. First, this is far from an exhaustive list of reasoning abilities, as the literature describes numerous others, particularly those abilities concerning the practice of biological experimentation such as designing experiments, testing hypotheses and using appropriate controls, or technical and practical skills (e.g., Quentin-Baxter & Dewhurst, 1992). Second, research has shown that some of the listed skills are at different levels of inherent difficulty for students. For example, students find memorization of information (see Table 2.1, A1) much easier than transfer and application of knowledge (A3) (Mayer, 2002), and decoding symbolism in a single diagram not as difficult as horizontal translation across multiple representations of the same phenomenon (Schönborn & Bögeholz, 2009). Third, clearly not all the skills (see Table 2.1) can be exclusively classified according to only one factor, as several of the skills may be applied both in the *mind's eye* (R-C) in the absence of an ER, and directly to an external representation (R-M). For example, experts can reason analogically (A4) both with or without an ER, whereas integration of knowledge (A2) can involve linking concepts

Table 2.1 Selected reasoning abilities classified according to the CRM model, central- to expert-level conceptual understanding and visualization of representations

A. Some examples of reasoning with concepts (classified as R-C)
Understanding a concept means the ability to:
1. Memorize knowledge of the concept in a mindful manner, as distinguished from rote learning
2. Integrate knowledge of the concept with that of other related concepts so as to develop sound explanatory frameworks
3. Transfer and apply knowledge of the concept to understand and solve (novel) problems
4. Reason analogically about the concept
5. Reason locally and globally about the concept (systems thinking)
6. Think metacognitively about the concept
B. Some examples of reasoning with ERs (classified as R-M)
Understanding a representation means the ability to:
1. Decode the symbolic language composing an ER
2. Evaluate the power, limitations, and quality of an ER
3. Interpret and use an ER to solve a problem
4. Spatially manipulate an ER to interpret and explain a concept
5. Construct an ER to explain a concept or solve a problem
6. Translate horizontally across multiple ERs of a concept
7. Translate vertically between ERs that depict various levels of organization and complexity
8. Visualize orders of magnitude, relative size, and scale

Adapted from Schönborn and Anderson (2010), Anderson and Schönborn (2008)

both in the *mind's eye* or while reasoning with a concept map. It is likely though, given the visual nature of biology, that even in cases where no ER is present, at least a mental model is involved in facilitating the reasoning process. Fourth, in some cases there is clearly a logical sequence for using reasoning skills. For example, knowledge cannot be integrated (A2) before key information has been memorized (A1), and both these reasoning processes need to precede higher-order reasoning such as problem solving (A3), analogical (A4) and systems thinking (A5), as well as any metacognitive activity (A6). Finally, and related to the above, it will become apparent, based on the examples of assessment tasks and student difficulties presented in this chapter, that more than one reasoning skill is always simultaneously engaged by biologists when ERs are being interpreted.

So the question arises: What is the purpose of dividing biological reasoning into separate skills? Why not study reasoning as an integrated process as it clearly occurs in this manner? The answer is simple—by distinguishing the different ways of reasoning, we are more easily able to identify the nature and source of specific reasoning difficulties and to devise ways of remediating them. In the following sections, we show how the CRM model, together with knowledge of the different reasoning abilities, can be used as an analytical tool for (1) guiding student interpretation of ERs, (2) identifying the unique nature and source of specific reasoning difficulties with ERs, and (3) devising approaches to remediate and develop student competence in these areas.

Using the CRM Model to Guide the Assessment and Interpretation of ERs

After the identification of the various cognitive skills that we considered central to biologists, the next step was to devise approaches to developing such competencies in students as part of formal biology curricula. In previous studies (Anderson, 2007; Schönborn & Anderson, 2008, 2010), we advocated the idea of assessment-driven development of conceptual understanding, including reasoning with concepts and representations. This idea stemmed from the crucial and reciprocal relationship that exists between the four key components of the educational process, namely, course objectives, teaching, learning, and assessment (Anderson, 2007). In line with this relationship, the *how* and *what* of assessment informs how and what students will focus on during learning—the idea of *learning to the test*! Based on this, we argue that specifically designed tasks, which focus on each of the reasoning abilities, as shown in Table 2.1, could be effective at both developing (formatively) and assessing (summatively) students' reasoning ability in biology. The approach involves giving students repeated practice at performing such tasks that specifically require them to use the particular visual skill that requires improvement.

To ensure that we developed sound assessment tasks—that specifically required students to reveal their conceptual understanding and reasoning ability with concepts and representations—we used (1) the guidelines presented in Anderson and Rogan (2010, p. 56), (2) the cognitive skills listed in Table 2.1 of this chapter, and (3) the CRM model to devise guidelines for assessment design. These guidelines are presented in Box 2.1. The guidelines provide criteria that correspond to each factor of the CRM model that instructors might wish to use to ensure that the tasks are both sound and focus specifically on assessing conceptual understanding and reasoning ability with representations. Establishing whether students have the necessary prior conceptual knowledge (factor C) that corresponds to the scientific propositional knowledge represented by the ER (C-M) is important because research has shown that one cannot assume that what students have studied in previous courses was necessarily learned. It is also essential to ensure that the ER is a sound representation (M) of the intended propositional knowledge (C-M). Also that such knowledge is appropriate for the course being taught and that it is of a suitable standard for the educational level so that it is neither too cognitively demanding for the students nor too easy for them (Anderson & Rogan, 2010). Finally, and most importantly for the present goals, each task must require students to use certain cognitive skills (R) so that a range of intended tasks can be designed to cover all reasoning abilities (see Table 2.1).

We are currently testing these guidelines by developing a wide range of tasks for use in various biological science disciplines, some examples of which are also included in this chapter in the section on student difficulties. We are also classifying and validating the tasks using the CRM model as an analytical tool. This is both from the perspective of expert opinion of what reasoning abilities are being tested and, most importantly, from a student perspective to ascertain if student response data can be coded for both R-C and R-M categories as well as for subcategories of reasoning abilities and any related reasoning difficulties. An example of such a task

Box 2.1 Guidelines for Designing and Analyzing Conceptual Assessment Tasks Involving Representations (ERs) Based on the CRM Model of Schönborn and Anderson (2009)

Factor C:

- Do students have the necessary prior conceptual knowledge to interpret the ER and answer the question?
- Will the task test and reveal evidence of both sound conceptual knowledge and any alternative conceptions in students?

Factor R:

- Will the task test and reveal evidence of students' reasoning skills and difficulties?
- See also subsets, R-C and R-M, below.

Factor M:

- How well or poorly does the ER represent the intended phenomenon?
- Do you think the ER and its constituent symbolism will be clear and not too complex for the students to understand?
- Do you think the ER will help the student to answer the question?

Factor R-C:

- Will the task test students' cognitive skills required for scientific reasoning?
- Will the task reveal evidence of students' cognitive difficulties?
- Which cognitive skills are being tested by the task?

Factor R-M:

- Will the task test students' visual skills (representational competence)?
- Will the task reveal evidence of students' visual difficulties?
- Which visual skills are being tested by the task?

Factor C-M:

- What propositional knowledge is represented by the ER and required for answering the question? That is, what specific concept(s) is the question designed to probe?
- Is the propositional knowledge appropriate for the educational level of the course? That is, is the extent and complexity of the required knowledge not too cognitively demanding?

Factor C-R-M (can students master the assessment task?):

- Does the task test students' conceptual understanding?
- Does the task allow for a range of scientifically correct (creative) answers?

(continued)

Box 2.1 (continued)

- Does the task probe students' ability to interpret, visualize, and learn from the ER?
- If the task reveals student difficulties interpreting the ER, check whether soundness of an ER (M), prior conceptual knowledge (C) or cognitive skill competence (R), is limiting.
- Is the instrument suitable as a formative task for promoting students' conceptual understanding and learning during the course?
- Is the instrument suitable for grading students' conceptual understanding?

is presented in Box 2.2 together with an analysis of the task using the CRM model to suggest, from an expert perspective, what reasoning abilities (see Table 2.1) might be required for students to answer the question.

On examining the example in Box 2.2, one is struck by the enormous amount of conceptual, symbolic, and strategic knowledge that we as instructors require students to master in order to merely interpret a single ER. This suggests the importance of clearly explaining ERs to students and giving them sufficient time to interpret them. As can be seen by the structure of the question in Box 2.2, the student is guided to link to all the critical concepts (C-M) that are important for interpreting the graph. Then, they are required to use Table 2.1 to identify which ways of reasoning (R) they think are necessary to use their conceptual knowledge (R-C) to make sense of the ER (R-M). In addition, they need to think of other representations of the kinetic experiment depicted by the graph (horizontal translation) (see Table 2.1, B6) in order to obtain greater insight into the nature and purpose of the experiment and the underpinning molecular processes. They also need to translate vertically (see Table 2.1, B7) (Schönborn & Bögeholz, 2009) to place the kinetic process being studied in the context of a living system. In so doing, they achieve a deeper analysis of the graph.

We have found that using the CRM model as an analytical tool to systematically and separately consider the various critical concepts, ways of reasoning (see Table 2.1) and related representations of relevance to the ER can significantly facilitate student interpretation of ERs. Although this remains to be confirmed by research, in our experience this approach gives students some sort of meaningful structure for making sense of an ER rather than the somewhat random manner used by some students. In this regard, our studies on secondary-level biology students' interpretation of a diagram of the thermoregulation process showed that students often completely ignored certain symbolism (e.g., arrows) or parts of an ER in attempting to interpret an ER (du Plessis, Anderson, & Grayson, 2003). In response to this problem and several other student difficulties with symbolism and ERs, we developed a strategy and tutorial for developing students' ability to interpret arrow symbolism in biology diagrams. Implementation of the strategy and tutorial in a small-scale study involving 18 grade 9 students resulted in significant improvement in the ability of some students

Box 2.2 An Example of the Use of the CRM Model as an Analytical Tool to Guide ER Interpretation

Interpret the graph below in as much detail as possible by doing the following:

1. List (C-M) and explain (C) the biochemical concepts related to the graph.
2. List and explain the experimental and mathematical concepts related to the graph.
3. List other ERs that represent the same phenomenon (e.g., equation, apparatus, models).
4. Use the supplied list of reasoning abilities (R; Table 2.1) to identify which:
   (a) Cognitive skills are required to make sense of the graph (R-C).
   (b) Visual skills are required to make sense of the graph (R-M).
   (c) Explain how you use each reasoning ability (a and b) to interpret the graph (C-R-M).
5. Describe the method a biologist would use to collect the data represented in this graph.

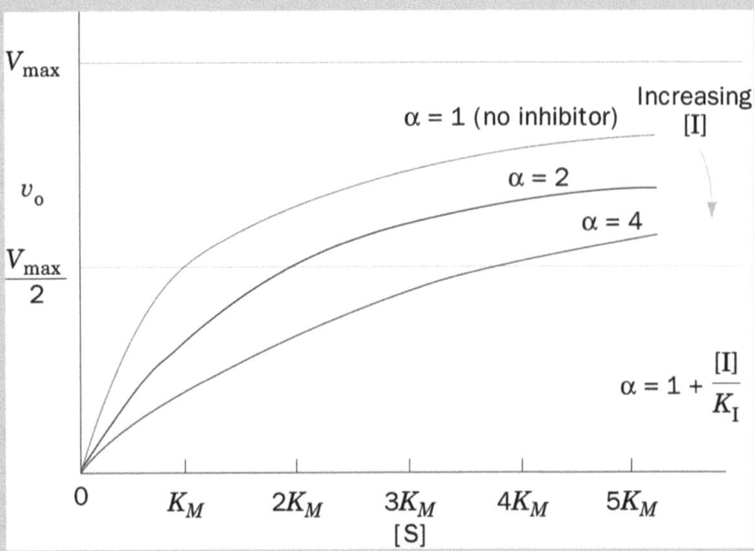

An example of a possible (brief) answer:

1. Biochemical concepts include enzyme, substrate, inhibitor, active/binding sites, and affinity.

(continued)

Box 2.2 (continued)

2. Mathematical/graphical concepts include V_{max}, K_m, K_{cat}, K_i, dependent and independent variable, constant, concentration, reaction velocity, and saturation curve versus linear relationship.
3. Other related ERs: experiments, equipment (macro level), double reciprocal plot, table of plotted data, Michaelis-Menten equation and formulas, visual competitive inhibition models, animation of enzyme substrate interaction, and qualitative illustration of near-equilibrium (reversible) reactions versus far-equilibrium (irreversible) reactions.
4. (a) Memorize, analyze, transfer, integrate, systems thinking, and analogical reasoning.
   (b) Decode, horizontal/vertical translation, construction, interpretation, transfer, and apply.
   (c) This is a graph depicting the effect of increasing concentrations of a competitive inhibitor (as compared to no inhibition) of an enzyme-catalyzed reaction occurring at constant enzyme concentration. The kinetics profile is typical of all competitive inhibition situations occurring in cells.
5. Set up the enzyme assay under optimal conditions of temperature, pH and ionic strength. Set up tubes with a range of concentrations of substrate up to 5 times the value of the enzyme's K_m for that substrate. Add a fixed concentration of enzyme to each tube, mix gently and incubate for a fixed time period. To determine the initial velocity, measure the disappearance of substrate, or the appearance of product, at two early time periods, for example at 15 and 45 secs. Plot the results on a Michaelis-Menten curve. Repeat the experiment but at 2 different inhibitor concentrations and plot these data on the same curve.

to interpret arrow symbolism in a nitrogen cycle diagram (du Plessis & Anderson, 2009). This strategy contained several similar elements of the proposed CRM-guided strategy in that students are required to systematically analyze each part of a diagram and identify and interpret the meaning of all the constituent symbolism.

Using the CRM Model to Analyze Student Difficulties for the Nature and Potential Source of Unsound Reasoning

In this section, we present some selected examples of student reasoning difficulties to provide further support for the importance of formally teaching scientific reasoning as part of all biology curricula. These examples were identified by our research group in different areas of biology and classified according to the CRM model and the reasoning abilities presented in Table 2.1.

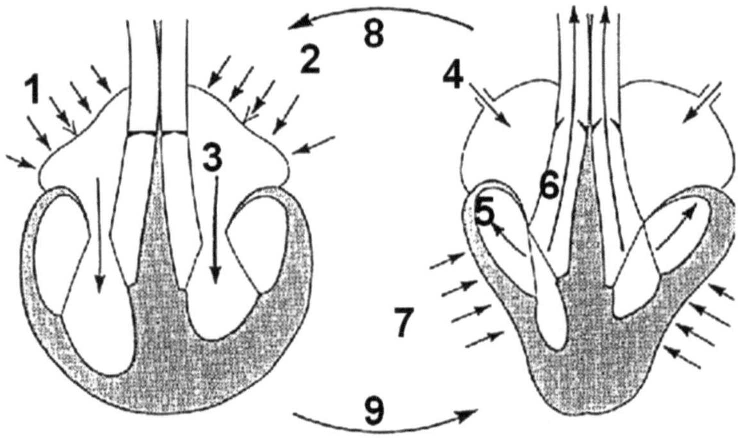

Fig. 2.2 A stylized diagram of the cardiac cycle (Wright, 1989, p. 55) (Reprinted with permission)

Reasoning Difficulties with an ER of the Cardiac Cycle

The following diagram (see Fig. 2.2; Wright, 1989) depicting the cardiac cycle was used in a study by our group to investigate secondary-level students' interpretation of arrow symbolism (du Plessis et al., 2003). The diagram, without its labels and caption, had previously been used in a biology examination at a secondary school in South Africa.

Extensive data obtained from open-ended and multiple-choice questions, as well as student-generated diagrams and clinical interviews, revealed evidence of a range of major student difficulties with their interpretation of the various arrows in the diagram. Regarding arrow 1, 39% of students interpreted it as blood entering the atrium rather than its intended purpose (as in the case of arrow 2) of indicating that blood could not flow into the closed atrium. In addition, 41% of students thought that the cluster of arrows on either side of arrows 1 and 2 represented pressure being applied to the outside of the atria causing them to contract, rather than simply indicating that the muscular wall of the atria was contracting. Regarding arrows 1 and 4, 36% of students did not see any difference in their intended purpose, suggesting that they thought both arrows show blood entering the atrium. Furthermore, many students did not recognize arrows 1 and 2 as being separate from their perceptual unit of similarly styled arrows. Whereas arrow 5 is intended to show blood pushing against and closing the tricuspid valve, 24% of students interpreted it instead as blood flowing out of the heart. Finally, 14% of students suggested that arrows 8 and 9 were part of blood flow.

Analysis of the above difficulties according to the CRM model suggests a problem with both the diagram or representation mode (M) and student reasoning (R). In the case of the diagram, the arrows are drawn in the same style but represent several purposes, including direction of flow (arrows 3, 4, and 6), direction of flow stopped by closed valves (arrows 1 and 5), alternating processes (arrows 8 and 9), and contraction (arrow groups 2 and 7). Similar problems have been noted by various authors (e.g., Ametller & Pinto, 2002) who reported that confusion can result when similarly styled arrows are used for different purposes (synonymy) or differently styled arrows for the same purpose (polysemy) (cf. Strömdahl, 2012). Thus, the issue of synonymy (corresponding to factor M of the model) as well as the number of arrows clearly contributes to the complexity of the diagram, and this was evident in various reasoning difficulties shown by students. Such difficulties probably included incorrect decoding of arrow symbolism (R-M; see Table 2.1, B1), incorrect interpretation of the ER (R-C; B3), inappropriate application of their knowledge of the cardiac cycle (A3), and inappropriate analogical reasoning (R-C and R-M; A4) about the ER—an analogical model of heart function. In addition, spatial reasoning (R-M; B4) might have been a problem in cases where students included arrow 1 together with the neighboring arrows as one perceptual unit.

Using the CRM model to classify the difficulty in the above manner leads to greater insight into the nature and possible source of the difficulty and permits the design of a more informed remediation strategy that specifically targets those reasoning abilities with which students have problems. Clearly in the above case this strategy would need to include ways of familiarizing students with the issue of synonymy and developing their ability to recognize and interpret diagrams with this problem, that is, to also improve students' ability to evaluate the quality and limitations (see Table 2.1, B2) of ERs. Alternatively, a different ER could be used to teach the cardiac cycle, but this will not solve the problem of the numerous other ERs with the same problem of synonymy.

Reasoning Difficulties with Symbolism in Molecular Biology

Gupthar and Anderson (2003) investigated student difficulties associated with DNA-strand symbolism and function. Double-stranded DNA is composed of two antiparallel strands which are complementary in terms of base sequence and *run* $5' \rightarrow 3'$ in opposite directions. The two strands are labeled either *coding* or *template*, depending on their respective function. The coding strand is the strand of DNA within a gene whose nucleotide sequence is identical to that of the transcribed RNA with the replacement of T by U in RNA. The template is defined as the strand of DNA within a gene whose nucleotide sequence is complementary to that of the transcribed RNA (Scism, 1996). During transcription RNA polymerase binds to, and moves along, the template in the $3' \rightarrow 5'$ direction, catalyzing the synthesis of RNA in a $5' \rightarrow 3'$ direction. In DNA replication, which occurs semiconservatively, each DNA strand serves as a template for complementary DNA synthesis. The result is two molecules of double-stranded DNA, each of which contains one of the template strands. A typical question given to biochemistry students to probe understanding of this topic is presented in Box 2.3.

The following difficulties, coded as R-C (with italics font) or R-M (with regular font), based on student interviews, revealed that some students interchanged the DNA-strand labels and thereby failed to differentiate between the functions of the template and coding strands:

Box 2.3 An Example of a Typical Probe for Symbolism in Molecular Biology

The following is representative of double-stranded (ds) DNA:

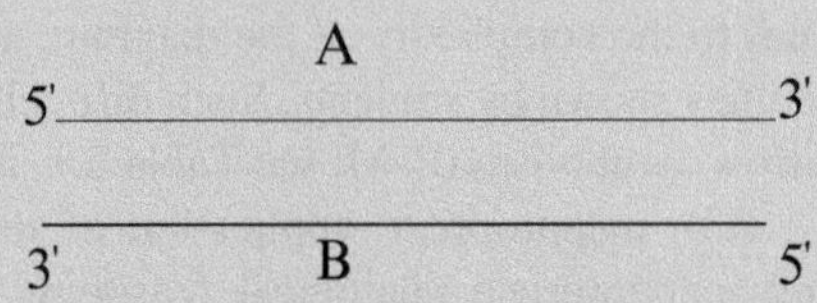

1. Name strands A and B and explain why you named them as such.
2. (a) Which strand(s) is/are implicated in:

   (i) Replication?
   (ii) Transcription?

   (b) Explain why in each case.

> A is the leading strand. *Replication occurs in a 5′→3′ direction within a replication bubble or fork.* There is a problem with the polarity of B, *resulting in the formation of Okazaki fragments*, thus B is the lagging strand.
>
> A is the leading strength [strand] *because nucleotides move from a 5′→3′ direction.* B is the lagging strand *because nucleotides move from a 5′→3′ direction.*
>
> A – leading strand. It begins from 5′→3′ left to right. B – lagging strand. It forms in the opposite direction to the leading strand and therefore it is from right to left in the 5′→3′ direction.

Analysis of these difficulties with the CRM model revealed various reasoning difficulties. First, the reference to *leading strand, lagging strand*, or *Okazaki fragments* clearly demonstrates a substitution of DNA-strand labels with nomenclature associated with DNA replication intermediates. This suggests a problem with decoding the symbolism (R-M; see Table 2.1, B1). Furthermore, students failed to transfer (R-C, A3) the appropriate knowledge to each strand to identify its function, thereby failing to correctly interpret (R-M, B3) the ER.

Reasoning Difficulties with an ER of the Structure of Immunoglobulin G (IgG)

We have reported elsewhere a wide range of difficulties shown by biochemistry students when interpreting textbook diagrams of immunoglobulin G (IgG), which included the following ER (see Fig. 2.3) (Schönborn, Anderson, & Grayson, 2002).

Fig. 2.3 Stylized diagram of the three-dimensional structure of an IgG antibody molecule (Reprinted with permission from Pearson Education, Inc., Upper Saddle River, NJ 07458, USA)

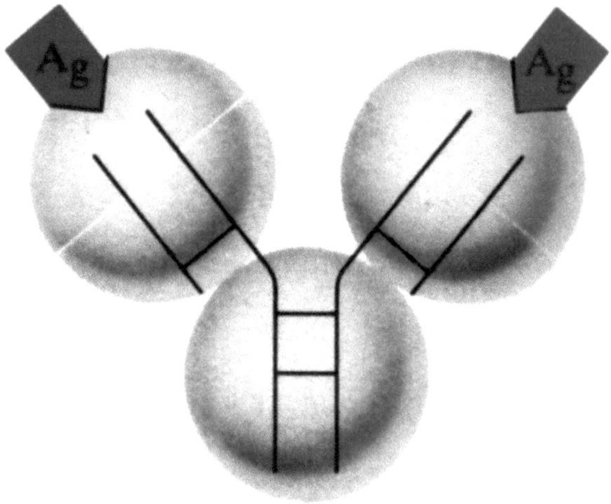

The following are selected examples of difficulties identified in interviews related to the interpretation of Fig. 2.3 which we coded in italics font for R-C and in regular font for R-M:

> *Heavy and light chains and [with] H-bonds between them.*
>
> Black lines [are] some form of bond or attachment holding the 3 cells together- *blood cells, biconcave type shape.*
>
> The colored (grey) region represents *different amino acid residues attached to the backbone* (black line) *of the antibody.*
>
> *Cell (C), cell division takes place, two cells (V) are formed. Cell C old mature structure attaches 2 cells with black lines or bonds. Young immature cells (V) are attacked by Ag.*
>
> This is meant to represent *a DNA molecule, leading strands and a lagging strand of DNA...*
>
> It looks like *a new replicating strand of DNA. Ja [yes]... it is nucleotide synthesis...*

Analysis of these difficulties using the CRM model as a guide suggests that the major problem was an incorrect decoding of the symbolism (R-M; see Table 2.1, B1) in the diagram, incorrect interpretation (R-M, B3), as well as inappropriate transfer and application (R-C, A3) of knowledge from biological domains concerning blood cells, cell division, and DNA replication (R-C, A3; and R-M, B6). In addition, there is also an analogical reasoning problem (R-C, A4) stemming from a diagram that poorly represents the intended protein structural information. Once again a remediation strategy would be designed to specifically address these reasoning difficulties so that students would improve their ability to evaluate the quality and limitations (R-M, B2) of ERs.

Reasoning Difficulties with Metabolic Pathways Occurring in Cells

Hull (2004) performed a study in our group on students' mental models of various biochemical processes. Data collection consisted of audiotaped interviews as well as

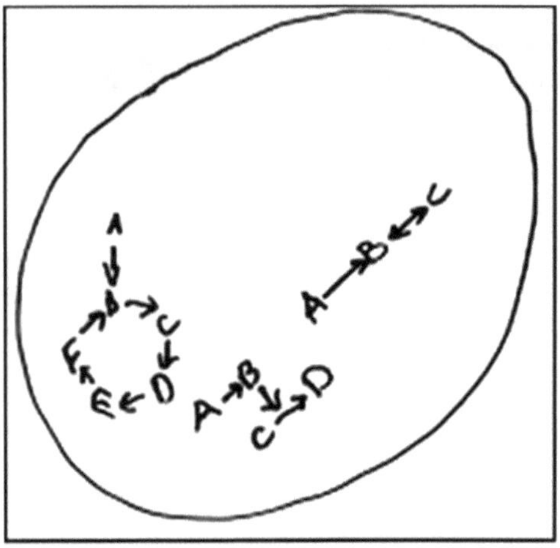

Fig. 2.4 A biochemistry student's representation of various biochemical processes occurring in vivo

student-generated diagrams in which students were asked to draw what they were visualizing. All interviews were in English and transcribed verbatim. The following are examples of such data which we have coded in italics for R-C and regular font for R-M:

I: Ok, let's say that we're sitting in the cytoplasm and we can see a cyclic process, for example the TCA cycle, happening in front of us, describe what you think that will look like.

S: Aah, I think they [metabolic constituents] would be going in a circle in front of me *and you'll have products and various substances going off into the rest of the cell* and ja [yes], it would be going round and round.

I: Ok, and what about a linear process?

S: Linear processes occur in a straight line. Linear processes occur at 180° in any direction... and occur vertically or horizontally.

I: Ok, let's come out of that cell and imagine we're looking at that same cell through a very powerful microscope, draw a rough outline of the cell and the processes you saw in the cytoplasm.

S: [draws cell outline in Fig. 2.4].

The above data represents a clear case of inappropriate horizontal translation (R-M; see Table 2.1, B6) from a typical textbook ER of metabolic pathways to how students imagine such processes would look in the cell. It is a typical case of literal interpretation (R-M, B3) and incorrect decoding (R-M, B1) of diagrams and demonstrates that students with this difficulty did not transfer (R-C, A3) their earlier acquired chemical knowledge of collision theory and kinetic energy of molecules to the cellular scenario. This led to the construction (R-M, B5) of an inappropriate ER based on an unsound mental model. Vertical translation (R-M, B7) was also a problem as students attempted to *move* from the molecular level to the cellular level. Thus, in summary, any remediation strategy would need to focus on developing a range of reasoning abilities in students—including the transfer and application of knowledge; the decoding, interpretation, and construction of ERs; and the horizontal and vertical translation across such ERs.

The above examples of student difficulties with representations, alongside numerous other examples in the literature, constitute strong evidence for the importance of addressing such difficulties, either through the devising of remediation strategies or by improving or replacing a specific ER. That is, in our view, course curricula, teaching and assessment approaches, learning activities, and pedagogical content knowledge need to be informed and shaped by the representations we use to educate biology students. Possible approaches are discussed in the next three sections.

Application of the CRM Model to the Design of Remediation Strategies

Since students in our studies showed such a wide range of conceptual, reasoning, and visualization difficulties with representations, there is clearly an urgent need to address the remediation and/or prevention of such difficulties in course curricula. In this section we present an example of three related difficulties in the context of metabolism and briefly show how we used the CRM model to both analyze them and design a remediation strategy that successfully addressed the difficulties.

Box 2.4 contains an example of a typical question which we gave to biochemistry students to probe their reasoning difficulties with metabolism (Grayson, Anderson, & Crossley, 2001). In this particular study, we also used more focused probes and interviews to delve deeper into the nature of the difficulties.

The expert response (Box 2.4) was analyzed by the CRM model and the results used to guide the coding of student responses with respect to the types of reasoning we could expect when answering the question. Clearly all questions require memory (R-C; see Table 2.1, A1) of numerous critical concepts concerning the functioning of metabolism which students need to transfer from various contexts (mainly chemistry) and apply (R-C, A3) to the context of metabolism. They also need to integrate (R-C, A2) such concepts in order to establish a sound explanatory framework for interpreting the ER (R-M, B3) and answering the question. In addition, question 2 requires systems thinking (R-C, A5) in that there is a need to consider the influence of the inhibition on other reactions in the pathway. Furthermore, questions 1 and 3 require horizontal translation (R-M, B6) from the equation of the inhibited reaction to an ER of its mechanism in order to fully understand the effect of enzyme inhibition on the reaction. Question 1 also requires horizontal translation (R-M, B6) to activation energy diagrams to realize the key function of the enzyme as a catalyst under cellular conditions. Finally, analogical reasoning (R-M, A4) is also important in that the diagram is an analogical model of the real process occurring in cells.

The expert response and the above classification, using the CRM model, were used as a standard to code student responses with respect to sound and unsound ways of reasoning. The following are selected descriptions (quotes not shown) of three difficulties, revealed by the question in Box 2.4, which we termed Essential

Box 2.4 An Example of a CRM-Guided Assessment Task

Consider the following part of glycolysis functioning in a cell:

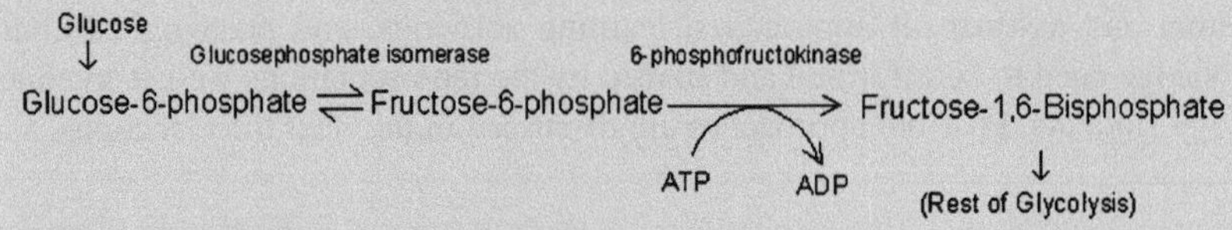

If 6-phosphofructokinase is totally and irreversibly inhibited by a toxic substance, explain what effect this would have on:

1. The conversion of fructose-6-phosphate to fructose-1,6-bisphosphate
2. The relative concentrations of intermediates before and after the inhibited reaction
3. The half-reaction for the conversion of ATP to ADP
4. The overall flux through glycolysis

Example of expert response:

1. The reaction will stop because the enzyme is an essential catalyst in the mechanism of the reaction, by stabilizing a high energy intermediate so that the reaction can occur under cellular conditions.
2. Both G-6-P and F-6-P will increase, while intermediates after the *point* of inhibition will deplete in concentration.
3. ATP will not be converted to ADP unless the enzyme facilitates the transfer of the phosphate from ATP to F-6-P in the active site.
4. The flux will decrease to zero as neither glucose is used nor pyruvate produced, because F-1,6-BP is no longer produced as a substrate for the next reaction.

Summary of CRM analysis of expert response:

R-C: Memory (A1), integrate (A2), transfer/apply (A3), systems thinking (A5), analogical reasoning (A4)

R-M: Decode (B1), interpretation (B3), horizontal (B6) translation

(E) Nature Difficulties due to students not being able to appreciate the indispensable nature of enzymes as key participants in the mechanism of metabolic reactions:

E_1: The inhibited reaction will proceed without enzyme, but at a slower rate.

E_2: One of a pair of half-reactions, coupled in parallel, can occur without the other.

E_3: An inhibited enzyme-catalyzed reaction will proceed because other factors override the effect of inhibition, such as whether the inhibited reaction is spontaneous (E_{3a}) in nature or is displaced from equilibrium (E_{3b}).

Analysis of students' written quotes that corresponded to the above descriptions revealed evidence of several different reasoning difficulties. First, students with E_1 difficulties had clearly rote learned (R-C, A1) the basic definition of an enzyme as a catalyst but did not remember its essential role in the mechanism of the reaction. Nor did they translate horizontally (R-M; see Table 2.1, B6) to activation energy diagrams to realize the key function of the enzyme. Thus, integration (R-C, A2) of the concept of an enzyme with other critical concepts—such as mechanism, kinetics and, in the case of E_2 with the concept of parallel coupling, bi-substrate reactions, and, for E_3 with equilibrium, Le Chatelier's principle, spontaneity, and exergonicity—was clearly poor, while their transfer and application (R-C, A3) of such concepts and principles to solving the problems was in many cases inappropriate. When using the diagram or representation mode (M) to answer the questions, some students incorrectly decoded the meaning of the straight arrow/curved arrow symbol used to depict parallel coupling and thought that ATP cleavage was not essential (E_2) for the reaction to occur. But the major reasoning difficulty across all three difficulties was a failure to translate horizontally (R-M, B6) to ERs concerning the enzyme catalytic mechanism of the reaction.

Thus, based on the above CRM-informed analysis of the difficulties, our remediation strategy was designed to specifically target the following reasoning difficulties: memory (see Table 2.1, A1), integration (A2,), transfer/application (A3), decoding of symbolism (B1), ER interpretation (B3), and horizontal translation (B6). The strategy was structured as a tutorial that included questions and tasks that specifically focused on the following:

- Critical concepts (e.g., spontaneity, chemical energy, chemical equilibrium)
- Integration of critical concepts composing an explanatory framework
- The essential nature of enzymes.
- The mechanisms of enzyme catalysis

In presenting the tutorial and the constituent tasks, we attempted to create a conceptual ecology and status that favored conceptual change as discussed by Duit and Treagust (2003) and others. In brief, we attempted to expose students to sound metabolism concepts and principles in the hope that they would find their new conceptions intelligible, plausible, and fruitful. Since students' lack of understanding and integration of the critical concepts was generic to all three difficulties, step 1 of the strategy was to address this problem with a concept-mapping task (cf. Schönborn & Anderson, 2008). The concept map (not shown) included the following concepts which we considered critical to the functioning of metabolism: *spontaneity*, *metabolic reactions*, *substrate*, *kinetics*, *coenzyme* or *cofactor*, *coupling*, *inhibitor*, *equilibrium*, *mechanism*, *thermodynamics*, *enzyme*, *energy*, and *ATP*.

Step 2 of the strategy was designed to specifically target the E_1-type difficulty by addressing integration (see Table 2.1, A2), transfer/application (A3), and horizontal translation (B6). This step required students to respond to tasks requiring them to:

- Determine which components (e.g., enzyme, coenzyme, cofactor, substrate) are essential for occurrence of metabolic reactions

Table 2.2 Results showing the effect of the remediation strategy on the incidence of student difficulties over a period of four consecutive years

	Percentage incidence and fraction of students showing each difficulty				
Type of difficulty[a]	No remediation		Before remediation	After remediation	Prevention[b]
Year	1	2	3		4
E_1	51%	48%	31%	2%	5%
	44/86	52/108	29/95	2/98	4/89
E_2	27%	53%	44%	1%	0%
	23/86	55/103	43/97	1/98	0/89
E_{3a}	30%	20%	34%	4%	1%
	26/86	23/118	32/94	4/98	1/89
E_{3b}	44%	16%	11%	1%	2%
	38/86	19/118	10/94	1/98	2/89

[a]See text for descriptions of each type of difficulty
[b]The remediation strategy was incorporated into the normal teaching process in an attempt to prevent the development of the student difficulties

- Determine what role each component plays in the mechanism of the reaction from analysis of various diagrams and an animation of an enzyme mechanism
- Use the kinetic graph (see Box 2.1) to compare the effect on reaction rate of reducing enzyme concentration to zero versus decreasing enzyme activity to zero by means of an inhibitor

Finally, step 3 targeted both E_2- and E_3-type difficulties by addressing reasoning concerning integration (A2), problem solving (A3), decoding (B1), and horizontal translation (B6) by requiring students to perform the following:

- E_2 tasks predicting the mechanism of reactions coupled in parallel (i.e., single mechanism)
- E_{3a} and E_{3b} tasks requiring application of knowledge of spontaneity, exergonicity, chemical energy, and equilibrium to metabolic reactions

As shown in Table 2.2, the revealed incidence of the difficulties was high for three consecutive years, whereas implementation of this strategy in the third year almost totally eliminated all the difficulties, while in the fourth year we were able to *prevent* the difficulties, rather than having to *cure* them.

In summary, our results suggest that the CRM model is a very useful analytical tool for identifying the nature of student reasoning difficulties and for developing more informed and better designed remediation and prevention strategies to address such difficulties. Since it might not always be feasible to design such a strategy for every difficulty, future work should focus on identifying more generic strategies that might be useful in addressing a range of related reasoning difficulties. Indeed, such strategies, if successful, could be incorporated into instructors' pedagogical content knowledge so that many of the difficulties are addressed in instruction rather than in remediation.

Conclusion

In this chapter we have shown that the CRM model can be extremely useful to biology education practitioners and researchers as a guiding framework and analytical tool for various aspects of the educational process. This includes using the model to guide the classification and assessment of reasoning abilities and to develop students' problem-solving strategies for interpreting ERs in biology. In addition, the CRM model is a valuable analytical tool for identifying the nature and potential source of students' reasoning difficulties with ERs and thereby for informing the design of remediation strategies for addressing the difficulties.

Like all models, the CRM model has limitations. In particular, the CRM-guided coding approach has revealed the following two problems concerning the analysis of quotes from student interviews: (1) the quotes do not reveal situations where students lack certain ways of reasoning and (2) the quotes do not always reveal all the types of reasoning being engaged by students, as this depends on the extent of their responses and therefore, to some degree, on the nature of probe design. Both these problems, though, can be minimized, respectively, by comparing student responses to multiple coded expert responses and by delving deeper into student reasoning during clinical interviews. The application of the presented examples of coding is also highly dependent on a complete list of reasoning abilities, whereas the nature of the reasoning displayed in students' quotes is not always lucid, which means that the coding is often subjective and requires validation by several experts.

Despite these limitations, we believe that the CRM model could become an important component of a biology education practitioner's and researcher's pedagogical toolkit, particularly in the area of scientific reasoning and visualization of external representations. Future work will focus on testing and validating reasoning tasks that could be used to both assess and develop reasoning in our students while at the same time yield data that enables instructors to monitor student progress. Ultimately, we believe that the teaching, learning, and assessment of reasoning ability should be integrated into all biology course curricula. Given that practical and technical skills are explicitly taught in all biology courses, there is no reason why we should not place the same emphasis on reasoning skills. This is because instructors cannot simply assume that these central skills will be automatically acquired through informal interactions with scientists and other students.

Acknowledgment We thank Dr. Nancy Pelaez for valuable feedback on an earlier draft of this manuscript.

References

Ametller, J., & Pinto, R. (2002). Students' reading of innovative images of energy at secondary school level. *International Journal of Science Education, 24*(3), 285–312.

Anderson, T. R. (2007). Bridging the educational research-teaching practice gap: The power of assessment. *Biochemistry and Molecular Biology Education, 35*(6), 471–477.

Anderson, T. R., & Rogan, J. M. (2010). Bridging the educational research-teaching practice gap: Tools for evaluating the quality of assessment instruments. *Biochemistry and Molecular Biology Education, 38*(1), 51–57.

Anderson, T. R., & Schönborn, K. J. (2008). Bridging the educational research-teaching practice gap: Conceptual understanding, part 1: The multifaceted nature of expert knowledge. *Biochemistry and Molecular Biology Education, 36*(4), 309–315.

du Plessis, L., & Anderson, T. R. (2009). Strategy and tutorial for the development of student's ability to interpret and learn from arrow symbolism in biology diagrams. In M. Hammann, A. J. Waarlo, & K. T. Boersma (Eds.), *The nature of research in biological education* (pp. 263–278). Utrecht, The Netherlands: Freudenthal Institute for Science and Mathematics Education, Utrecht University.

du Plessis, L., Anderson, T. R., & Grayson, D. J. (2003). Student difficulties with the use of arrow symbolism in biological diagrams. In J. Lewis, A. Magro, & L. Simonneaux (Eds.), *Biology education for the real world: Student, teacher, citizen* (pp. 89–103). Toulouse, France: Ecole Nationale de Formation Agronomique.

Duit, R., & Treagust, D. F. (2003). Conceptual change: A powerful framework for improving science teaching and learning. *International Journal of Science Education, 25*(6), 671–688.

Grayson, D. J., Anderson, T. R., & Crossley, L. G. (2001). A four-level framework for identifying and classifying student conceptual and reasoning difficulties. *International Journal of Science Education, 23*(6), 611–622.

Gupthar, A. S., & Anderson T. R. (2003, January 11–15). Student difficulties associated with DNA-strand nomenclature and function. In *Proceedings of the eleventh annual meeting of the Southern African Association for Research in Mathematics, Science and Technology Education* (Vol. 1, pp. 361–365), Mbabane, Swaziland.

Hull, T. L. (2004). *Students' use of diagrams for the visualisation of biochemical processes.* Unpublished M.Sc. thesis, University of KwaZulu-Natal, Berea, South Africa.

Mayer, R. E. (2002). Rote versus meaningful learning. *Theory into Practice, 41*(4), 226–232.

Quentin-Baxter, M., & Dewhurst, D. (1992). An interactive computer-based alternative to performing a rat dissection in the classroom. *Journal of Biological Education, 26*(1), 27–33.

Schönborn, K. J., & Anderson, T. R. (2006). The importance of visual literacy in the education of biochemists. *Biochemistry and Molecular Biology Education, 34*(2), 94–102.

Schönborn, K. J., & Anderson, T. R. (2008). Bridging the educational research-teaching practice gap: Conceptual understanding, part 2: Assessing and developing student knowledge. *Biochemistry and Molecular Biology Education, 36*(5), 372–379.

Schönborn, K. J., & Anderson, T. R. (2009). A model of factors determining students' ability to interpret external representations in biochemistry. *International Journal of Science Education, 31*(2), 193–232.

Schönborn, K. J., & Anderson, T. R. (2010). Bridging the educational research-teaching practice gap: Foundations for assessing and developing biochemistry students' visual literacy. *Biochemistry and Molecular Biology Education, 38*(5), 347–354.

Schönborn, K. J., Anderson, T. R., & Grayson, D. J. (2002). Student difficulties with the interpretation of a textbook diagram of immunoglobulin G (IgG). *Biochemistry and Molecular Biology Education, 30*(2), 93–97.

Schönborn, K. J., & Bögeholz, S. (2009). Knowledge transfer in biology and translation across external representations: Experts' views and challenges for learning. *International Journal of Science and Mathematics Education, 7*(5), 931–955.

Scism, A. J. (1996). *The biochemistry student companion.* Upper Saddle River, NJ: Prentice Hall.

Strömdahl, H. R. (2012). On discerning critical elements, relationships and shifts in attaining scientific terms: The challenge of polysemy/homonymy and reference. *Science Education, 21*(1), 55–85. doi:10.1007/s11191-010-9335-y.

Tsui, C.-Y., & Treagust, D. F. (2003). Genetics reasoning with multiple external representations. *Research in Science Education, 33*(1), 111–135.

Wright, D. (1989). *Human biology.* Oxford, UK: Heinemann Educational.

Chapter 3
Pictures in Biology Education

Wolff-Michael Roth and Lilian Pozzer-Ardenghi

Introduction

We live in a visual culture and images abound not only in the media generally but also in scientific texts and scientific popularizations more specifically. The adage "a picture is worth a thousand words" renders the perception of the power of the image in communication. Although communication is a social phenomenon, the general and dominant approach to the study of images today is cognitive psychology, which investigates mental capacities for processing images. Research in this area increasingly focuses on multimedia contexts, where, in the context of ever-more-powerful computers and software, images may be interactive (Rasch & Schnotz, 2009). This line of research shows a mitigated effect of images on learning, where some scholars claim that they assist learning (Meyer, 2001), whereas others suggest that there is an interaction effect so that the copresence of pictures and text may not necessarily lead to differential effects on learning (Rasch & Schnotz, 2009; Sweller, 2005).

One aspect that all these studies assume is that the perception of the images/pictures is unproblematic and that the participants in the studies have developed the competencies to read images in the same way they also read text. The lack of attention to requirements and competencies of reading images also is prevalent in teaching, where teachers and textbooks appear to assume that students do read pictures in the way intended by the authors/presenters (Pozzer-Ardenghi & Roth, 2005a). Our own research showed that pictures are read in different ways, and there

W.-M. Roth (✉)
Applied Cognitive Science, University of Victoria, Victoria, BC, V8W 7G6, Canada
e-mail: wolffmichael.roth@gmail.com

L. Pozzer-Ardenghi
Department of Curriculum, Teaching and Learning, Faculty of Education
University of Manitoba Winnipeg, Canada
e-mail: Lilian.Pozzer-Ardenghi@ad.umanitoba.ca

D.F. Treagust and C.-Y. Tsui (eds.), *Multiple Representations in Biological Education*,
Models and Modeling in Science Education 7, DOI 10.1007/978-94-007-4192-8_3,
© Springer Science+Business Media B.V. 2013

are simple reasons for images/pictures to give rise to many different figure/ground constellations independent of and prior to any interpretive issues.

A very different approach to inscriptions generally and pictures more specifically can take reading these visual representations as aspects of social practices, which require forms of apprenticeship for their acquisition (Roth & Bowen, 1994; Roth & McGinn, 1998). In contrast to seeking competencies somewhere in the mind and its multiple ways of coding, social practice research—anthropology, social psychology, phenomenology, pragmatic philosophy—focuses on the way in which pictures (inscriptions) are used. The methodological implications of the practice approach reading inscriptions are as follows:

> Before attributing any special quality to the mind or to the method of people, let us examine first the many ways through which inscriptions are gathered, combined, tied together, and sent back. Only if there is something unexplained once the networks have been studied shall we start to speak of cognitive factors (Latour, 1987, p. 258).

This method, though developed independently in anthropology/philosophy, actually is consistent with the social-psychological approach that we sketch and exemplify below. The social-psychological approach fundamentally locates all higher cognitive functions in social interactions both historically—reading and imaging evolved as cultural practices—and ontogenetically, in the development of the individual child who participates in reading books with naturalistic features, watching TV shows on natural history, or looking up information on the Internet.

The purpose of this chapter is to articulate a social-psychological approach to the role of inscriptions (pictures including photographs, lifelike drawings, and diagrams) in biology education. We begin by situating the general approach, which focuses on the public and therefore objective and shared nature of inscriptions generally and pictures more specifically. We then exemplify and review the existing research on the social nature of reading pictures. We provide exemplary analyses that show how reading pictures emerges from social interactions. We conclude by drawing pedagogical implications for biology education.

Pictures in the Continuum of Inscriptions

Pictures are counted among *inscriptions*, a category that includes all forms of representing information other than language. However, the resemblance between a particular phenomenon and the various inscriptions that can represent this phenomenon differs. Some contain more contextual information and are more specific compared to others. For example, the photograph of a lotus flower in Fig. 3.1 bears a strong resemblance with the original flower. We see its colors, some of its leaves, and part of the lake in which it was found. On the other hand, a diagram of a lotus flower retains only the most outstanding characteristics and, therefore, may represent any generalized lotus flower of a particular species. An equation showing the height of a lotus flower over time does not bear any iconic relationship with the

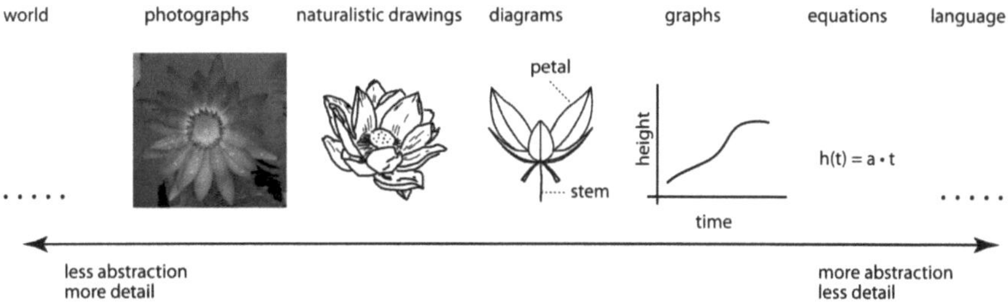

Fig. 3.1 Inscriptions representing a lotus flower fall along a continuum of abstraction

lotus flower even though it expresses a characteristic of a flower or—if many flowers have been sampled—the mean height. Therefore, inscriptions can be thought to exist along a continuum of different types of representations (Roth & Bowen, 1999b; Latour, 1993). Photographs and pictures (e.g., naturalistic drawings, paintings) are at one extreme and contain a lot of other often-irrelevant, gratuitous information. To read the transformation between any two neighboring inscriptions requires learning (social practices).

The gratuitous information comes both as an advantage and as a shortcoming. On the one hand, this information makes photographs and paintings more lifelike. On the other hand, precisely because the inscription is more lifelike, it is also more specific and interferes with the learning of concepts. Thus, for example, a study among bird watchers shows that field guides with figures that lie somewhere between naturalistic drawings and diagrams in Fig. 3.1 lend themselves more to the learning of bird identification than guides that contain photographs (Law & Lynch, 1990).

It is therefore not surprising that in biology journals, there are less photographs than in high school biology textbooks (Roth, Bowen, & McGinn, 1999). Moreover, journals focusing on theoretical or molecular aspects of biology tend to have no photographs at all, whereas journals such as the *Journal of Natural History* include many photographs, naturalistic drawings, and diagrams. For example, an article on the morphology of the mouthparts of a crustacean includes 8 plates with a total of 41 photographs, some of which are composed of 2–5 individual photos (Nickel, Atkinson, & Pinn, 1998). Furthermore, the photographs that are used in scientific publications may be different from those used in textbooks for popularization and teaching because in the sciences, their veridical nature is of importance, whereas in the popularized version, the aesthetic value is emphasized (Lynch & Edgerton, 1988).

The gratuitous information actually allows readers of photographs to focus on aspects that are present but irrelevant to the phenomenon or argument, and therefore, this results in alternative readings (Bastide, 1990). Thus, it is not surprising that when scientists publish photographs, these come with extensive captions that exhibit just what is to be seen (Roth et al., 1999). This also points us to the potential need to develop specific pedagogies for reading, interpreting, and deconstructing photographs in the biology classroom—pedagogies that are currently underdeveloped (Roth & Bowen, 1999a).

Pictures in Printed and Online Media

Over the past 15 years, we have conducted different studies concerning photographs accompanying text in informal and formal print and online media. These included North American high school (Roth et al., 1999), Brazilian high school (Pozzer & Roth, 2003), Korean middle school textbooks (Han & Roth, 2006), and online media (Roth, 2010a, 2010b). As part of this work, we developed a framework that articulates the resources that textbooks and online media make available for reading generally and for reading (photographic) inscriptions more specifically.

The resources are only one part of the social practices of reading photographs but also the work of reading. Together, work and resources constitute a pair of the structure of *doing [notational particulars]* (Garfinkel & Sacks, 1986). Here *doing* refers to the lived work of reading, whereas *[notational particulars]* refers to resources/structures mobilized in and by reading. Thus, in reading, the words and image aspects are notational particulars that are mobilized in/by reading. This work tends to be invisible—which is possibly why it is so little attended to and accepted as going without saying. But it is precisely by participating in such work that human beings learn the practices of reading generally (Livingston, 1995) and the practices of reading inscriptions more specifically (Roth, 2003). This is precisely why anthropological methods are well suited for the study of reading photographs and pictures and why the framework proposed by Vygotsky (1989) is well suited for theorizing the learning of reading practices.

Analysis of Photographs in High School Textbooks

In our analysis of high school biology textbooks, we found four different categories of photographs according to their functions: decorative, illustrative, explanatory, and complementary (Pozzer & Roth, 2003). Decorative photographs appear without a caption and are deictic references in the text; that is, these photographs stand on their own and their content is not further specified or explicitly related to the main text. Illustrative photographs include captions that name the object or phenomenon depicted but no other information is provided. Explanatory photographs include a caption that names an object or phenomenon and, in addition, provides further contextual information about the object or phenomenon. Finally, complementary photographs include captions that name the object/phenomenon, provide explanation or classifications of this object/phenomenon, and contain new information about it not available in the main text.

The absence of text directly associated with the decorative photographs makes them more difficult to interpret. In our study of students reading pages of textbooks with different categories of photographs (Pozzer-Ardenghi & Roth, 2005a), we showed how students attempted to connect the decorative photograph with the text available on the same page. Even though students were able to name the object depicted in the photograph, their interpretation of what the photograph was representing in terms of scientific concepts varied widely. When reading a page where the representations in a photograph and the text were explicitly associated through the caption and indexical reference, students easily identified the topic of the photograph and their conceptual interpretations were overall very similar. Moreover, students did use the indexical reference to switch attention between reading the text and reading the photograph, and their interpretation of what to see in the photograph changed according to the amount and type of text provided. Thus, readers interpret photographs with different functions differently; the amount and type of text accompanying each photograph alters the readers' interpretation of what is to be seen in the photograph.

Our study also showed that students always try to associate photographs and text, even when this association is not explicitly available, as in the case of decorative photographs. Yet, when textual information is not available, most readers rely on conventions of perspective (focus, size, and color) to interpret the photographs. When texts are provided, however, their interpretation changes and they more confidently name or point out the object represented in the photograph. Thus, photographs and texts mutually inform each other in the readers' interpretation while they are reading texts and pictures in science textbooks.

Our work on photographs in Korean science textbooks showed that additional text and inscriptions may be layered upon photographs (Roth, Pozzer-Ardenghi, & Han, 2005). In this situation, the photograph constitutes a resource for making a link between a scientific inscription such as a table, graph, or formula, on the one hand, and students' experiences in the everyday lifeworld (*Lebenswelt*), on the other hand. However, these additional resources also require additional work, such as *structuring* the inscriptions (including the differentiation between layers), *transposing* between inscriptions, and *translating* between the different layers of inscriptions. A simple arrow superposed in a photograph is one kind of additional resource that scientists often use as an aid to their readers for identifying the phenomenon of interest. Lettering of parts, information about the scale size, and even the energy and image number from the apparatus used to produce the picture may be provided in scientific publications to aid readers in their interpretive work. Without these aids—which teach the reader how to *read* the picture—students reading textbooks and attending lectures (where images are presented) might look at these images in ways not intended by the authors/lecturers. This might undermine the potential pedagogical value of these inscriptions.

Pictures in Lectures

When pictures are used in lectures, gestures provide a particular resource for identifying relevant features in the inscription, thereby becoming part of the pedagogy that helps students learn how to read images. Thus, these gestures can be understood as resources similar to captions and other layered materials that assist in describing the

Fig. 3.2 In lectures, gestures and body position provide resources for perceiving relevant information and learning how to read images

content and serve as pedagogy for helping students to identify what is relevant (Roth & Bowen, 1999a). Moreover, the gestures also add other dimensions to the two-dimensional, static pictures, aiding in explaining details that are not visible in the picture or drawing and also providing a dynamic feature of the concept or object represented in the inscription.

In particular, drawings are intrinsically related to the sequential presentation of content during lectures; and they are also part of an integrated unit—of speech, gesture, and inscription, which evolve simultaneously with the lesson on a scientific concept—so that the final imagery produced is quite complex and represents several sequential and interdependent stages of communicating a scientific concept. Consider, for example, Fig. 3.2. During a lesson on the excretory system, the teacher is lecturing about glomerular filtration; he draws on the board a schematic drawing of a nephron and then identifies its structures and their functions in the filtration process. This identification occurs through the simultaneous employment of multiple resources: speech, through which he names the structures and their functions; drawing, which he completes as he moves along with the lecture, adding complexity by drawing new structures and words that label them; and through his gestures, he is pointing toward the structures drawn on the board and referred to in his speech. He is also performing iconic gestures that represent the same structures from a different perspective, which is not available in the two-dimensional drawing.

The leftmost image on Fig. 3.2 shows the teacher performing an iconic gesture that *represents* a feature or property of the object drawn on the board, which is not available in the drawing itself; the teacher uses both his hands to represent the three-dimensional shape of the glomerulus while saying, "this glomerulus is surrounded by a capsule, sort of a, I've only drawn it in two-dimensions but obviously it gets three-dimensions, so it is enclosed by this capsule."

While representing the three-dimensional shape of the glomerulus, the teacher is positioned toward the students, and the drawing of the nephron where the glomerulus is represented in two dimensions stays in the background. After acknowledging through his speech that the drawing presents only two dimensions of the structure, the teacher moves closer to the board and gestures over the drawing, *emphasizing* the glomerulus and its three-dimensionality with both his hands. By performing these gestures, the teacher is in fact teaching students how to look at and interpret the drawing: as a two-dimensional (and, therefore, inherently, incomplete) representation of a three-dimensional structure. Moreover, his gestures are connecting his speech—and the scientific names he is introducing to students for the structures of the nephron—with the drawing, which is another way of representing (i.e., another *signifier* of) these structures. Thus, words, drawings, and gestures represent the same structure in different ways and are all connected within the same meaning unit to communicate a scientific concept to students during the lecture. As the teacher continues to develop the concept, introducing new information and new names and new structures, the drawing also evolves, becoming more and more complex. However, because speech, gestures, and drawings are integrated, this complexity reflects the sequential unfolding of the communication (i.e., *teaching*) of the concept.

Drawing on lectures in Canadian middle schools, universities, and research centers, we derived a classification of the functions gestures have with respect to reading and understanding images used in lectures (Pozzer-Ardenghi & Roth, 2005b). Our classification includes eight different functions of gestures and body orientations produced as semiotic resources for interpreting pictures: (a) representing, (b) emphasizing, (c) highlighting, (d) pointing, (e) outlining, (f) adding, (g) extending, and (h) positioning. These functions vary according to the lecturer's position in relation to the photograph and the specificity of the gesture performed.

(a) Representing gestures include those used to represent objects/phenomena not directly available in the picture; usually, lecturers use gestures to represent three-dimensional aspects of the object depicted in the two-dimensional picture, like in the first example in Fig. 3.2. Because the entity is not available in the picture, the lecturer is usually positioned toward the audience when performing representing gestures.
(b) Emphasizing gestures focus on a particular feature of the image by generally mimicking the shape of the object of interest, thus bringing this aspect to the foreground and turning everything else to the background; when the teacher in Fig. 3.2 gets closer to the board and gestures over the drawing, he is emphasizing a particular area on the drawing (the glomerulus) and a particular feature of it (its three-dimensionality).
(c) Highlighting gestures are also general gestures made over the pictures to call students' attention to an area or object(s) on it.
(d) In contrast, pointing is a very specific gesture that pinpoints a specific aspect of the object/phenomenon represented in the picture.
(e) Outlining is also specific and refers to tracing the outline of the object to make it salient.
(f) Adding gestures use the picture as the background upon which a phenomenon not available in the photograph is brought to bear. It is usually used to represent dynamic processes, rather than static structures.
(g) Extending occurs when the gestures are performed beyond the visible limits of the image, thus effectively transcending the borders of the printed or projected image; in this sense, these gestures add another dimension to the pictures, bringing the depicted environment/larger object to the classroom.

(h) Finally, positioning makes the role of the human subject in the production of the image evident, by showing the angle and particular view depicted in the image and therefore teaching how to *see* the picture.

Our research program as a whole therefore shows that in addition to being the topic of talk—in the case of which making something salient or emphasizing it is the main function—a picture may also be the background that supports the addition of other forms of information not directly available in it but relevant to understanding the phenomenon of interest. The visible aspects of the phenomena might be augmented or contextualized by information available only in a third dimension—that would happen outside the current frame—or even by dynamic processes that cannot be represented in the static image but that are added to it through associated words and gestures (Roth & Bowen, 1999a).

Social Origin of Picture Content

The question about how we learn to read generally and how we come to read (photographic) images more specifically ought to be of more interest to biology educators than it currently is. When the question is approached from a constructivist or cognitive perspective, then the problem of learning to read is relegated to the mind. It takes the concrete sensorimotor schema and abstracts from it patterns of action that are used at a formal level. In both instances, at the concrete and formal level, the mind has to test what it does. A radically different approach was developed in Marxist social psychology in the Soviet Union. It is based on the fact that

> any higher psychological function was external; this means that it was social; before becoming a function, it was the social relation between people. The means of acting upon oneself is first a means of acting on others and the action of others on one's personality (Vygotsky, 1989, p. 56).

This fact was beautifully illustrated in the work with deaf-blind children in the Soviet Union, who, contrary to the contentions of scholars such as Jean Piaget and Ernst von Glasersfeld, did not have an investigatory reflex and did not construct anything of any order. They vegetated and resisted even the touch of other people (Meshcheryakov, 1979). Mediated by social interaction, these children were literally awakened to life and, in some cases, even became university professors. Here, too, reading was something that exists for the child as social relation first, generally with the parent, and something that is observed as enacted by the child on its own only subsequently. General pedagogical practices concerning images in the biology classroom appear to assume children and older students to read images on their own.

Not only does children's or students' reading of images emerge as a social relation but adults' reading of images also is socially related; we have been able to document these social relations during a 5-year ethnographic study of one science laboratory (Roth, 2009). Those new to the laboratories engage in social interaction with older members; and it is precisely from these social interactions that competent reading emerges. The following example is from this ethnographic study.

Scientists Learn to Read Photographic Images

During this ethnographic study of a biology laboratory investigating the absorption of light in the retina of salmonid fishes, the composition of laboratory members changed repeatedly. Initially, two physicists joined the investigation focusing on data collection, apparatus, and the processing of the data. There were also a post doctoral fellow, a PhD student, and an MSc student who participated at various stages and to varying extent in this research. Each time a new person joined, those already familiar with the experimentation introduced others on every aspect, including how to read both the photographic images of the material on the microscopic slides and the graphs derived from the absorption measurements. In this particular transcript fragment alone, there are three instances where the professor, who had done already more than 20 years of research on fish vision, and the two newcomers enacted social relations that pertained to the reading of photographic materials in the following transcript fragment:

01 C: <<p>it's an air bubble or something in there>
02 (23)
03 T: <<p>nothing for photo .h ha>
04 C: uh no here we go we'll change the sample * pretty soon I don't know this looks a little
05 T: yea
06 03

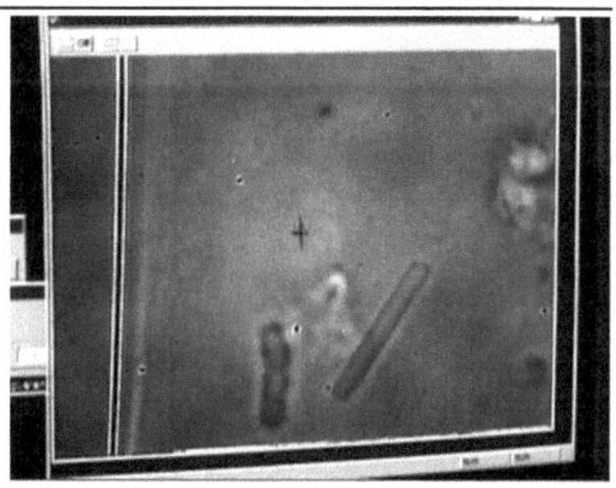

07 T: pretty good for a first try
08 C: yea it's beautiful okay yea it's the camera
09 T: yea
10 (6)
11 C: yea () big blob up there
12 (14)
13 C: I didn't get really good bushes so I am wondering*

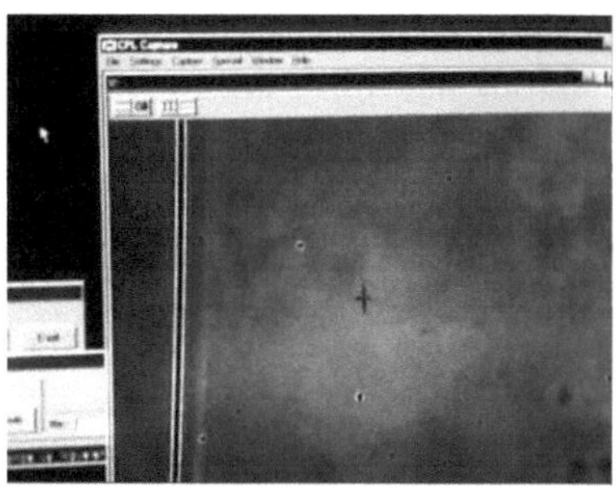

(continued)

(continued)

14 T: yea
15 C: * there is an air bubble ((focuses in and out))
16 (24) ((continues to move the microscopic slide))

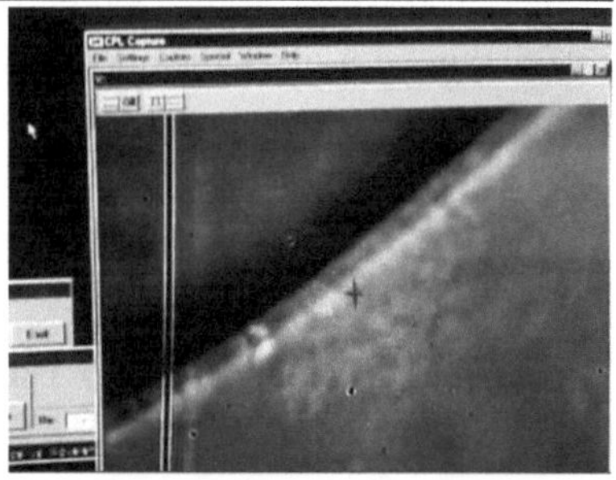

17 M: do you think it has to do with remember that one blob that came so (0.3) off so difficult from the epithelium
18 (1.5)
19 C: yea (1) well (1.5) I think (2.0) one thing that happened with this sample is * this is a cone right here a double cone * you see
20 M: uh hm

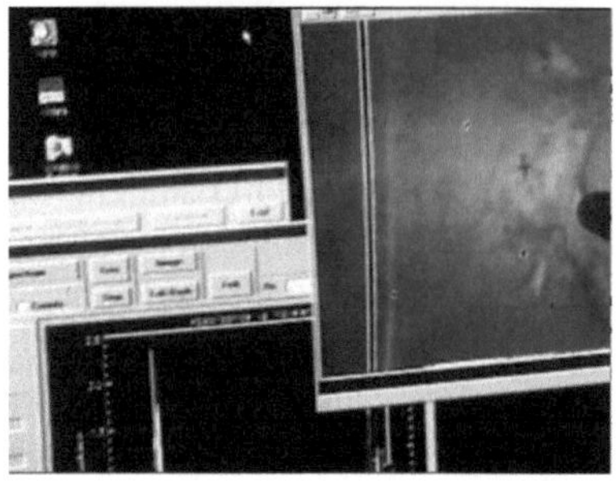

21 C: the dividing * ((moves back and forth on the "dividing line"))
22 M: uh hm
23 (5)
24 C: not very good shape

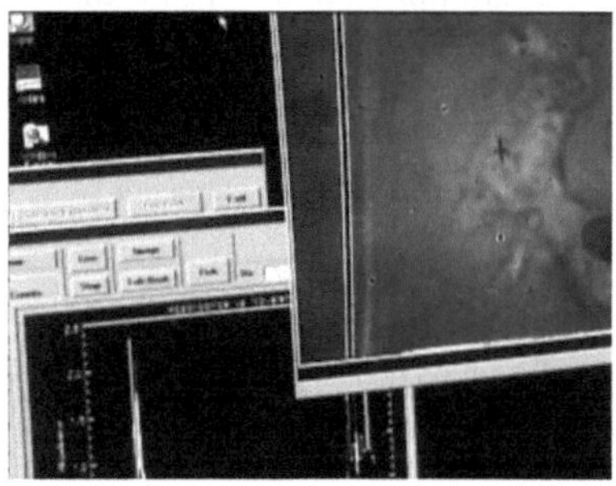

In this transcript fragment, we observe very long speaking pauses. At the same time, the images on the monitor are continually changing. In fact, one study showed that the members of this laboratory only tended to speak when there was something relevant and when there was the potential for misunderstanding (Roth, 2004). As long as the members could presuppose intersubjectivity with respect to their assessment of the objects/processes apparent on the display, no talk was recorded. The moments when they were talking therefore indicate moments of significance. That is, the very fact that the professor articulates the names of these objects—as they come into focus and do not name other features—is a sign that they are significant and those present ought to note them.

The professor suggests changing the sample and then begins a sentence about how "this" looks. T suggests that "pretty good for a first try" (turn 07). The professor at first appears to accept this assessment ("yea"), then articulates something about a "blob up there," and then makes a negative assessment ("I didn't get really good bushes"). In this instance, the positive assessment that the physicist research associate provides contrasts with the negative assessment that the director of the laboratory articulates. The "yea" (turn 11) does not in fact create a tension, for it can be heard as an acknowledgment of attention; and having heard the preceding assessment immediately followed by a reframing/rearticulation of the assessment is more suited to the situation at hand. There are other indicators that the situation is not as favorable as T's assessment makes it appear. The professor talks about air bubbles, a big blob, and about changing the sample (which he only does when there are no more suitable photoreceptors for making measurements). M in fact formulates an earlier occurrence, where some "blob" "came off so difficult" "the epithelium" (turn 17).

Biology professor C continues by formulating that he is wondering about something and subsequently describes an air bubble to be there just as the image on the monitor changes and shows a sharply defined but slightly bent entity. That is, the thing that just at that time comes into focus is named an "air bubble" (turn 15). Although it appears as if this corresponds to the illustrative function, where a name and a thing/phenomenon come to be associated with each other, this is not so. In this instance, the image changes from fuzzy (turn 13) to a sharply defined image (turn 15), which subsequently disappears again. Something has come into sharp focus, clearly defined; and it is this something that therefore stands out as salient. Coming into sharp focus, therefore, is equivalent to the indexical and iconic functions of emphasizing, pointing, and outlining gestures. It is an opportunity where physicists M and T learn to identify the image of an air bubble on the microscopic slide.

In this instance, biology professor C teaches the two physicists how to see and recognize double cones as distinct from single cones and rods. His gestures are an integral part of this communicative interaction, whereby he not only points but actually traces along the feature that he names the dividing line that we learn to see as the separation between the two cones. Here, the gesture has a double function of both pointing to and outlining the dividing line between the two members of a double cone, thereby emphasizing what might have escaped the attention of the two physicists in the room. The identification may actually not be so easy depending on the way in which the phenomenon is placed on the slide. For example, when the double cone is on its side, it looks like a single cone. The professor may actually be heard initially to articulate seeing a cone and then he revises the description to a double cone. Moreover, it is not just that he perceives and then interprets the photo to depict a double cone but rather, the difference between perception of the feature and thinking about the feature is undecidable (Wittgenstein, 1958). That is, in such social interactions, newcomers not only find out about what to see in a particular image, but their perception is changed—in contrast to the constancy hypothesis that Piaget upheld (Gurwitsch, 2010)—and with it, their forms and content of thinking.

Implications for Biology Curriculum

In the previous sections, we observe how salience, objects/phenomena, and orientation are the results of social interaction. What is to be seen in (photographic) pictures generally and what a particular participant sees more specifically is the result of social interactions rather than something that can be given. Similarly, assessments of the content of photographic images and the nature (name) of the salient objects are the result of social interactions. In this brief fragment alone, the professor reframes an assessment and then points to and names particular features on the display while leaving others that pass as he scans across the slide without commenting on their nature. We have reported very similar aspects in the classification of scientific objects in very different situations, including fish hatcheries, fieldwork in ecology, and everyday work of fish biologists (Roth, 2005). In each case, what is to be seen emerges from and is the result of a social relation.

Perception actually is a complex process involving the movements of the eyes back and forth between a figure and its background, on the one hand, and along particular features of the image, on the other hand (Roth, 2011). Perception therefore is not something that is merely constructed in our heads but is a process that emerges from the relation between the eyes and the world outside of the organism (Marion, 2004). To us this implies that biology teachers ought not to assume that their students see in photographic images in the way intended by the curriculum. Moreover, rather than assuming that students learn to read by constructing something in their minds or by abstracting from their first attempts in looking at images, the focus on social relations attributes a major role to student-student and student-teacher interactions as the very source for perceiving images in didactically useful and appropriate ways.

We have shown in this chapter that gestures play an important role in the social relations from which emerges the reading of pictures as a social practice. There are multiple relations between the possible features of a photograph. What can be seen is not identical with what ought to be seen in the relevant instance. Sometimes, what ought to be seen may be absent in didactic materials (biology textbooks), and students need to learn to interrogate these features. For example, there is no way of ascertaining the size of a small insect; students ought to learn to ask questions about it. Furthermore, they need to learn asking questions as to the nature of the habitat, which in part can be taken from the context in which the small insect appears—the leaf. Other questions that may be asked concern the overall frequency of occurrence. Our theoretical framework suggests that students learn such social practices by engaging in them with others, most notably their teacher but also with their peers.

Social interactions are important because the work of reading is not otherwise available to children and newcomers. That is, although the resources that pictures and the accompanying forms of text provided may be thought of as being out there and therefore available for picture-learner interactions, their significance is available only in the social practice itself. What is to be communicated essentially is a definite mode of perception, which, as in sociological research parlance, consists of "a set of principles of vision and di-vision" (Bourdieu, 1992, p. 222). These principles cannot be learned other than by seeing or experiencing them in practical operation while doing what is to be done alongside another person.

Acknowledgments The research in this chapter was made possible by grants to both authors from the Social Sciences and Humanities Research Council of Canada.

Appendix: Transcription Conventions

We used the transcription convention below, following common usage in conversation analysis:

(3)	Time in seconds, including accuracy
=	Two phonemes are linked
((points))	Transcriber's descriptions of actions are enclosed in double parentheses
*	Asterisk aligns verbal transcript and images
; . , ?	Punctuation marks intonation movement toward the end of the utterance as slightly down, strongly down, slightly up, and strongly up, respectively
∧ ˇ	Movement of intonation within a word up-down and down-up, respectively
<<p>> <<f>>	Speech intensity is piano or forte, respectively, that is, lower or higher (louder) than normal
[ ]	Square brackets align simultaneous features
guee:::n	Colon marks lengthening of phoneme, here "ee," by about 0.1 seconds per colon
.h	In-breath

References

Bastide, F. (1990). The iconography of scientific texts: Principles of analysis. In M. Lynch & S. Woolgar (Eds.), *Representation in scientific practice* (pp. 187–229). Cambridge, MA: MIT Press.

Bourdieu, P. (1992). The practice of reflexive sociology (The Paris workshop). In P. Bourdieu & L. J. D. Wacquant (Eds.), *An invitation to reflexive sociology* (pp. 216–260). Chicago: University of Chicago Press.

Garfinkel, H., & Sacks, H. (1986). On formal structures of practical action. In H. Garfinkel (Ed.), *Ethnomethodological studies of work* (pp. 160–193). London: Routledge & Kegan Paul.

Gurwitsch, A. (2010). *The collected works of Aron Gurwitsch (1901–1973) vol. III – The field of consciousness: Phenomenology of theme, thematic field, and marginal consciousness.* Dordrecht, The Netherlands: Springer.

Han, J., & Roth, W.-M. (2006). Chemical inscriptions in Korean textbooks: Semiotics of macro- and microworld. *Science Education, 90*, 173–201.

Latour, B. (1987). *Science in action: How to follow scientists and engineers through society.* Milton Keynes, UK: Open University Press.

Latour, B. (1993). *La clef de Berlin et autres leçons d'un amateur de sciences* [The key to Berlin and other lessons of a science lover]. Paris: Éditions la Découverte.
Law, J., & Lynch, M. (1990). Lists, field guides, and the descriptive organization of seeing: Birdwatching as an exemplary observational activity. In M. Lynch & S. Woolgar (Eds.), *Representation in scientific practice* (pp. 267–299). Cambridge, MA: MIT Press.
Livingston, E. (1995). *An anthropology of reading*. Bloomington, IN/Indianapolis, IN: Indiana University Press.
Lynch, M., & Edgerton, S. Y. (1988). Aesthetics and digital image processing: Representational craft in contemporary astronomy. In G. Fyfe & J. Law (Eds.), *Picturing power: Visual depiction and social relations* (pp. 184–220). London: Routledge.
Marion, J.-L. (2004). *The crossing of the visible*. Stanford, CA: Stanford University Press.
Meshcheryakov, A. (1979). *Awakening to life: On the education of deaf-blind children in the Soviet Union*. Moscow: Progress.
Meyer, R. (2001). *Multimedia learning*. Cambridge, UK: Cambridge University Press.
Nickel, L. A., Atkinson, R. J. A., & Pinn, E. H. (1998). Morphology of thalassinidean (Crustacea: Decapoda) mouthparts and periopods in relation to feeding, ecology, and grooming. *Journal of Natural History, 32*, 733–761.
Rasch, T., & Schnotz, W. (2009). Interactive and non-interactive pictures in multimedia learning environments: Effects on learning outcomes and learning efficiency. *Learning and Instruction, 19*, 411–422.
Pozzer, L. L., & Roth, W.-M. (2003). Prevalence, function, and structure of photographs in high school biology textbooks. *Journal of Research in Science Teaching, 40*, 1089–1114.
Pozzer-Ardenghi, L., & Roth, W.-M. (2005a). Making sense of photographs. *Science Education, 89*, 219–241.
Pozzer-Ardenghi, L., & Roth, W.-M. (2005b). Photographs in lectures: Gestures as meaning-making resources. *Linguistics & Education, 15*, 275–293.
Roth, W.-M. (2003). *Toward an anthropology of graphing*. Dordrecht, The Netherlands: Kluwer Academic Publishers.
Roth, W.-M. (2004). Perceptual gestalts in workplace communication. *Journal of Pragmatics, 36*, 1037–1069.
Roth, W.-M. (2005). Making classifications (at) work: Ordering practices in science. *Social Studies of Science, 35*, 581–621.
Roth, W.-M. (2009). Radical uncertainty in scientific discovery work. *Science, Technology & Human Values, 34*, 313–336.
Roth, W.-M. (2010a). An anthropology of reading science texts in online media. *Semiotica, 182*, 409–442.
Roth, W.-M. (2010b). A social psychological reading multimodal scientific texts in online media. *Reading Psychology, 31*, 254–281.
Roth, W.-M. (2011). *Passibility: At the limits of the constructivist metaphor*. Dordrecht, The Netherlands: Springer.
Roth, W.-M., & Bowen, G. M. (1994). Mathematization of experience in a grade 8 open-inquiry environment: An introduction to the representational practices of science. *Journal of Research in Science Teaching, 31*, 293–318.
Roth, W.-M., & Bowen, G. M. (1999a). Complexities of graphical representations during lectures: A phenomenological approach. *Learning and Instruction, 9*, 235–255.
Roth, W.-M., & Bowen, G. M. (1999b). Digitizing lizards or the topology of vision in ecological fieldwork. *Social Studies of Science, 29*, 719–764.
Roth, W.-M., Bowen, G. M., & McGinn, M. K. (1999). Differences in graph-related practices between high school biology textbooks and scientific ecology journals. *Journal of Research in Science Teaching, 36*, 977–1019.
Roth, W.-M., & McGinn, M. K. (1998). Inscriptions: A social practice approach to "representations. *Review of Educational Research, 68*, 35–59.

Roth, W.-M., Pozzer-Ardenghi, L., & Han, J. (2005). *Critical graphicacy: Understanding visual representation practices in school science*. Dordrecht, The Netherlands: Springer-Kluwer.
Sweller, J. (2005). Implications of cognitive load theory for multimedia learning. In R. E. Mayer (Ed.), *The Cambridge handbook of multimedia learning* (pp. 19–30). Cambridge, UK: Cambridge University Press.
Vygotsky, L. (1989). Concrete human psychology. *Soviet Psychology, 27*(2), 53–77.
Wittgenstein, L. (1958). *Philosophische Untersuchungen* [Philosophical investigations] (3rd ed.). New York: Macmillan.

Chapter 4
Possible Constraints of Visualization in Biology: Challenges in Learning with Multiple Representations

Billie Eilam

Introduction

Learning with multiple representations involving visual representations has long been proven empirically in particular conditions to promote students' construction of knowledge, understanding, and transfer of the represented information (Mayer, 2005). Accordingly, such learning involves the processing of the visual external representations—a stimulus perceived from the multiple representations—for the construction of internal representations. These internal representations may include mental or visual imagery, internal mental models, memory, or knowledge representations that are broader than the mere description of the perceived stimuli and include individuals' prior knowledge concerning the represented information. The value of these constructed internal representations lies in the individual's ability to store them in memory for future use and application (Hegarty, 2004; Rapp & Kurby, 2008).

Displays of external representations in the visual modality alone, such as texts and images, constitute physical representations that are created and exist outside an individual's mind, on a paper or computer screen or other materials, with the aim of augmenting and enhancing human cognition (Tversky, 2005). For example, for learning purposes, external visual representations may free mental resources for processing information because when learners can rely on such an external image, they do not need to maintain the image in working memory. These physical, external entities may be shared among individuals for purposes of learning, discussion, communication, and so forth, by making thoughts and abstract ideas visible (Rapp & Kurby, 2008).

B. Eilam (✉)
Head of the Department of Learning, Instruction and Teacher Education, Faculty of Education, University of Haifa, Mount Carmel, Haifa 31905, Israel
e-mail: beilam@edu.haifa.ac.il

D.F. Treagust and C.-Y. Tsui (eds.), *Multiple Representations in Biological Education*, Models and Modeling in Science Education 7, DOI 10.1007/978-94-007-4192-8_4,

© Springer Science+Business Media B.V. 2013

Hegarty (2004) suggested the existence of the interplay between the processes involved in the comprehension of an external visual representation and the processes involved in the construction of an internal representation of the external one. Accordingly, learning with multiple representations—involving visual representations—is based on the combination of perceptions of external visual representations, internal visualization processes, and storage of constructed internal representations for future applications and visual thinking. It is therefore important to develop students' abilities related to both internal visualization and external representations. One way to accomplish this would be to teach students to read, comprehend, manipulate, and create external visual representations, and this would require an effective pedagogy (Hegarty, 2004; Reed, 2010).

Despite the reported advantages of visual representations, the complex learning environment of multiple representations may hinder students' learning. Difficulties are usually attributed to one or more of four main sources: (a) the *learners' characteristics* such as students' cognition—prior knowledge and abilities—regarding the content represented, the specific representation or multiple representations, and the technology used (e.g., Mayer & Massa, 2003; Molinari & Tapiero, 2007); (b) the *representation's characteristics* like complexity, structure, abstractness, cognitive load, attention, spatial arrangement, and ease of processing (e.g., Ainsworth, 1999; Mayer, 2005; Sweller, 1988); (c) the *characteristics of the pedagogy* applied to the visual representations, such as interactive versus passive learning, instructional approach, explicitness, instructional support, and training mode (e.g., Berthold & Renkl, 2009; Eysink et al., 2009); and (d) the difficulty least studied, the *contextual characteristics*, which may affect further learning processes, including aspects like students' sociocultural background, students' and teachers' related attitudes and beliefs, and wider local and global sociocultural contexts (e.g., Deregowski, 1989; Maes, Foesenek, & Hoogwegt, 2008). With regard to difficulties related to context, researchers indicated that learning and instruction are culturally embedded, and that to achieve successful learning outcomes, instructional practices and beliefs must entail *translation* into students' local culture and alignment with their beliefs and attitudes for learning (Hatano & Inagaki, 1994). I suggest an additional, fifth source of difficulty: (e) a particular visual representation's *status and placement in the horizontal and vertical school curriculum*, which can affect learning and interpretation; this difficulty refers to students' transfer of knowledge regarding a particular visual representation to the interpretation of another, which is inherently different (Eilam & Ben-Peretz, 2010).

Need for Effective Visualizations in the Biology Domain

Biological phenomena are complex—involving the concrete and abstract; the spatial, systemic, and temporal; the micro- and macrolevels; as well as two-dimensional (2-D) and three-dimensional (3-D) representations of life forms. Moreover, biological phenomena require examination at different points in time to show the temporal development of organisms or stages of processes as well as from different viewpoints such as longitudinal or cross-sectional view. Hence, biology learning and teaching may be expected to receive ample support from the well-documented advantages of external visualization—such as ways to complement text information, enable a holistic spatial view of phenomena, represent sequential stages and changes in a process, and deepen understanding. Indeed, learners commonly encounter visualizations within all four biology learning environments: (a) the natural world, (b) school lessons and laboratories (i.e., formal curricula and learning materials), (c) technology (e.g., both static and dynamic formal learning programs like simulations and visual information on the Internet), and (d) individuals' everyday lives and home environment (e.g., health brochures, media, books). Students gradually construct formal and informal biological knowledge through interactions with these learning environments.

These characteristics of the biology domain call for extensive use of visualization in schools, together with improvement in computer designs and development of more diverse and effective external visualizations (to include models, graphs, diagrams, schema, timelines, and more). In addition, biology's characteristics call for enhancement of teachers' visual literacy to efficaciously utilize these visual representations as well as greater consideration of teachers' knowledge in cognitive psychology to better target students' cognition, thinking, and knowledge processing.

In this chapter, I discuss the outcomes of empirical studies on the uses of static visualization in biology as well as data that I collected over several years from classroom observations of Israeli biology students and teachers and from learning with teaching aids and other learning materials. For each case of the studies and classroom observations, I discuss a specific type of representation and then relate it to its constraints for use within a specific context in the biology domain. I argue that only teachers' and curriculum developers' awareness of the problems may initiate processes to avoid the difficulties related to the aforementioned constraints. To help these professionals develop the systematic, critical perspective for designing effective static visual representations, I start by discussing illustrative or decorative representations and then present models, representations of processes, and referents of different scales. Next, I discuss students' tendency to process common representations superficially or by erroneously transferring knowledge about representations of similar shape and structure. Finally, I discuss confusing visualizations that teachers sometimes present to the class on the classroom board.

Decorations

Textbooks and learning materials are richly loaded with illustrative or decorative elements. These may initiate students' affective and cognitive responses such as aesthetic pleasure, drawing their attention, or raising their curiosity, but such elements play no role in promoting cognition. Some may contend that advantages (pros) like drawing learners' attention and raising their curiosity constitute adequate reasons for inserting such representations into learning materials. However, at times, certain unintentional disadvantages (cons) occur due to the presence of such representations in the materials, which may negatively affect students' learning. I next describe two such cases where the *cons* may outweigh the *pros*, both from the same junior-high biology textbook on feeding (in Hebrew)—Chapters on *Feeding in Humans and Plants* (Sivan et al., 1992).

Diet and Cholesterol

A chapter discussing organic materials presents information under the heading "Cholesterol – A Necessary Fat or a Harmful Material?" The information discusses cholesterol's characteristics, functions in the body, and hazards when present in large amounts in the blood. The information is presented within a box (a convention intended for emphasis), which also contains a humorous schematic drawing of an overweight woman standing on a scale checking her weight. Unfortunately, as indicated by the results in a test that seventh-grade students took on this learning unit, about 68% of students inaccurately linked cholesterol to being overweight. Although nowhere in the text were such relations mentioned, this datum suggests that the humorous drawing drew most of the students' attention and was easily but erroneously interpreted as indicating that only overweight people suffer from high cholesterol. This erroneous conclusion could have stemmed from (a) the perceptual contiguity (Mayer, 2005) between text and image (both in the same box), leading to their interpretation as directly related elements, and/or (b) the effortless interpretation of the depictive-iconic representation (Schnotz, 2005), which could have initiated only surface processing of the textual information instead of deeper processing. A third factor that could have affected students' test responses was their prior knowledge gained in the home environment through media, health brochures, or family discussions about dieting. Although the information of these images were intended to alert students to the hazards involved in eating too much cholesterol, many students ended up believing that these hazards should alarm overweight people only.

Opening Page of Book Chapters

In many textbooks, the first page of a new chapter is covered with different representations that are frequently related to the chapter's content. However, these representations activate students' related prior knowledge and initiate expectations regarding the upcoming chapter's content. More specifically, as students encounter the chapter's content, they may interpret it in the framework of the knowledge activated by the opening page or may construct interpretations not intended by the textbook developers.

Fig. 4.1 Feeding or color blindness?

In the same textbook about feeding described above, the opening page for a chapter discussing considerations for choosing healthy foods (*Smart Feeding*) shows a colorful, attractive photograph of fruits and vegetables (see Fig. 4.1). However, the presented photograph is one used for diagnosing red-green color blindness because the red items (darker items in Fig. 4.1) forming the number 25 are distinguishable from all the other items (mostly yellow and green – brighter items in Fig. 4.1) by color alone. Because most students are familiar with the idea of color-blindness tests, knowledge concerning this topic was activated, and students formed some expectations unrelated to food choices. Many students asked, confused, how color blindness was related to feeding, and a few coerced relations between the topics (e.g., commenting that there are 25 essential minerals and vitamins required by the human body). As can be seen, such a visual representation is far from being merely a decoration. It carries with it the possibility of hindering students' learning.

Models

A 3-D Model of Human Anatomy

Models are frequently used in biology. They may be iconic or conventional. For example, a conventional model of the human respiration system would exhibit

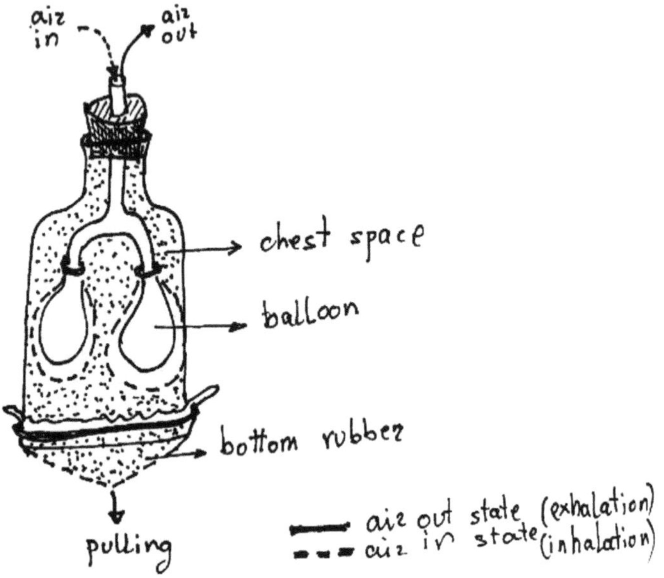

Fig. 4.2 A common 3-D bell jar model of the human respiratory system to demonstrate its function (inhalation and exhalation)

this system's parts for demonstrating its function, whereas an iconic model would exhibit the various inner organs as well as their relative size, exact location, and proportion in the body space as related to the size and locations of other organs. Such a model would constitute a *3-D map* of the body space. These models may be taken apart to examine the different organs and reassembled into the whole; hence, they may also serve to assess students' knowledge of inner anatomy. However, one aspect of the anatomy model may mislead learners, especially in the elementary and junior high school grades where it is most used—the colors and textures. Students erroneously perceive the colors used to represent the various parts of the model and the model's uniform texture to be as authentic as the model's 3-D structure. Frequently, the color conventions used in a model reflect considerations regarding salience or distinctiveness rather than authenticity.

A 3-D Model of Respiratory System Function

A conventionally used classroom model—which aims only at teaching the function of respiration (*inhalation* and *exhalation*) but not the lung structure—can lead to its own set of misconceptions on the part of learners. In one observed sixth-grade biology class, the teacher demonstrated the function of the human respiration using a common 3-D model of a bell jar (see Fig. 4.2). By pulling the bottom rubber (diaphragm) downward demonstrates inhalation—the space inside the bell jar (the chest) increases, its pressure decreases, and air is drawn into the balloons (the lungs) and fill them up, returning to the initial chest volume. Following the release of the bottom rubber it returns to its horizontal position demonstrates exhalation.

Unlike the iconic model of human anatomy, which resembles a real body in shape and proportion, this common bell jar model of the lungs is intended to teach only the function of the human respiratory system. To represent how pressures in the chest volume operate the lungs, this model only partially resembles the structure of the chest. For example, differences are evident in the structure's characteristics (e.g., the balloons are made of a single smooth unit, whereas the lungs are made of small bubbles—the alveoli; the tubes are stiff and inflexible, unlike the trachea). Two of the model's features may cause misunderstanding. First, the bell jar's side walls cannot move (unlike the ribs), suggesting erroneously that the increase in chest volume is caused only by the diaphragm. Second, the model's *diaphragm* is flat (horizontal) and must be pulled down to cause inhalation, whereas in reality the diaphragm is convex in shape and becomes horizontal through muscle contraction, causing inhalation by increasing the chest volume.

In a test for sixth graders ($n = 33$) about the respiratory system's function, the majority seemed to base their responses on the bell jar model, rather than using more accurate information that they discussed in the classroom and learned from the text in their textbook. In their test answers, 87% described the diaphragm as flat during exhalation and concave during inhalation as shown in the model, and 58% did not mention the ribs at all, relating the increase in chest volume to the diaphragm alone. On the other hand, students made fewer errors regarding other structural differences (e.g., trachea, alveoli), possibly due to the information from the media, iconic pictures, and brochures that they frequently encounter in their everyday lives.

Students' misinterpretation of information may be attributed to several factors. First, because visual perception is so effective in promoting memory, it may have greatly impacted students' construction of a mental model of the phenomenon at hand. Regarding the diaphragm and ribs, many students constructed a mental model of the artifact model instead of the real system. Second, the dissimilarities of the model from the real system may have contributed to a high cognitive load. Students had to hold in their working memory both the model as perceived and its differences from what they had learned from other sources (the textbook, prior knowledge, and so on). The model may have activated stored *mental images* of the respiratory system constructed during past (formal or informal) encounters with the system characteristics and the descriptions and pictures of the system in their textbooks, and students had to map them to the various artifact model parts.

Many students' low performance and erroneous knowledge could probably have been prevented if the teacher had carried out explicit classroom discussion regarding the differences between the model and the real, live system. Teachers often express concern about devoting time to such discussions when limited by severe time constraints; however, by involving individual students in an activity aiming to identify differences between reality and the model, students may be able to slowly make better sense of the external model and to gradually refine their mental model at their own pace.

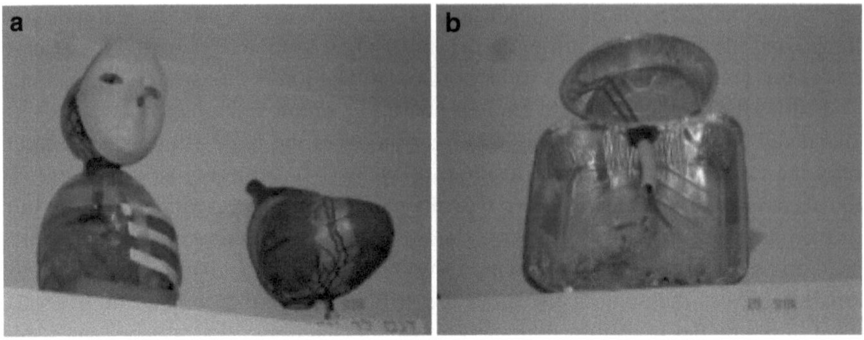

Fig. 4.3 Two models of the anatomical structure of the respiratory system (**a**) and (**b**) created by two forth graders. (note, the heart seen on the right in Fig. 4.3a is a ready-made model)

Self-Generated Model of Respiratory System Structure

A common task intended to activate students while learning biology is to ask them to build a model of their own (e.g., of various systems or of the cell). The advantages of such activities are not limited to the physical activation and students' involvement. To build the model, students must carefully examine all details of the structure, decide what must be left out of their model due to the scale used, identify all its components (e.g., tissues), and so forth. Such preparation requires reading about the structure and examining both photographs and schematic drawings (including photomicrographs, if needed, for models of cells). However, very frequently students are not taught explicitly about uses and shortcomings of models or about their possible relations with the represented information. Consequently, students conceive models as merely smaller or larger copies of reality, and this results in frequent modeling errors, which in turn enter students' stored mental models, thus hindering their learning.

For example, I observed at least two problems in the 3-D models of the respiratory system structure constructed by fifth graders and displayed on a shelf in their science classroom (see Fig. 4.3). As can be seen, students' chosen materials were straws, rubber tubes, aluminum plates, transparent bottles, bubble paper, and so on. First, the proportions among the different components of the model were not kept relative to reality, thus distorting the structural features because of students' deficient knowledge of models as scaled representations and/or because the specific chosen materials affected the proportions (e.g., the diameter of a straw or of the selected tube to represent the trachea). Second, the same parts were represented by different materials and colors—unlike the representations of the ribs and the diaphragm in Fig. 4.3a—because of their deficient understanding of models and/or from restrictions imposed by the limited quantity of specific materials for building the model.

Being actively involved in this learning activity of building models, students probably remember their models much better than the pictures. Therefore, I suggest that students should acquire an understanding of models and other representations before using them; otherwise, models may become obstacles rather than affordances for their learning.

A Live Ecosystem Model: The Aquarium

While studying ecology through a long-term inquiry, ninth graders built an aquarium as a live model for an ecosystem (Eilam, 2002)—a very effective model to enable students to collect different kinds of data and follow changes in real time regarding the aquarium's conditions (e.g., changes in pH, turbidity, temperature and mortality, reproductive, and growth rates) for understanding the freshwater ecosystem. Other advantages are that these models can elicit students' inquiries about the represented ecosystem, their search for solutions to problems raised while building the model, and their awareness of the model's differences from the represented ecosystem—the differences that they might account later on for the outcomes of the inquiry (Eilam, 2012a).

However, the main problem I observed in this example was that students had difficulties recognizing that the aquarium itself was both a limited living environment and a small model for a freshwater ecosystem like a lake. For example, sometimes students seemed to ignore some necessary components of the represented phenomena (e.g., plants as producers or light) or to ignore the feeding relations by adding some unrelated components (e.g., introducing predators). In such cases, a rapid collapse of the system often occurred, which either prevented students' learning from it or resulted in erroneous knowledge. Students had to think about abstract elements that were not directly observable components in the ecosystem (e.g., gases in the water) along with processes and causes that could only be inferred. My observations indicated that they needed more time and a serious teacher's consistent guidance for understanding of an ecosystem model (Eilam, 2002, 2012a).

Scales

Topics in biology involve both size scales (e.g., molecular referents, microscopic cells, the biosphere) and temporal scales (e.g., molecular processes in parts of a second, bacterial reproduction in minutes, blooming in hours, growth in years, and evolution in millions of years). Lacking an understanding of scale may seriously hinder learning. For example, research findings showed that elementary school students and older students lacked the awareness of causality particularly when temporal and spatial gaps were evidenced (e.g., Grotzer & Basca, 2003).

Size Scales

A well-known problem in biology is the use of different scales of referents within the same picture, page, or textbook or over various textbooks, teaching aids, and other learning materials, including those found in the Internet. Frequently, the scale used is not indicated (e.g., a drawing of DNA double helix structure, a photomicrograph of a cell, and a photograph of the organism in the same textbook figure). Difficulties with scale conception were reported for students of all ages from the elementary school to university and also for teachers of all experiences (Jones & Taylor, 2009; Jones, Tretter, Taylor, & Oppewal, 2008). These studies reported that both novice and expert teachers as well as students of all ages were more accurate with large scales than with smaller ones and were most accurate regarding the human scale (1 m).

Understanding many macroscopic phenomena in biology requires the understanding of their microscopic and ultramicroscopic structures and the ability to shift between different scales. Students' difficulties in relating macroscopic observations to microscopic explanations were reported by many researchers (e.g., Liu & Lesniak, 2006). To students, both photomicrographs and electron micrographs—microscopic pictures of biological phenomena—are perceived as macroscopic. Hence, even if the magnitude is indicated, individuals have difficulties in perceiving phenomena as microscopic. Some current textbooks attempt to enhance students' understanding of scale by presenting some relevant proportions between different objects. Such scaffolding may be used as prior knowledge for assessing scales in other cases.

Temporal Scales

Any representation of processes inherently involves the temporal dimension. These may be presented in several ways: a linear schema with arrows (e.g., molecular reactions or growth from a seed to a tree), different tree diagrams (e.g., *pedigree*s in genetics or *cladograms* in evolution), cycles (e.g., biochemical cycles or production of ATP in cellular respiration), or a series of pictures of the same referent changing over different points in time (e.g., a series of animal embryo pictures at different stages of development). However, all these forms for representing temporal processes may pose problems for students.

Whereas linearity may be the easiest for students to interpret, understanding the equilibrium state as represented by two arrows going in opposite directions was found to be highly difficult, probably because students often use unidirectional thinking and views of physical processes as irreversible (e.g., Eilam, 2012b). Evidence also showed that students have difficulties interpreting tree diagrams despite their simple form (e.g., Ainsworth, Matuk, Uttal, & Rosengren, 2010). Accordingly, visual structure greatly influences interpretation independent of content, and content beliefs influence representation; hence, interpretational errors evolve from conceptual and perceptual biases. Likewise, scaffolding activities for tree diagram manipulations were shown to significantly improve undergraduate students' reasoning.

Catley, Novick, and Shade (2010) found that university students erroneously interpreted *noncladogenic diagrams*, where one species turns into a different one over time, whereas students succeeded in interpreting cladograms, which depict the splitting of a species into two species that are then subject to different environmental and selection pressures. Hence, although scientists use cladogenic diagrams only, many of the textbooks present noncladogenic diagrams, which were found to result in students' conception of evolution as *anagenic* (one species evolves directly into another without branching events) and *teleological* (purpose driven); in other words, these diagrams led to students' misconception that organisms turn into a new species by changing in order to adapt to their environment (Catley et al., 2010). Other studies reported students' difficulties in understanding evolution due to its complex characteristics, citing difficulties such as evolution counterintuitiveness, time frames, causality, religious beliefs, prior learning based on anatomy rather than on evolution, and teleological thinking or an intuitive Lamarckism view (e.g., Blackwell, Powell, & Dukes, 2003; Demastes, Good, & Peebles, 1996; Sinclair, Pendarvis, & Baldwin, 1997). Considering that evolution is a central organizing theory in biology (Hatano & Inagaki, 1994), instruction of tree thinking and related skills should be introduced from an early age in schools.

Temporal cycles, which are common representations in biology, may also pose some difficulties for students of all ages. Cyclical representations are used for describing repetitive processes involving materials or organisms (e.g., water cycle in nature; oxygen, carbon, or nitrogen cycles; biochemical cycles in cells; and life cycles of organisms). Eilam (2012a) reported on students' difficulties in understanding that large parts of the different matter cycles constitute the abiotic and biotic, as well as the inorganic and organic relationships in the biosphere and all ecosystems in it. Students fail to relate feeding relations with these cycles. In addition, most students have difficulty in comprehending the idea that an infinite number of cycles—occurring simultaneously in the biosphere—actually pass through each single organism in the biosphere. Studies have suggested that these difficulties are due to students' lack of the ability to think of simultaneous processes (Eilam, 2012a; Hmelo-Silver & Azevedo, 2006). This deficient understanding of cycles hinders students' understanding of the functions and structure of ecosystems—a topic currently considered the most central in learning biology. However, this difficulty evolves not only from lack of representation-related skills and pedagogies applied but also from the difficulties inherent to the content.

Last, while presenting students with a series of pictures of a process transpiring over time, learning materials often fail to emphasize the temporal scale to prevent students' misperception of the process. For example, a series of pictures—depicting a plant from embryo to adult—may be presented along an axis, adjacent to one another and spaced by equal distances, but each of them represents growth across a different number of days (sometimes even hours and days mixed). My observations revealed that when exposed to such series, 82% of the classroom students erroneously misinterpreted the equal distances to be equal time units (as they were found in growth graphs), thus constructing erroneous knowledge regarding that organism's growth process (Eilam, 2007). This phenomenon leads us to the next section.

Temporal Changes in Structures

A difficult task is the demand to construct a 3-D mental structure (sometimes even construct a real artifact) from a series of 2-D pictures of cross/longitudinal sections of that structure in order to understand its inner structural organization or how a process transpires along time within a 3-D space. The following examples from my observations show how such difficulties may constitute an obstacle to understanding different issues in biology. First, a deficient understanding of the many layers of organisms in different habitats of an ecosystem—presented by a set of longitudinal sections—may hinder understanding of the system (e.g., causality, changes, and relations between abiotic and biotic components). Second, a cross section of the human chest fails to show the location of the esophagus as behind the trachea and thus prevents students from understanding the trachea's structure—which contains horseshoe-shaped cartilage rings with only a membrane closing the trachea in its rear—allowing food in the esophagus to pass by. Third, comprehension of growth areas in plants and animals (specifically, the inner structure of root tips) may be hindered by viewing a series of sections showing the location of the initially dividing cells (*meristem*) as always remaining close to the tip (under the root cap), while the cells produced are pushed up or toward the cap. Fourth, deficient understanding of the processes of *endocytosis* or *exocytosis* in cells may result from exposure to a series of pictures representing bubbles forming and disappearing over time. These shortcomings may cause difficulties for students in comprehending many biological processes (e.g., cellular transmissions of materials, eliminating wastes, feeding, or the passage of materials between *axons* and *dendrites* of nerves in the course of stimuli transfer).

To help students with such tasks, learning materials are sometimes designed to present both the 3-D structure and its cross sections and longitudinal sections. However, at least for young and junior high school learners, these schematic drawings are too complex. Such ideas require the allocation of much time and effort in biology lessons to enable students' training and acquisition of the required perception abilities. Figure 4.4 shows a simple training exercise for constructing a 3-D structure of a biological specimen from a set of 2-D sections or vice versa. To achieve maximum efficiency in the translation of these representations, students' attention to measures and proportions in the constructed 3-D object or its cross sections should be raised. Unfortunately, this is seldom done, and teachers constantly complain about the lack of time to *cover* the materials.

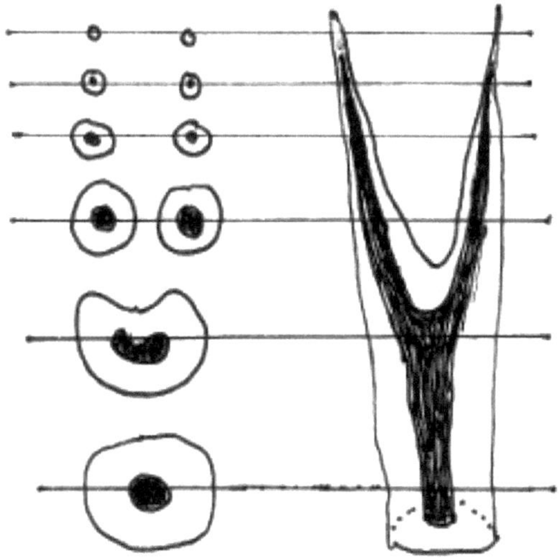

Fig. 4.4 A series of cross sections of a 3-D objects

Superficial Interpretation of Familiar, Common Representations

Students encounter a large number of different external representations not only in diverse biology topics but also in other disciplines and in the media. If students process them superficially, as they are often shown to do, they automatically perceive the surface features of the representation (e.g., an ascending line graph, a pyramid, or a tree diagram) as conveying certain qualities, and then they transfer this interpretation to other representations of similar surface features, resulting in erroneous interpretations. For example, students' misinterpretation of tree diagrams, as described above, probably evolved from the human tendency to perceive higher objects as better ones. Similarly, their reading of equal time intervals on the axis of plant development despite clear indication of the number of days, as described above, was probably due to what students had learned about the structure of graphs.

A more problematic case of misinterpretation due to superficial processing of familiar representations was found in a study when students transferred ideas from one pyramidal representation to another (Eilam & Ben-Peretz, 2010). The ninth-grade national curriculum for students in Israel contains three different pyramidal representations, respectively, in three subjects: the first one in *biology*, portraying energy loss along food chains (see translation from Hebrew to English in Fig. 4.5b); the second one in *English as a foreign language*, portraying recommended amounts of food materials to be eaten (see translation from Hebrew to English, Fig. 4.5a); and the third one in *history*, representing layers of society in Europe (tenant, farmer, aristocrat, and king). The understanding of energy loss in an ecosystem poses a great difficulty for students of all ages, whereas the other two pyramids are easier to comprehend. Therefore, students are often found to transfer their understanding of the other pyramids to the biology one. When they erroneously transfer the notion of amounts that underlies the English and history pyramids to the ecosystem pyramid,

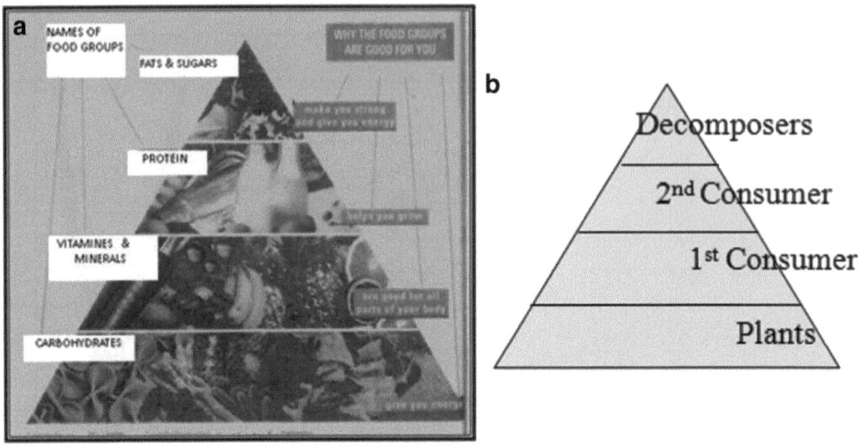

Fig. 4.5 Different pyramids: (**a**) pyramid of food amounts from English class and (**b**) pyramid of energy loss from biology class

students misperceive the layers of the biology pyramid as showing the amounts of producers (plants), consumers, and decomposers. Moreover, students tend to interpret the perceived proportions between the different layers literally; in other words, students interpret pyramids drawn with dissimilar angles differently. Initiating a conceptual change concerning these mistaken beliefs—to help students understand the complexity of the energy issue in the biospheres—is a difficult task requiring teachers' analysis of these three pyramids.

There is another problem in grasping temporal scale concerning life cycles. Cycles describing transformations of materials in nature indeed start with the same material, which transforms along the cycle and returns to its original form, whereas life cycles are intended for describing communities of organisms rather than a single one. However, for the ease of representation, visualizations generally portray only a single organism in the cycle, thus often conveying the misperception that at the end of the life cycle, we return to the same organism with which the cycle began. That is, the same old corn is redeveloped by its own seeds (see Fig. 4.6b), or the same exact butterfly redevelops from its own pupa (see Fig. 4.6a). This may be a simple realization for adults, but my interviews of school students showed that they were frequently unable to explain this paradox, erroneously transferring their knowledge of the water cycle, for example, to such life cycles, and merely ignoring the differences.

Teachers' Representations on the Classroom Board

Most teachers draw many representations on the classroom board in the course of their lessons. Some of these representations summarize the lesson content in one way or another. This can serve as a helpful device for students in organizing the knowledge acquired in that lesson. However, students rely heavily on these

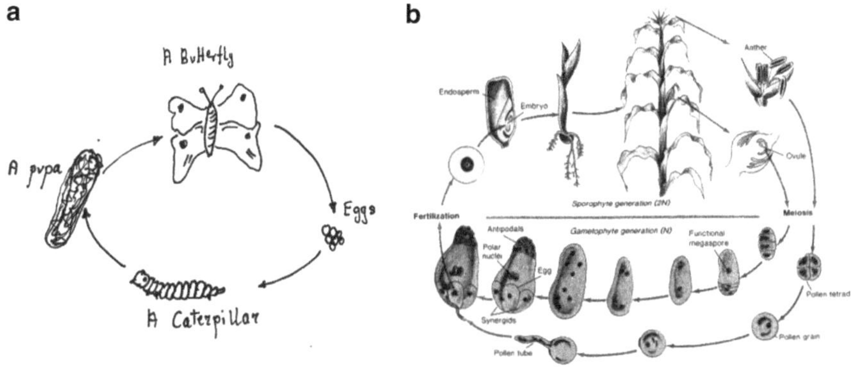

Fig. 4.6 Life cycles of two organisms: (**a**) a butterfly and (**b**) *Zea mays* – corn

representations and frequently ignore teachers' accompanying explanations. They copy the visualizations from the board (hopefully without mistakes) and learn them as it is for their examinations, applying no checking mechanisms or critiques. This may result in students' erroneous knowledge. Moreover, this adherence to the one visualization—as the ultimate kind of organization for the phenomenon at hand—impedes students' development of the cognitive flexibility (Spiro, Feltovitch, Jacobson, & Coulson, 1991) that would otherwise emerge if other suitable organizations for students' knowledge structures were available for enabling students to examine the phenomenon from several different points of view.

In one of my observations in a sixth-grade biology classroom, I witnessed some of these trends (Eilam, 2012b). The lesson topic was about the different body materials transferred by blood. The teacher drew a simple diagram on the board, depicting the two double circulation cycles as a horizontal number eight, created by four arrows (see translation from Hebrew to English in Fig. 4.7a). One arrow (labeled CO_2) exited the heart and led toward the lungs, another arrow (O_2) exited the heart and led toward body cells, the third returned to the heart from the lungs (O_2), and the fourth (CO_2) returned to the heart from body cells.

Next, building on this diagram, the teacher intended to differentiate between the two main bodily wastes—from the bowel and from the kidneys (see translation from Hebrew to English in Fig. 4.7b). She added two more arrows and their labels: One exited the digestive system, carrying food materials and O_2 to the blood and the body cells, and the other left the body cells carrying waste (and CO_2) from the blood toward the kidneys and digestive system. Students copied the entire diagram into their notebooks.

However, this diagram was misleading the students. Waste leaving the digestive system is excreted without ever entering the double circulation cycle. The waste that passes through cells and is brought to the kidneys constitutes harmful material that the body secretes through cells' metabolism. Thus, the diagram introduced erroneous information, possibly hindering students' understanding. A diagram for the representation of this correct biological content is shown in Fig. 4.7c (not drawn in the classroom).

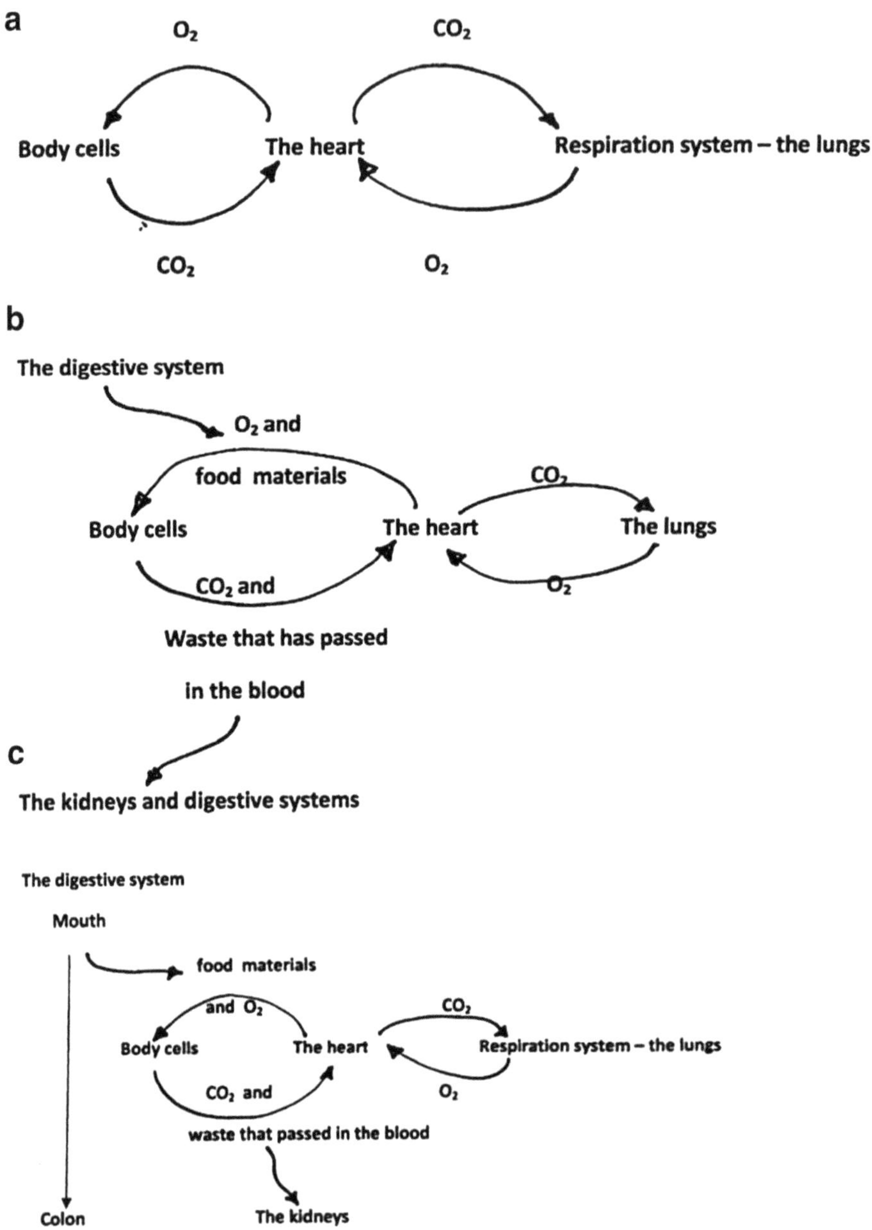

Fig. 4.7 Two-step diagram (drawn on the classroom board) and a suggested error-free version (not drawn). (**a**) Teacher's initial diagram of double circulation cycle. (**b**) Teacher's continuation of diagram to include wastes carried by blood. (**c**) Suggested correct version

Some Final Words

The many examples presented in this chapter concerning possible constraints of utilizing static visual representation in biology involve the impacts of several factors: students' available representational skills, students' prior knowledge of the biological content, students' ages and experiences (related to their available skills and prior knowledge), the characteristics of the task involved, as well as the pedagogies applied. In this chapter, by no means do I attempt to warn teachers against the use of such representations; rather I cannot imagine teaching and learning biology without them. However, in light of the research and direct observations described here, as well as the extensive use of visualization in the biological and scientific research and the globalization processes spreading visual culture internationally, I argue that two aspects of instruction are in urgent need of change: curriculum planning and teacher training (Eilam, 2012b).

First, curricula must be designed to explicitly enhance students' awareness of the diverse characteristics of different representations and students' ability to compare and contrast familiar or novel representations. This should become a prerequisite set of skills for biology learning through visualizations; therefore, biology curriculum developers should consider and emphasize in teachers' guides the potential interpretational difficulties whenever a representation is at hand. Second, teacher preparation programs, both preservice and in-service, must increase teachers' awareness of students' difficulties in interpreting external representations and must train teachers in the dual task of visual literacy both as teachers and as lifelong learners themselves (Eilam, 2012b). Because the domain of biology is based on theories and principles that are shared by many organisms and phenomena, such considerations and awareness may decrease students' numerous misconceptions and errors and advance their ability to learn not only the represented content but also many broad principles that may be applied for understanding other biological phenomena.

I hope this chapter is useful in promoting teachers' and policy makers' awareness not only of the benefits of visual representations to student learning but also of their inherent dangers in creating student difficulties and misconceptions if such representations are not appropriately designed and implemented. In this regard, this chapter should contribute toward the development of relevant learning materials and the resolution of some of these teaching challenges.

References

Ainsworth, S. (1999). The functions of multiple representations. *Computer & Education, 33*(2/3), 131–152.

Ainsworth, S., Matuk, C. F., Uttal, D. H., & Rosengren, K. (2010). Learning to understand the tree of life. In *Proceedings of the 9th international conference of the learning sciences* (Vol. 2, pp. 221–226). Chicago: The International Society of the Learning Sciences.

Berthold, K., & Renkl, A. (2009). Instructional aids to support a conceptual understanding of multiple representations. *Journal of Educational Psychology, 101*(1), 70–87.

Blackwell, W. H., Powell, M. J., & Dukes, G. H. (2003). The problem of student acceptance of evolution. *Journal of Biological Education, 37*(2), 58–67.

Catley, K. M., Novick, L. R., & Shade, C. K. (2010). Interpreting evolutionary diagrams: When topology and process conflict. *Journal of Research in Science Teaching, 47*(7), 861–882.

Demastes, S. S., Good, R. G., & Peebles, P. (1996). Patterns of conceptual change in evolution. *Journal of Research in Science Teaching, 33*(4), 407–431.

Deregowski, J. B. (1989). Real space and represented space: Cross cultural perspectives. *Behavioral and Brain Science, 12*, 51–119.

Eilam, B. (2002). Strata of comprehending ecology: Looking through the prism of feeding relations. *Science Education, 86*, 645–671.

Eilam, B. (2007). *Temporal scales and plants growth: Nine graders' perceptions.* Unpublished manuscript.

Eilam, B. (2012a). System thinking and feeding relations: Learning with a live ecosystem model. *Instructional Science, 40*(2), 213–239.

Eilam, B. (2012b). *Teaching, learning, and visual literacy: The dual role of visual representation in the teaching profession.* New York, NY: Cambridge University Press.

Eilam, B., & Ben-Peretz, M. (2010). Revisiting curriculum inquiry: The role of visual representations. *Journal of Curriculum Studies, 42*(6), 751–774.

Eysink, T. H. S., de Jong, T., Berthold, K., Kolloffel, B., Opfermann, M., & Wouters, P. (2009). Learner performance in multimedia learning arrangements: An analysis across instructional approaches. *American Educational Research Journal, 46*(4), 1107–1149.

Grotzer, T. A., & Basca, B. B. (2003). How does grasping the underlying causal structures of ecosystems impact students' understanding? *Journal of Biological Education, 38*(1), 16–29.

Hatano, G., & Inagaki, K. (1994). Young children's naïve theory of biology. *Cognition, 50*, 171–188.

Hegarty, M. (2004). Diagrams in the mind and in the world: Relations between internal and external visualization. In A. Blackwell, K. Mariott, & A. Shimojima (Eds.), *Diagrammatic representation and inferences: Lecture notes in artificial intelligence* (Vol. 2980, pp. 88–102). Berlin, Germany: Springer.

Hmelo-Silver, C. E., & Azevedo, R. (2006). Understanding complex systems: Some core challenges. *The Journal of the Learning Sciences, 15*(1), 53–61.

Jones, M. G., & Taylor, A. R. (2009). Developing a sense of scale: Looking backward. *Journal of Research in Science Teaching, 46*(4), 460–475.

Jones, M. G., Tretter, T., Taylor, A., & Oppewal, T. (2008). Experienced and novice teachers' concepts of spatial scale. *International Journal of Science Education, 30*(3), 409–429.

Liu, X., & Lesniak, K. (2006). Progression in children's understanding of the matter concept from elementary to high school. *Journal of Research in Science Teaching, 43*(3), 320–347.

Maes, A., Foesenek, K., & Hoogwegt, H. (2008). Visual health communication: Why and how do literate and low literate South Africans differ in their understanding of visual health messages? In P. Swanepoel & H. Hoeken (Eds.), *Adaptive health communication to cultural needs: Optimizing documents in South African health communication on HIV/AIDS prevention* (pp. 151–170). Amsterdam: Benjamines.

Mayer, R. E. (Ed.). (2005). *The Cambridge handbook of multimedia learning.* New York: Cambridge University Press.

Mayer, R. E., & Massa, L. J. (2003). Three facets of visual and verbal learners: Cognitive ability, cognitive style, and learning preferences. *Journal of Educational Psychology, 95*(4), 833–846.

Molinari, G., & Tapiero, I. (2007). Integration of new domain-related states and events from texts and illustrations by subjects with high and low prior knowledge. *Learning and Instruction, 17*, 304–321.

Rapp, D. N., & Kurby, C. A. (2008). The "ins" and "outs" of learning: Internal representations and external visualizations. In J. K. Gilbert, M. Reiner, & M. Nakhleh (Eds.), *Visualization: Theory and practice in science education* (Section A, pp. 29–52). Dordrecht, The Netherlands: Springer.

Reed, S. K. (2010). *Thinking visually*. New York/Hove, UK: Taylor & Francis Group/Psychology Press.

Schnotz, W. (2005). An integrated model of text and picture comprehension. In R. E. Mayer (Ed.), *The Cambridge handbook of multimedia learning* (pp. 49–69). New York: Cambridge University Press.

Sinclair, A., Pendarvis, M. P., & Baldwin, B. (1997). The relationships between college zoology students' beliefs about evolutionary theory and religion. *Journal of Research and Development in Education, 30*, 118–125.

Sivan, A., Orvi, N., Kave, D., & Abulafia, N. (1992). *Chapters on feeding in humans and plants.* Jerusalem: Maalot (in Hebrew).

Spiro, R. J., Feltovitch, P. J., Jacobson, M. J., & Coulson, R. L. (1991). Cognitive flexibility, constructivism, and hypertext: Random access instruction for advanced knowledge acquisition in ill-structured domains. *Educational Technology, 31*(5), 24–33.

Sweller, J. (1988). Cognitive load during problem solving: Effects on learning. *Cognitive Science, 12*, 257–285.

Tversky, B. (2005). Functional significance of visuospatial representations. In P. Shah & A. Miyake (Eds.), *Cambridge handbook of visuospatial thinking* (pp. 1–34). Cambridge, UK: Cambridge University Press.

Chapter 5
Promoting the Collaborative Use of Cognitive and Metacognitive Skills Through Conceptual Representations in Hypermedia

Lei Liu and Cindy E. Hmelo-Silver

Introduction

Conceptual representations can serve as frameworks that can guide learners to organize their knowledge. These representations are important in constructing knowledge, engaging in inquiry, and helping understand phenomena (Liu & Hmelo-Silver, 2009). Suthers and Hundhausen (2003) found that using different representations could guide the learning process and alter the course of collaborative learning conversations. In addition, hypermedia can be an effective representational aid for supporting individual understanding (e.g., Jacobson & Archodidou, 2000). This study addressed a gap in research on how learners' cognitive and metacognitive skills are developed in a hypermedia context in collaborative learning settings. Specifically, in this study we used alternative conceptual representations to organize hypermedia. In hypermedia, students are given access to a range of nonlinear information. Such nonlinear organization provides new possibilities for teaching about the structure of a domain but also provides challenges for students' self-regulated learning (Azevedo, 2005). By self-regulated learning (SRL), we take the definition by Azevedo, Johnson, Chauncey, and Burkett (2010) that "SRL is an active and constructive process whereby learners set learning goals and then attempt to monitor, regulate, and control their cognitive and metacognitive processes in the service of those goals" (p. 229). In this chapter, we focus, in particular, on knowledge co-construction and the metacognitive aspect of

L. Liu (✉)
Educational Testing Service, Princeton, NJ, USA
e-mail: lliu001@ets.org

C.E. Hmelo-Silver
Department of Educational Psychology, Graduate School of Education, Rutgers, the State University of New Jersey, 10 Seminary Place, New Brunswick, NJ 08901-1183, USA
e-mail: chmelo@rci.rutgers.edu; cindy.hmelo-silver@gse.rutgers.edu

D.F. Treagust and C.-Y. Tsui (eds.), *Multiple Representations in Biological Education*, Models and Modeling in Science Education 7, DOI 10.1007/978-94-007-4192-8_5,
© Springer Science+Business Media B.V. 2013

self-regulated learning. We take a situated cognitive perspective in noting how a technology-rich learning environment can provide affordances and constraints for cognitive and metacognitive processing (Lajoie, 2008).

Hypermedia by itself may provide challenges to cognitive and metacognitive processing. Research has shown that a browsing task, due to its lack of focus, yields low levels of both factual and conceptual learning (Jonassen & Wang, 1993). This finding suggests that learning in a poorly designed hypermedia environment may add additional cognitive complexity to self-regulated learning, because it requires learners to generate inferences and connections between and among elements of the database. These challenges may be further magnified during collaborative learning. In the study reported in this chapter, we hypothesized that embedding conceptual representations to help organize hypermedia should support learners' deployment of metacognitive skills during their collaborative learning as they explore and co-construct their understanding of the information presented. The major goal of our study was to examine how hypermedia structure affects students' metacognitive strategies and knowledge co-construction.

Theoretical Framework

Hypermedia as Representational Tools

Hypermedia provides an environment for active learning about facts, concepts, and principles of a domain. However, understanding also requires integration of knowledge into a framework that includes the relations among concepts and principles (Newton & Newton, 2000). The organization of hypermedia systems can help learners develop such a framework as the links and cross-references can promote navigation patterns that model expert knowledge organization (Nunes & Fowell, 1996). In other words, the nodes and links in hypermedia can be used to provide explicit conceptual representations to learners. It is the organization embedded in hypermedia—rather than the media and modalities in which they are presented—that is important for comprehension of complex systems. Conceptual representations refer to frameworks or models that people can use to organize their knowledge (Davis, Shrobe, & Szolovits, 1993). Such representations can highlight common organization across different systems and can promote developing a deep understanding by highlighting key aspects of a complex system, such as the relationship among structural, functional, and behavioral levels of a system (Liu & Hmelo-Silver, 2009; Novick & Hmelo, 1994). One approach to help students learn about complex systems is to provide instruction that focuses on the functional aspects of the system. This chapter extends our previous work (to be described below) to collaborative learning contexts and investigates effects of alternative representations on knowledge co-construction and metacognition.

Structure-Behavior-Function (SBF) Representations

Using Structure-Behavior-Function (SBF) as a conceptual representation for instructional design builds on what we have learned about how experts represent complex systems in terms of structures, behaviors, and functions (Hmelo-Silver, Marathe, & Liu, 2007). Goel et al. (1996) proposed that systems could be understood in SBF terms. Structures refer to the basic elements that compose the system; behaviors refer to the mechanisms that allow structures to achieve their functions; functions are the roles or outputs of an element of the system or the system as a whole. For example, the lungs are one structure within the human respiratory system. The behavior of the lungs is to exchange oxygen and carbon dioxide by the differences in the gasses' concentrations at the alveoli. The function of the lungs is to bring air into the body and help get rid of wastes. Previous studies have shown that novices focus on the perceptually available structures of a system, whereas experts integrate structures, behaviors, and functions (Hmelo-Silver et al., 2007). Such results suggest that SBF representations have implications for teaching and learning. Specifically, SBF representations can be embedded in hypermedia and used for teaching and learning about complex systems.

In an experimental study, Liu and Hmelo-Silver (2009) compared the effects of alternative versions of hypermedia, function-centered and structure-oriented hypermedia, in the domain of the human respiratory system. The function-centered version emphasized the interrelationships within the system. This version was organized around how and why questions in a way similar to experts' mental models (Graesser, 1999; Narayanan & Hegarty, 1998). The structure-oriented hypermedia used an organization similar to traditional textbooks. In this version, the learners had a picture of the respiratory system with structures that participants could select to get more information. The findings showed that students who used the function-centered hypermedia gained a deeper understanding of the system than did students using the structure-oriented version. This effect was pronounced at the microlevel, which is not perceptually salient, and was more difficult for the novices to understand (Hmelo-Silver et al., 2007). These studies demonstrated the power of embedding conceptual representations in hypermedia, especially their effects on deeper understanding.

Cognitive and Metacognitive Processes in Learning from Hypermedia

Early research on hypermedia systems was critiqued for focusing on the technology rather than cognitive and learning issues and for lack of theoretically grounded design (Jacobson & Azevedo, 2008). One conclusion was that students benefit when a hypermedia system encourages learners to exercise their metacognitive skills. Regulatory learning emerged in the process of knowledge co-construction

(Hickey & McCaslin, 2001). In a classroom study of collaborative learning using hypermedia, Azevedo, Winters, and Moos (2004) demonstrated that collaborative outcomes were related to the use of regulatory behaviors. In their follow-up study, Azevedo, Moos, Greene, Winters, and Cromley (2008) found that students made qualitative shifts in conceptual understanding and frequently deployed regulatory and metacognitive processes with hypermedia prompted by external facilitation. Although Azevedo et al. (2008) used *soft* scaffolds—adaptive scaffolding provided by a human facilitator—in this study we used conceptual representations embedded in hypermedia to prompt students' co-regulatory learning in dyads. We investigated how using hypermedia for collaborative learning can support learner's use of cognitive and metacognitive skills. The conceptual representations embedded in hypermedia as a cognitive tool may help students develop metacognitive and regulatory skills such as setting learning goals, sequencing exploration to seek and collect information to construct (or co-construct in collaborative learning environment) meanings, and monitoring and evaluating if the learning goals are reached. Shapiro (1998) suggested that one effective strategy to improve learners' regulatory skills when using hypermedia may be to attach notations to links that help learners understand the relationships between different levels of conceptual knowledge. The hypermedia used in this study incorporated these ideas by embedding SBF representations into the design of the hypermedia systems.

Function-Centered (F-Hypermedia) and Structure-Centered Hypermedia (S-Hypermedia)

This study used two different versions of hypermedia (Liu & Hmelo-Silver, 2009) to teach about the human respiratory system: the function-centered version (F-hypermedia) and the structure-centered version of hypermedia (S-hypermedia) on the human respiratory system. Both versions presented similar content; the major difference was in the underlying conceptual representations. In the F-hypermedia, the information was organized in a way that emphasizes the functional and behavioral perspective on the respiratory system. Specifically, learners first viewed the two major functional-behavioral questions: "Why do we need oxygen?" and "How does oxygen get into our body?" (see Fig. 5.1). These two questions led them to study the function of the whole system first and then its behaviors and structures. Hence, the F-hypermedia makes the functional aspect salient. In the S-hypermedia, learners started with a diagram of the human respiratory system with links to each component in the system (see Fig. 5.2). In this way, the structural aspects were salient to learners. This organization moved learners from isolated structures to their respective behaviors and functions.

As noted earlier, the use of conceptual representations in hypermedia affects what students learn individually (Liu & Hmelo-Silver, 2009). This study explored how different forms of hypermedia affect students' cognitive and metacognitive processing

Fig. 5.1 Opening screen of the function-centered hypermedia (F-hypermedia)

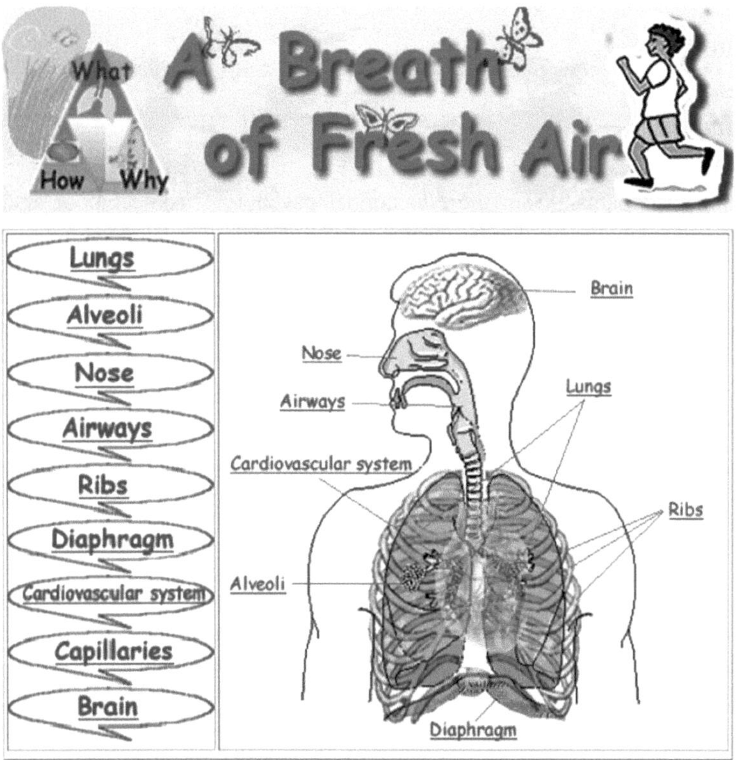

Fig. 5.2 Opening screen of the structure-oriented hypermedia (S-hypermedia)

emerging during knowledge co-construction. The conceptual representations set different goals for student learning, which is regarded as a powerful intervention with regard to learners' use of hypermedia (Gall, 2006). The functional conceptual representations helped students to set a goal of understanding how various components function together within the system, whereas the structural conceptual representations stressed the importance of the components of the systems and how each one worked. Although both types of hypermedia contained the same information, the organization differed in terms of the goals that were promoted. Furthermore, the functional and behavioral questions in F-hypermedia afford the deployment a myriad of cognitive and metacognitive strategies to monitor and evaluate whether sufficient information is collected and constructed to reach the learning goals as learners try to tease out the interrelationships in the F-hypermedia. Because each structure is presented in a more isolated way in the S-hypermedia, learners do not need to engage in as much cognitive and metacognitive processing. By making the F-hypermedia focus on the interrelatedness in the system, we have rocked the "cognitive boat"—as Reiser (2004, p. 288) suggested—that may sometimes be important to get learners to engage in deeper cognitive and metacognitive processing. Therefore, we hypothesized in this study that in collaborative learning the F-hypermedia leads to more advanced use of cognitive and metacognitive skills.

Method

Participants

Twenty participants—from the educational psychology subject pool at a large public university in the United States—participated in a single 2-hour session. Participants received course credits for participation. In each session, students were grouped into dyads and randomly assigned to explore either the function-centered hypermedia (F-dyads) or the structure-centered hypermedia (S-dyads) to learn about the human respiratory system. Therefore, five dyads explored the function-centered hypermedia and the other five dyads explored the structure-centered hypermedia.

Procedure

Each session lasted around 2 hours. Initially, the students in each dyad were asked to take a pretest on their conceptual understanding of the human respiratory system. Then, they were asked to explore the hypermedia collaboratively starting from the main page. The participants were told that collaboration meant that the participants needed to explain to their partner what the content meant to them and how it related to what they had already known about the human respiratory system. Following the exploration, each of the participants completed an individual posttest and completed a questionnaire on their attitude toward using the software and the collaborative learning activities. The session was both video- and audiotaped.

Coding and Analysis

The tapes were transcribed verbatim. All transcriptions were marked with conversational turns. The main interest of this study lies in how conceptual representations affect collaborative regulatory behaviors. To address this issue, we compared the conversations of the dyads in the two conditions using qualitative and quantitative analyses. The coding and analysis procedures are shown in Fig. 5.3.

In the pilot study (Liu, Hmelo-Silver, & Marathe, 2005), we developed three hypotheses regarding what might characterize the difference across the conditions: (1) when negotiation takes place in their conversations, (2) what kind of knowledge is negotiated, and (3) how the F- and S-hypermedia promote differential knowledge co-construction. To address these issues, we applied the same three-pass coding method to capture emerging trends in the negotiation segments that occurred in the collaborative interactions. Reliability was greater than 90% agreement between our coding of the transcripts with that of an independent coder who coded 20% of all the transcripts blind to the conditions.

The first pass of coding focused on identifying topics, such as cellular respiration, the lungs, transport, and movement of air into the body, on which the dyads were working by reviewing the videotape as well as their concentration in the discourse. In addition, the coder also identified salient topics (the more macrolevel structures) from nonsalient topics (the microlevel of a system) which are particularly important for a deep understanding of the system (Hmelo-Silver & Pfeffer, 2004).

In the second pass, we divided the entire discourse into five categories of functional episodes: social talk, tool-related task talk, reading, negotiation, and quizzing. The purpose of the social talk was to provide a common ground or to create a friendly environment for collaboration. Conversations such as "Which major are you in?" or "Which subject is your favorite?" were coded as social talk. Tool-related task talk, used to set goals for navigation, usually occurred after completing one topic. A reading episode occurred when there was only verbatim reading of the content on the computer. Negotiation episodes occurred when students exchanged ideas as they attempted to reach common understanding. These segments might include referring to prior knowledge, asking questions, paraphrasing, elaborating, or reasoning and making meaning of the information in the hypermedia. Quizzing occurred between the students as they tested each other's understanding or memorization of the learned facts.

The last pass of coding was our focus for this chapter—the negotiation episodes. A coding scheme was constructed to identify indicators of constructive processing that are associated with learning (Chi, Siler, Jeong, Yamauchi, & Hausmann, 2001),

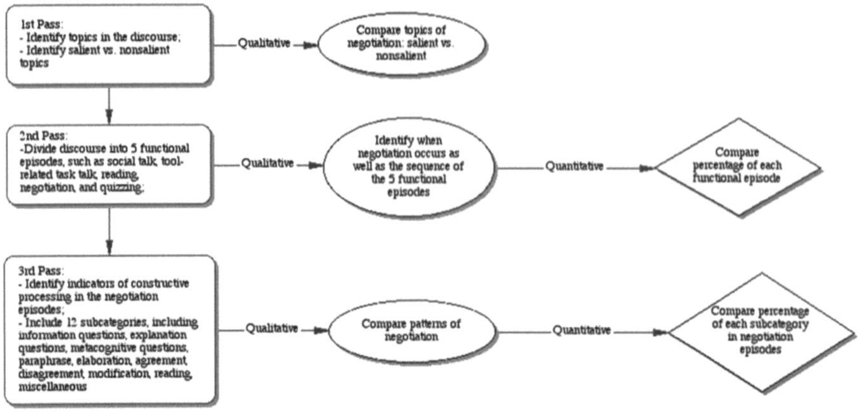

Fig. 5.3 Coding and analysis procedures

including questions (classified into information question, explanation question, and metacognition question), prior knowledge, paraphrase, elaboration, agreement, disagreement, modification of ideas, and metacognition (monitoring progress and understanding). A summary and coding examples of subcategories under negotiation episodes are provided in Table 5.1. Prior knowledge is an important indicator of learning because people usually learn better when they connect the current learning material with their previous knowledge. Asking questions indicates peers mediating each other's learning and is a key aspect of social knowledge construction (King, 2002). Paraphrasing and elaboration allow the partners to clarify their ideas, reorganize and integrate their understanding, and thus to construct their knowledge. Agreement, disagreement, and modification are verbal indicators of the cognitive process associated with negotiation. The final indicator, metacognition, demonstrates learners' awareness of their own cognitive processing, which is essential for knowledge transfer. Besides identifying all the above indicators of constructive processing, the coding scheme also included reading within negotiation, which refers to the purposeful reading to support one's idea. For all the other turns that could not be classified into any of the above categories, they were identified as miscellaneous categories, such as meaningless "ok" or "yeah."

Results

As reported elsewhere (Liu & Hmelo-Silver, 2010), students using the F-hypermedia showed gains in understanding structures and behaviors, whereas students using the S-hypermedia did not. In this chapter, we report quantitative and qualitative results related to these learning processes. The quantitative results

Table 5.1 Summary and examples for the third coding pass

Categories	Definition	Example
Information question	Simple questions; expected answers include "yes/no"	Is this a blood cell?
Explanation question	Asks for additional detail or causal information	How does the lung work?
Metacognitive question	Asks participant to think about their thinking, levels of understanding, planning	How do we do that?
Prior knowledge	Knowledge not in the hypermedia	It's like how ATP worked when turning food into energy. This was learned 10 years ago
Paraphrase	Rewording the hypermedia content for the purpose of clarification	"Alveoli make the area in the lungs bigger" (not exact wording)
Elaboration	Adding extra material or illustration to clarify the hypermedia content or to build idea	So the properties of the lungs help cellular respiration...
Agreement	To agree with the partner's understanding	Oh, yeah
Disagreement	To disagree with the partner's understanding	I don't really agree
Modification	Making something different that builds on or change previously mentioned ideas	It's not just a place for oxygen exchange. It also increases the area inside the lungs
Metacognition	Thinking about the process of knowing, thinking, planning, and monitoring	I did not know...
Reading	Reading in negotiation to find evidence in hypermedia to support one's idea	Verbatim reading
Miscellaneous	Cannot be categorized into other categories	Sentence fragments

presented here include the statistics for the functional episodes as well as the subcategories of negotiation episodes for both the F- and S-conditions. Because the number of dyads ($n = 10$) was small, we present these results only descriptively. We present in the following section the qualitative results of two case studies, respectively, from the two conditions for qualitative comparison.

Quantitative Results

Percentage of Functional Episodes

The percentage distribution of the five functional episodes—social talk, tool-related task talk, reading, negotiation, and quizzing—are presented in Table 5.2. The dyads in the F-condition (5.29%) made fewer verbatim reading statements than did the dyads in the S-condition (12.89%). This indicated that the dyads in the S-condition

Table 5.2 Mean percentage of functional episodes of dyads in F- and S-conditions

	Social talk	Tool-related task talk	Reading	Negotiation	Quizzing
Dyads ($n = 5$) in F-condition	6.12	24.21	5.29	53.55	10.86
Dyads ($n = 5$) in S-condition	2.79	22.02	12.89	53.00	9.98

engaged in more rote repetition in their collaborative exploration than did the dyads in the F-condition. The percentage of negotiation in both conditions was quite similar. However, the quantity of negotiation cannot determine the quality of the negotiation. Because of this, we complemented the quantitative analysis with a more in-depth analysis of the negotiation episodes in the following section.

Percentage of Subcategories Under Negotiation Episodes

The subcategories under negotiation episodes—that showed clear differences across the two conditions—are displayed in Table 5.3. The dyads in the F-condition (F-dyads) asked more explanation questions, made more connections to their prior knowledge, presented more elaborations, and made more metacognitive statements than did the dyads in the S-condition (S-dyads). In contrast, dyads in the S-condition asked more questions that only needed simple answers and conducted more verbatim reading in their negotiation. These results provided further evidence supporting the hypothesis that the F-hypermedia stimulated more constructive and metacognitive processing than did the S-hypermedia.

Qualitative Results

To compare the discourse of the dyads across the two conditions, we selected two dyads that collaborated best—one from each of the two conditions. The selection was based on the quantity and quality of the indicators of learning, such as elaboration and metacognition that occurred in dyads' transcripts. The qualitative analyses were centered around three research questions based on the aforementioned three hypotheses in our pilot study: (1) When does negotiation take place in their conversations? (2) What kind of knowledge is negotiated? (3) How do F-hypermedia and S-hypermedia promote (or fail to promote) knowledge co-construction and metacognition?

Table 5.3 Mean percentage of subcategories under negotiation episodes of dyads in F- and S-conditions

	Information question	Explanation question	Prior knowledge	Paraphrase	Elaboration	Metacognition	Reading
Dyads ($n = 5$) in F-condition	7.72	4.46	5.18	12.66	17.87	21.79	5.38
Dyads ($n = 5$) in S-condition	11.94	2.98	2.32	19.61	11.69	12.11	10.31

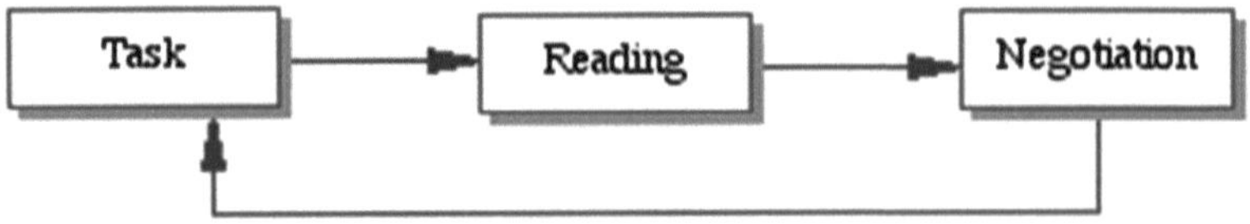

Fig. 5.4 Possible sequence of functional episodes of F-dyads

Fig. 5.5 Possible sequence of functional episodes of S-dyads

Sequence of Functional Episodes

First, the dyads in the F-condition tended to present a more complicated sequence of the discourse functions in the second pass of coding than did the dyads in the S-condition. Typically, the F-dyads produced multiple cycles of functional tasks. In contrast, the S-dyads presented much simpler sequences. For example, when the discussion topic was about the diaphragm, there were great differences in the pattern of negotiation between students R and G (a dyad in the F-condition) and students S and J (a dyad in the S-condition) when they discussed the topic. The typical sequences of both dyads' turn-taking discussions are shown in Figs. 5.4 and 5.5. Both these dyads realized the difficulty of understanding the function and behavior of the diaphragm. The difference was that R and G made great effort to make meaning of the content in the hypermedia by producing loops of content reading and negotiation. They kept coming back and forth between reading and negotiation, which indicated their engagement in thinking and trying to understand. The following excerpt illustrates how R and G in the F-dyad co-constructed knowledge during their hypermedia exploration:

12. G: I am interested, what is the, what exactly is the diaphragm?
13. R: I have no idea. . . .
14. R: Muscles in the chest consist of diaphragm and intercostals muscles, sheet-like muscles that raise the chest through a cavity. . . .
15. G: Oh is this, is this just a separation basically, or is this, what is it?
16. R: It is, yeah, it is this thing on the bottom. . . when it relaxed it is dome shaped, the top of the dome is inside the rib cage, when it is contracts it is flat and lies below the rib cage, primary muscle for pulling in air (reading website).
17. G: So right now it is relaxed or contract?
18. R: Yeah. . . it is relaxed I think, yeah it is dome shaped.
19. G: It is relaxed. . . ok, and it comes flat once like contact.
20. R: Yeah, when you breathe in (reading).
21. G: Ok (reading website)....

22. G: So that is how we pull in air.
23. R: Um uh.

R and G together co-constructed knowledge about the role of the diaphragm and how it supports the mechanics of ventilation. G started out by monitoring her understanding and noting that she was unsure of how the muscle worked (turn 12), and so did R (turn 13). Eventually they figured out collaboratively how the shape changed as the muscle contracted and relaxed and how that pulled in air through R's goal-oriented information collection (turns 14 and 16) and G's monitoring questions (turns 15 and 17). There was a give-and-take process between the partners as they converged on an understanding. R evaluated that convergence positively in turn 23.

In contrast, the typical pattern of S and J's discussion was that they simply read the content on the computer without a specific goal orientation, engaged in brief negotiation, and S ended up confused, saying "I still didn't get those." This indicated that they, too, engaged in monitoring their understanding but they did not work through their comprehension difficulties and instead jumped to another topic. The cycles of functional episodes illustrated that the F-dyad showed more effort and engagement in understanding the content than did the S-dyad even though they were dealing with identical content. This pointed to the differential function of the underlying conceptual representations in the two versions of hypermedia—the F-hypermedia (but not the S-hypermedia) models how to set learning goals with functional questions which require complex and interrelated system knowledge to understand the answer (see Figs. 5.4 and 5.5).

Knowledge Construction and Metacognition

The dyad's discourse in the F-condition constructed knowledge involved multiple aspects of the system as well as the interrelations within the system. According to the first pass of coding, in the F-condition, negotiation occurred in most of the topic-based episodes particularly those about the nonsalient structures—such as alveoli, capillaries, blood, and red blood cells—and their interactions with other components in the respiratory system as they participate in cellular respiration. Whereas in the S-condition, negotiation was focused on salient structures—such as airways, intercostal muscles, and diaphragm—and the interrelations among these structures were rarely mentioned. We compared examples of negotiation excerpts from the aforementioned selected dyads that collaborated best in the two respective conditions. The following excerpt illustrates a negotiation episode from the F-dyad, R and G, discussing how energy was produced in the respiratory system, in which various structures involved were discussed:

45. G: So basically air is just to keep the process of really going, of creating energy.
46. R: Pretty much.

47. G: It is sort of like, the unmoved, mover, if you will, if you are into the whole St. Francis of Assi, oxygen goes into the body and goes into the vein, alveoli goes in the blood, blood goes to all the cells, cells then have the energy needed to convert the food into energy.
48. R: Let me get a piece of paper, I will write on the back of this thing, the way ATP works, you got water that goes around in a cycle pretty much, an easy way to think about it, oxygen comes in, gets picked up by the water molecules because it is a polar molecule.
49. G: Um, yeah. . ..
50. R: Comes over to the other side of the thing, got $C_6 H_{12}O_6$, it doesn't matter, it can be a little bit different, it is just basically sugar, gets in here, breaks down, breaking of molecules creates energy, that energy, not only does it like, keep the body at a certain temperature but the, it breaks this apart, ahh so the addition of oxygen breaks it apart and creates more water, so you end up with CO_2 and like H over here and that is how that works.
51. G: Yeah, and you just store it to like where we need it.
52. R: Yeah, and it just goes back around here, the water stays, CO_2 leaves, need more oxygen.
53. G: Excellent. . . so that is it.
54. R: I hope I did that right; it has been about four years since I touched that material.

The above excerpt shows that both G and R actively made meaning out of the content on the computer by monitoring their understanding (e.g., turns 45 and 46), elaborating (e.g., turn 48), connecting to their prior knowledge, and accommodating old and new knowledge (e.g., turn 47). In particular, the negotiation between G and R showed that they had noticed that it was the coordination of the various structures that brought about the process of energy production. In addition, this negotiation example also illustrated that this dyad had noticed the functions of nonsalient structures, which are usually ignored by novices, and their discussion went deep into the molecular level. For comparison, we present another negotiation example from the S-dyad, S and J. This example shows a discussion focused on a salient structure—the airway:

28. S: So this is like the nose itself, like basically. . ..
29. J: Yeah. . . ahh trachea, that is how you spell it. . . there is also nose hairs that filter out.
30. S: Yeah, I don't remember that one.
31. J: So maybe it should say that, because that is pretty important, because you should inhale through your nose and exhale through your mouth because of those nose hairs, it is a lot more healthy.
32. S: We have two nostrils and that goes to the pharynx which is divided into and it goes to the eso (reading).
33. J: Esophagus (reading).
34. S: Esophagus, right?
35. J: That is the food pipe, that. . ..

36. S: And that is just the airpipe.
37. J: There is a little valve, that when you eat closes up, that is why you shouldn't talk when you eat because... that opens.
38. S: It might go down.
39. J: Yeah, cause you are going to need to breathe obviously while you talk, and it confuses things and food goes down the airpipe.
40. S: And that is why you cough to get it up.
41. J: Um uh....

Like the previous negotiation example from the F-dyad, this negotiation example from this S-dyad also included monitoring, elaboration, connecting to prior knowledge, and other indicators of knowledge construction. However it looked different from the F-dyad's negotiation in two ways. First, the negotiation in the S-dyad was centered on one unique structure, the airpipe, but failed to make connections to other structures. Second, the negotiation stayed at the structural level and did not involve much talk about the functions of these structures. Therefore, the students in this S-dyad monitored their understanding of the individual structures (e.g., nose, airpipe, etc.) in contrast with the F-dyads who monitored more complex processes.

The qualitative analyses comparing the two dyads, respectively, in the two conditions showed that there were differences in the patterns of discourse functions, as well as constructive and metacognitive processing. First, the students in the F-dyad produced more complicated sequences than did those in the S-dyad, and their negotiation reoccurred on the same topic. Second, there was also a difference across the conditions as to the kind of knowledge negotiated. The F-dyad tended to put their focus on making meaning of the nonsalient microlevel phenomena, whereas the S-dyad focused their negotiation on macrolevel structures. Moreover, the deep negotiation between the dyadic students in the F-condition did promote the knowledge co-construction and brought about shared understanding. Both conditions provided opportunities for metacognitive processing, such as monitoring, to occur. However, the F-dyad students were able to set explicit learning goals and evaluate their progress as the modeling effects of the function orientation of the conceptual representations, whereas the S-dyad students were less likely to persist on the kinds of ideas that would lead them to productively construct convergent knowledge and evaluate their understanding in the structure-oriented learning context.

Discussion

The results of this study suggest that the underlying conceptual representations affect students' collaborative knowledge co-construction and metacognitive processing. The quantitative results showed that the dyads in the F-condition engaged in less verbatim reading than did the dyads in the S-condition. In addition, the F-dyads were more elaborative as they asked explanatory questions and made connections to their prior knowledge, as well as demonstrating more metacognitive processing compared to the dyads in the S-condition when reviewing the hypermedia. The qualitative results illustrated that the navigation of the F-dyads was more purposeful than that of the S-dyads. The comparison of the two dyads in different conditions provided evidence that although both dyads showed evidence of monitoring their understanding, the F-dyad tended to set explicit goals for their exploration by trying to answer specific questions discovered through the monitoring process. The focus on functions and behaviors in F-hypermedia provided a context for students to plan a purposeful information-seeking sequence and evaluate their progress toward the learning goals. On the other hand, the S-dyad was constrained by the isolated structures of the conceptual representations in S-hypermedia and engaged in less monitoring and exploration compared to what we observed in the F-dyad. Although the students in the S-dyad identified where they lacked understanding, they did not deploy sophisticated cycles of metacognition to successfully co-construct deep understanding. Instead they ignored the deficiency in understanding and jumped to another new topic. In addition, the F-hypermedia directed the attention of the F-dyad to the hard-to-understand nonsalient phenomena, necessary for deep understanding of the system, because the function-centered conceptual representations led the students to make connections among nonlinearly related micro- and macrolevels of phenomena in the system. In contrast, the S-dyad focused their discourse on isolated structural knowledge and rarely discussed the functions and behaviors of the system. Understanding these functions and behaviors are critical for deep understanding of the system. This implies that the S-hypermedia failed to provide opportunities for students to co-construct deeper understanding of the system and to deploy advanced metacognitive skills such as planning and evaluating to lead to productive learning.

Our findings in this study suggest that the way information is organized in hypermedia and the navigational guidance provided can have powerful effects on cognitive and metacognitive processing. Conceptual representations can afford and constrain how learners set goals for exploration, monitor what they are learning, and evaluate what they have accomplished as they co-construct a shared understanding. We have used the Structure-Behavior-Function (SBF) representations as an example of conceptual representations grounded in research on expertise. As Collins and Ferguson (1993) suggested, there are likely other *epistemic forms* that might be equally successful in affording deep cognitive and metacognitive processing. Moreover, collaboration creates the need to make thinking visible and provides further opportunities for *co-regulated learning* (Liu & Hmelo-Silver, 2010). Together, these SBF representations suggest that hypermedia can be a powerful tool to promote cognitive and metacognitive processing. Nevertheless, we need to have a better understanding of how best to organize such media to take advantages of the affordances of hypermedia as well as recognize those limitations that require additional scaffolding.

References

Azevedo, R. (2005). Computer environments as metacognitive tools for enhancing learning. *Educational Psychologist, 40*, 193–197.

Azevedo, R., Johnson, A., Chauncey, A., & Burkett, C. (2010). Self-regulated learning with Met Tutor: Advancing the sciences of learning with metacognitive tools. In M. S. Khine & I. M. Saleh (Eds.), *New science of learning: Cognition, computers and collaboration in education* (pp. 225–247). New York: Springer.

Azevedo, R., Moos, D. C., Greene, J. A., Winters, F. I., & Cromley, J. G. (2008). Why is externally-regulated learning more effective than self-regulated learning with hypermedia? *Educational Technology Research and Development, 56*, 45–72.

Azevedo, R., Winters, F. I., & Moos, D. C. (2004). Can students collaboratively use hypermedia to learn about science? The dynamics of self- and other-regulatory processes in an ecology classroom. *Journal of Educational Computing Research, 31*(3), 215–245.

Chi, M. T. H., Siler, S., Jeong, H., Yamauchi, T., & Hausmann, R. G. (2001). Learning from tutoring. *Cognitive Science, 25*, 471–533.

Collins, A., & Ferguson, W. (1993). Epistemic forms and epistemic games: Structures and strategies to guide inquiry. *Educational Psychologist, 28*, 25–42.

Davis, R., Shrobe, H., & Szolovits, P. (1993). What is a knowledge representation? *AI Magazine, 14*, 17–33.

Gall, J. (2006). Orienting tasks and their impact on learning and attitudes in the use of hypertext. *Journal of Educational Multimedia and Hypermedia, 15*, 5–29.

Goel, A. K., Gomez de Silva Garza, A., Grué, N., Murdock, J. W., Recker, M. M., & Govinderaj, T. (1996). Toward designing learning environments: Exploring how devices work. In C. Fraisson, G. Gauthier, & A. Lesgold (Eds.), *Intelligent tutoring systems: Lecture notes in computer science* (pp. 493–501). New York: Springer.

Graesser, A. C. (1999, April). *How do adults comprehend the mechanisms of everyday devices? Texts, illustrations and breakdown scenarios*. Paper presented at the annual meeting of the American Educational Research Association, Montreal, Quebec, Canada.

Hickey, D. T., & McCaslin, M. (2001). A comparative sociocultural analysis of context and motivation. In S. Volet (Ed.), *Motivation in learning contexts: Theoretical advances and methodological implications* (pp. 33–55). Amsterdam: Elsevier Science Ltd.

Hmelo-Silver, C. E., Marathe, S., & Liu, L. (2007). Fish swim, rocks sit, and lungs breathe: Expert-novice understanding of complex systems. *The Journal of the Learning Sciences, 16*, 307–331.

Hmelo-Silver, C. E., & Pfeffer, M. G. (2004). Comparing expert and novice understanding of a complex system from the perspective of structures, behaviors, and functions. *Cognitive Science, 28*, 127–138.

Jacobson, M. J., & Archodidou, A. (2000). The design of hypermedia tools for learning: Fostering conceptual change and transfer of complex scientific knowledge. *Journal of the Learning Science, 9*, 145–199.

Jacobson, M. J., & Azevedo, R. (2008). Advances in scaffolding learning with hypertext and hypermedia: Theoretical, empirical, and design issues. *Educational Technology Research and Development, 56*, 1–3.

Jonassen, P. M., & Wang, S. (1993). Acquiring structure knowledge from semantically structured hypertext. *Journal of Computer-Based Instruction, 20*, 1–8.

King, A. (2002). Structuring peer interaction to promote high-level cognitive processing. *Theory into Practice, 41*, 33–39.

Lajoie, S. P. (2008). Metacognition, self regulation, and self-regulated learning: A rose by any other name? *Educational Psychology Review, 20*, 469–475.

Liu, L., & Hmelo-Silver, C. E. (2009). Promoting complex systems learning through the use of conceptual representations in hypermedia. *Journal of Research in Science Teaching, 46*, 1023–1040.

Liu, L., & Hmelo-Silver, C. E. (2010). Conceptual representation embodied in hypermedia: An approach to promoting knowledge co-construction. In M. S. Khine & I. M. Saleh (Eds.), *New science of learning: Cognition, computers and collaboration in education* (pp. 341–356). New York: Springer.

Liu, L., Hmelo-Silver, C., & Marathe, S. (2005). How representation matters: Comparing collaborative learning with alternative versions of hypermedia. In T. Koschmann, D. D. Suthers, & T.-W. Chan (Eds.), *Proceedings of computer supported collaborative learning (CSCL) 2005* (pp. 398–402). Mahwah, NJ: Erlbaum.

Narayanan, N. H., & Hegarty, M. (1998). On designing comprehensible interactive hypermedia manuals. *International Journal of Human Computer Studies, 48*, 267–301.

Newton, D. P., & Newton, L. D. (2000). Do teachers support causal understanding through their discourse when teaching primary science? *British Educational Research Journal, 26*, 599–613.

Novick, L. R., & Hmelo, C. E. (1994). Transferring symbolic representations across nonisomorphic problems. *Journal of Experimental Psychology: Learning, Memory, and Cognition, 20*, 1296–1321.

Nunes, J. M., & Fowell, S. P. (1996). Hypermedia as an experimental learning tool: A theoretical model. *Information Research News, 6*, 15–27.

Reiser, B. J. (2004). Scaffolding complex learning: The mechanisms of structuring and problematizing student work. *The Journal of the Learning Sciences, 13*, 273–304.

Shapiro, A. (1998). Promoting active learning: The role of system structure in learning from hypertext. *Human Computer Interaction, 13*, 1–36.

Suthers, D. D., & Hundhausen, C. D. (2003). An experimental study of the effects of representational guidance on collaborative learning processes. *Journal of the Learning Science, 12*, 183–218.

Part II
Implications for Biology Teaching and Teacher Education with Multiple Representations

Chapters in Part II address approaches by different researchers using a range of external representations for teaching biology.

Chapter 6
Learning and Teaching Biotechnological Methods Using Animations

Hagit Yarden and Anat Yarden

Rationale

It was recently suggested that school science should play a major role in the development of a citizenry that is capable of dealing with the scientific developments and changes in the vital field of biotechnology and their influence on our everyday lives (Steele & Aubusson, 2004). Biotechnology can be defined in the broadest sense as any technological application that uses biological systems, living organisms or derivatives thereof, to make or modify products or processes for specific use. Biotechnology is also an aspect of science in which its content is rich with opportunities for applying the knowledge, understanding, and attitudes gained from the study of science to everyday life (Lock, Miles, & Hughes, 1995). Indeed, the importance of biotechnology education has been recognized in a number of international curriculum frameworks around the world (Dori, Tal, & Tsaishu, 2003; Falk, Brill, & Yarden, 2008; Steele & Aubusson, 2004). Although biotechnology education has gained significant recognition, less has been published about how to effectively teach and learn this aspect of science.

One of the most problematic issues to comprehend while learning biotechnology concerns the methods involved (Falk et al., 2008). Molecular biology methods are completely unfamiliar to most students because these methods are remote from the everyday lives of the students' who usually have no opportunity to experience them hands-on in the school laboratory (Olsher & Dreyfus, 1999; Steele & Aubusson, 2004). In addition, the methods are based on the understanding of molecular

H. Yarden
Department of Education in Technology and Science, Technion Institute of Technology, Haifa 36006, Israel
e-mail: hagit_y@oranim.ac.il

A. Yarden (✉)
Department of Science Teaching, Weizmann Institute of Science, Rehovot 76100, Israel
e-mail: anat.yarden@weizmann.ac.il

D.F. Treagust and C.-Y. Tsui (eds.), *Multiple Representations in Biological Education*, Models and Modeling in Science Education 7, DOI 10.1007/978-94-007-4192-8_6,
© Springer Science+Business Media B.V. 2013

processes which are known to be an intellectual challenge for high school students (Falk et al., 2008; Marbach-Ad, 2001). According to Malacinski and Zell (1996), students' difficulties in understanding molecular concepts and processes are especially attributed to the emphasis on minute details and abstract concepts. Indeed, even though teachers regard this topic as important and interesting to students, most of them choose not to teach it, due to its subject matter difficulties (Steele & Aubusson, 2004). Thus, there is a strong need for a more concrete and accessible means of demonstrating and visualizing the course of action and applications of molecular processes.

Multimedia instructional environments in general, and animations in particular, have a great potential for improving the way people learn (Kelly & Jones, 2007; Mayer & Moreno, 2002; Stith, 2004; Williamson & Abraham, 1995). When an animation simulates real processes which include, for instance, motion, it allows learners to execute virtual experiments that would be costly, dangerous, or otherwise not feasible in a school laboratory. The idealization of complex laboratory experiments, as in simulations, is helpful in reducing errors and focusing students' attention on particular abstract concepts or isolating variables that are normally combined (Hennessy, Deaney, & Ruthven, 2006).

The studies presented in this chapter aimed to identify the cognitive as well as the pedagogical factors involved in using animations while learning and teaching biotechnological methods in high school. Specifically, we aimed to (1) explore how the use of animations affects high school students' comprehension of biotechnological methods and (2) characterize the pedagogical characteristics of enacting animations in class while teaching biotechnological methods. This chapter is divided into two parts; each part focuses on one of the above two aims, and a general discussion follows.

How the Use of Animations Affects High School Students' Comprehension of Biotechnological Methods

Cognitive Basis of Learning Using Visualization Tools

In designing multimedia presentations involving animations, instructional designers base their decisions on theories of how students learn from words and pictures. Those theories are relevant for learning and teaching in general, and they appear to be most relevant in biology education in particular. One of those theories is the cognitive theory of multimedia learning (Mayer & Moreno, 2002) which is based on three fundamental assumptions. According to the first assumption, the dual-channel assumption (Paivio, 1986), humans have separate channels for processing visual and verbal representations. Therefore, information encoded in both channels will be better remembered than information encoded in only one of the channels. Because pictures, whether they are dynamic or static, may be coded both visually and verbally, they are more likely to be remembered than words. There is a strong empirical evidence that learning outcomes are improved by presenting the learner with verbal and pictorial information in a coordinated fashion (Hoffler & Leutner, 2007). In biology education, where we are dealing with phenomena that are for the most part abstract, the integration between verbal and concrete pictorial information seems to be most significant.

The second assumption is the limited-capacity assumption (Baddeley, 1997) which postulates that only a few pieces of information can be actively processed at any one time in each of the two separate channels (for processing visual and verbal representations). This assumption goes together with the cognitive load theory (Sweller, 1994) in that the working memory's capacity sets very narrow limitations. This aspect is particularly relevant in biology education where there is a burden of diverse concepts and processes, most of which are totally new to the learners (Yarden, Marbach-Ad, & Gershony, 2004), as well as a requirement to generate large conceptual frameworks (Trowbridge & Wandersee, 1996). In this situation, cognition in general and memory in particular are faced with a considerable challenge. Hence, there is a need for tools that will assist in reducing the inherent cognitive load as well as relieving the limited organic capacities for processing information.

The third assumption, the active-processing assumption, states that meaningful learning (Ausubel, 1968) occurs when the learner engages in active cognitive processes such as selecting relevant material, organizing it into a coherent representation, and integrating it with existing knowledge (Mayer, 1996; Wittrock, 1974). This active processing is most likely to occur when the learner has corresponding pictorial and verbal representations in his/her working memory simultaneously, and thus this theory predicts that multimedia presentations, such as narrated animations, are most likely to lead to meaningful learning.

According to the information delivery theory of multimedia learning (Mayer, 1996), the computer is an information delivery system for learners. When the information is presented in words (such as narration), the learner stores the information in his or her memory. According to this theory, adding multimedia (such as animation) to the verbal information should have no effect on what is learned if the pictures contain the same information as the words. Thus, according to this theory, multimedia presentations should not result in better learning than single-medium presentations. However, in a mixed situation with learners who favor visual presentations and others who favor verbal ones, a multimedia presentation might be equally effective in delivering information to both groups of learners. We are most familiar with students' multiplicity of learning styles (Tobias, 1990); therefore, tools such as animation, which can be effective for visual as well as verbal learners, could be extremely valuable.

In distinguishing between static and dynamic visualizations, multimedia may be a relatively new technology, but the addition of images to text in order to facilitate learning has a much longer history. Pictures can be used to accompany texts in order to improve their comprehensibility and memorability (Large, 1996). However, Tversky and Morrison (2002) found no advantage of animations over static graphics in 20 primary studies that they reviewed. In contrast, a more recent meta-analysis indicated a statistically significant advantage in favor of animations over static pictures (Hoffler & Leutner, 2007).

Obviously, there are some significant differences in the interpretation of information from dynamic versus static displays, which are not consistently in favor of the dynamic ones. Some of those differences can be explained from the perspective of the cognitive load theory (Sweller, 1994). For example, when viewing an animation, "one views one frame at a time, and once the animation or video has advanced beyond a given frame, the previous frame is no longer available to the viewer" (Hegarty, 2004, p. 346). This situation may place a heavy demand on the working memory, especially in cases when information presented earlier in the animation should be integrated with information that is presented later. In contrast, when viewing a static display, viewers can reinspect different parts of the display as often as they wish (Ainsworth & van Labeke, 2004). An alternative point of view is that the ability to introduce each step independently in animations reduces the clutter of static illustrations, in which all of the steps are shown at once (Stith, 2004). Individual differences, such as spatial ability (Yang, Andre, Greenbowe, & Tibell, 2003) or prior knowledge (ChanLin, 2001), can also influence whether static pictures or animations are superior within a specific domain. In the case of low prior content knowledge, learning from molecular representations can be a difficult process (Cook, 2006). Students who have little or no knowledge of the domain depend heavily on observable phenomena to construct understanding (Seufert, 2003), that is, they use what can be easily observed. For that reason, some educational practices favor the use of dynamic visuals over static illustrations because they provide the learners with a ready-made, explicit, and dynamic representation of the phenomena (Williamson & Abraham, 1995). On the other hand, static displays require the learner to construct a dynamic mental model using the static information provided. For instance, students who are expected to learn about changes in matter or motion using static visuals have reported that they had to visualize those changes using static information, whereas when learning from dynamic visuals, the corresponding changes were apparent (Ardac & Akaygun, 2005). Still, students with low levels of prior knowledge may have difficulty extracting information from complex animations. Blissett and Atkins (1993) reported that individuals with less prior knowledge or lower achievers tended to find the learning demands confusing when animation is used.

From the cognitive load perspective, the preference of the visualization format can be conflicting. Although dynamic visuals may reduce the load of cognitive processing by directly supporting the construction of a mental model, their transitory nature may cause higher cognitive load because learners have less control of their cognitive processing (Lewalter, 2003). In addition, although animations can provide learners with explicit dynamic information that is unavailable in static graphics, the inclusion of a temporal change in visual displays introduces additional information-processing demands (Lowe, 2003).

Even though there is no obvious cognitive advantage to dynamic over static media, dynamic media are considered to have enormous potential for instruction (Hegarty, 2004). In the next part of this chapter, we attempt to determine what conditions or what learning terms may enable dynamic visualizations to be effective in learning biotechnological methods.

Examples of Animations of Biotechnological Methods

At the molecular level, biotechnological methods are completely invisible and intangible to students. To demonstrate the mechanisms behind those methods, we developed animations which accompany a textbook which we developed in genetic engineering (Michael & Yarden, 2007). Each animation introduces, sequentially, the procedure of the biotechnological method being demonstrated—using restriction enzymes to digest DNA, cloning a gene into a plasmid, creating a DNA library, and the polymerase chain reaction (PCR) (Falk et al., 2003; Yarden & Yarden, 2007).

One of the most helpful and effective features of animations is their interactive use (Hegarty, 2004; Stith, 2004). Stopping, starting, and replaying an animation can allow reinspection, focusing on specific parts and actions. Animations that allow close-ups, zooming in, alternative perspectives, and speed control are even more likely to be facilitative to learners (Tversky & Morrison, 2002). Thus, two alternative versions were developed for each animation: a continuous version, showing the whole procedure of the biotechnological method continuously, and a sequential version, showing the process gradually, or step by step. The animations were divided into steps according to the way in which the various biotechnological methods are carried out in the laboratory, that is, whenever a new stage is encountered such as heating, a new step is demonstrated in the animation. In addition, the steps were selected according to transitions from macro to micro perspectives and vice versa.

Each animation includes a written text which appears in close proximity to the animation and describes what is being shown—according to the spatial contiguity principle (Mayer & Moreno, 2002). In addition, each animation is accompanied by components of active learning in the form of computerized tasks—according to the cognitive theory of multimedia learning (Mayer & Moreno). The tasks are aimed at identifying students' attention to key issues in the biotechnological methods being demonstrated as well as to understanding the symbols and images which appear in the animations themselves. Those animations were used as a context to the study that is described in the following sections.

Students' Comprehension of PCR Using Animation and Still Images

In our study (Yarden & Yarden, 2010) using pre- and post-intervention questionnaires, we identified the differences between a group of students (12th graders, biology majors; $n = 90$) who used the PCR animation in order to visualize the PCR method and a comparison group ($n = 83$) who used equivalent still images to visualize the PCR method. We found a statistically significant advantage for the animation group over the still images group using a t test, t (171) = 4.64, and $p < 0.0001$. Since no significant differences were found between students' prior knowledge, we concluded that the use of the PCR animation as a visualization tool provided an advantage to learners of the PCR method. In addition, regression analysis indicated a positive correlation between students' prior knowledge and their understanding of the PCR method in the still images group ($R^2 = 0.412$). Students with a low prior content knowledge achieved low scores in the post-intervention questionnaire, while students with a high level of prior knowledge achieved high scores in the post-intervention questionnaire. In contrast, for the animation group, the level of students' prior content knowledge seemed to have no noteworthy effect on their success in the post-intervention questionnaire, namely, on their understanding of the PCR method ($R^2 = 0.091$). Thus, prior content knowledge was found to be an important factor for students who learned PCR using still images, whereby low prior knowledge could serve as an obstacle to learning the PCR. In contrast, the same variable had no noticeable effect on students who learned PCR using animation.

Using the conceptual status framework (Hewson & Lemberger, 2000; Tsui & Treagust, 2007) for analyzing students' discourse while learning about the PCR, we also found that the use of the animation was advantageous in understanding the mechanistic aspects of the method compared to students who learned using still images. Students from the animation group and from the still images group had reached the kind of understanding reflected by the conceptual status of intelligibility, indicating that they knew what the concepts of PCR mean and they could represent them using images, language, or examples. However, the next level of understanding—which is reflected by holding the plausibility conceptual status—appeared to be available only to the students who watched the animation. As expressed in their conversations, the students who used the animation were able to understand the causal relationships between different molecules in the PCR method, as well as the ontological function of those molecules. Regarding the third and highest conceptual status of fruitfulness, it appeared that neither group reached this level of understanding: They did not reveal significantly in their conversations that they had found the concepts of the PCR method useful in solving problems or in suggesting new possibilities and directions (Yarden & Yarden, 2010).

Students' Comprehensions of Restriction Enzyme Digestion of DNA Using Animation

In an additional study (Yarden, 2010), concept maps were used as a tool for identifying students' (12th graders, biotechnology majors, $n = 38$) understanding of the process of restriction enzyme digestion of DNA using animation. Students were asked to construct concept maps before and after watching an animation from a written list of eight concepts, namely, DNA, restriction enzyme, restriction site, nucleotides, sticky ends, DNA strands, phosphodiester bonds, and palindromic sequence. Students were instructed to think about as many connections as possible between those eight concepts, to draw lines between any two concepts, and to write

Table 6.1 Analysis of students' propositions in their pre-watching and post-watching concept maps

	Sample B1[a] ($n = 15$)			Sample B2[a] ($n = 23$)		
Factors that were tested	Pre-watching concept maps	Post-watching concept maps	Significance of the difference between the paired maps[b]	Pre-watching concept maps	Post-watching concept maps	Significance of the difference between the paired maps[b]
Average number of propositions	10.66	16.4	($p < 0.0001$)	5.21	7.43	($p < 0.0001$)
Average percent of correct propositions	84.66	90.2	($p < 0.0001$)	81.91	92.91	($p < 0.0001$)
Average percent of structural propositions	61.26	56.06	($p < 0.0001$)	75.21	65.47	($p < 0.0001$)
Average percent of functional propositions	38.74	43.94	($p < 0.0001$)	24.79	34.53	($p < 0.0001$)

[a]Samples B1 and B2 represent two 12th grade classes from two different high schools
[b]The Wilcoxon signed rank statistical test was used to test whether the differences identified between the pre-watching and the post-watching maps in each subgroup are significant. A *t* test was not used here because of the small sample size

on the line a sentence which reflects a proposition between those two concepts. After the students had watched the restriction enzyme animation, they were asked to build another concept map from the same eight given concepts.

As can be seen in Table 6.1, the number of propositions was significantly greater in students' post-watching concept maps than in their pre-watching maps. A closer look at the nature of the propositions in both student samples reveals that besides the significant increase in the number of propositions between the pre-watching and the post-watching maps in general, there was also a significant increase in the percentage of the correct propositions in students' post-watching concept maps in both groups. Thus, as reflected in the accuracy of the propositions made in students' post-watching concept maps, it seemed that watching the animation demonstrating restriction enzyme digestion of DNA had made this biotechnological method clearer and more coherent to the students,

After classifying the propositions that students had written in terms of structural versus functional type of propositions, we observed a significant decline in the number of propositions with a structural nature between the pre-watching and the post-watching concept maps of both groups. Accordingly, there was a significant increase in the number of propositions that were classified as functional. Within the pre-watching concept maps, structural propositions—such as "restriction site is composed

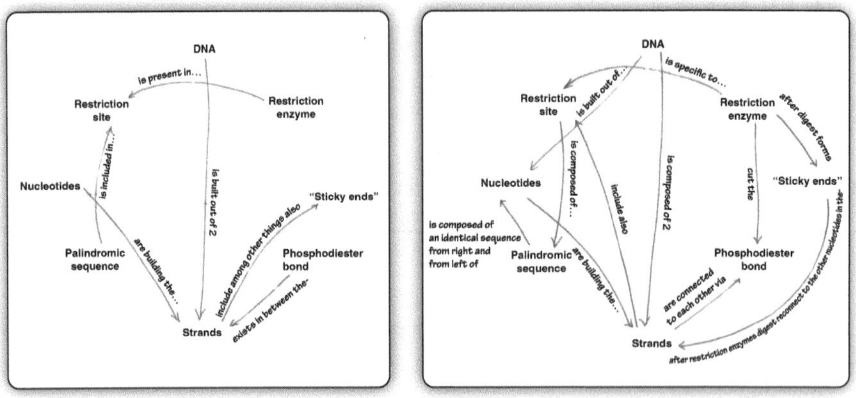

Fig. 6.1 An example of student's paired pre-watching (*left*) and post-watching (*right*) concept maps (subsample of Sample B2)

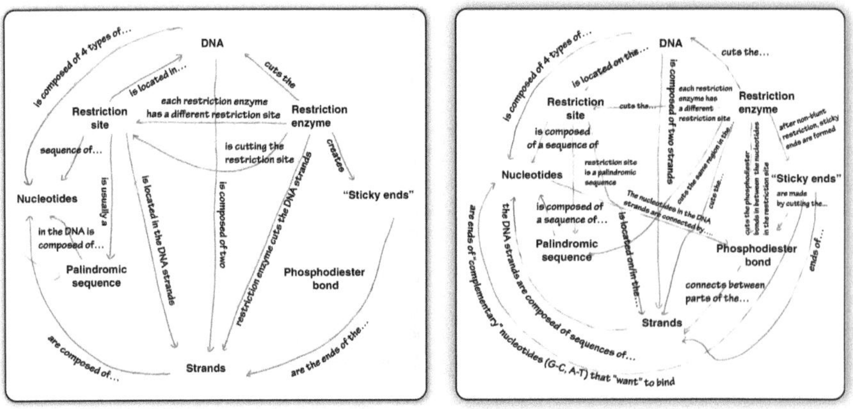

Fig. 6.2 Another example of student's paired pre-watching (*left*) and post-watching (*right*) concept maps (subsample of Sample B1)

of nucleotides," or "palindrome sequence is inside the DNA strands"—were most common. In the post-watching concept maps, most of the propositions dealt with the functions or configuration through action of molecules such as "the restriction enzyme cuts the phosphodiester bond" or "sticky ends are being configured as a consequence of a graded digestion by the restriction enzyme." Two examples of paired pre-watching and post-watching concept maps are shown in Figs. 6.1 and 6.2.

As can be seen in Fig. 6.1, there are more propositions and more correct propositions, as well as more propositions that can be classified as functional in the post-watching concept map, compared to the paired pre-watching map. For example, it can be seen that the concepts sticky ends and phosphodiester bond are more connected to other concepts in the post-watching concept map. In the pre-watching concept map, those concepts appeared in the propositions—"[strands] include among other things also [sticky ends]" and "[phosphodiester bond] exists in between the [strands]" (structural nature of propositions)—whereas in the paired post-watching concept map, they appeared in new propositions such as "[restriction enzyme] cuts the [phosphodiester bond]" or "after restriction enzyme digest reconnect to the other nucleotides in the [sticky ends]."

In Fig. 6.2, it is also observable that there is an increase in the number of propositions from the pre-watching concept map to the post-watching concept map. On looking deeper into the nature of the propositions, it can be noticed that also here the concepts sticky ends and phosphodiester bond are more connected to other concepts in the post-watching concept map, with propositions whose nature can be classified mostly as functional, for instance: "[restriction enzyme] cuts the phosphodiester bond in between nucleotides in the restriction site." However, in the paired pre-watching concept map, the concept sticky ends appeared only in two propositions, and only one of them can be classified as functional ("[restriction enzyme] creates [sticky ends]").

Thus, the use of the restriction enzyme animation enabled the students to increase the number of propositions they could write between concepts in general and the correct ones in particular. Additionally, the use of animation while learning about restriction enzyme digestion promoted the students' understanding about the functional relationships between molecules that participate in this biotechnological method in terms of its mechanistic aspects in a way similar to the case of learning the PCR method using animation (cf. Yarden & Yarden, 2010).

The Pedagogical Characteristics of Enacting Animations in Class While Teaching Biotechnological Methods

The Role of the Teacher While Enacting Animations in Class

Using animations alone does not ensure learning. Animations are occasionally linked with unquestionable, sometimes simplified, models of a scientific process that give students the impression that every variable is easily controlled (Hennessy et al., 2006). It seems that students tend to attribute a great deal of authority to the computer and accordingly may develop misconceptions by taking animations and images of abstract concepts too literally (Wellington, 2004).

In some studies, students were reported to be engaged in unplanned, inefficient, and inconclusive experimentation while learning with simulations (de Jong & van Joolingen, 1998), and sometimes they missed essential features while watching animations alone (Kelly & Jones, 2007). Productive learning requires staged, structured tasks and systematic experimentation (Linn, 2004). Hence, it is most important to make implicit reasoning explicit to highlight any inconsistencies

(Hennessy et al., 2006). For students to be able to learn new concepts and processes which they encounter in a meaningful way, students must also relate new knowledge and information with concepts and claims they have already held (Ausubel, 1963). Since learning is viewed by this perspective as an accumulating process, it is also most important to construct the knowledge being learned gradually, as well as to organize it under main principles (Chi, Feltovich, & Glaser, 1981).

In view of the above perspectives about learning, it seems that the teacher plays a crucial role while learning from animations. There is a strong necessity for the teacher's coaching together with the software supports to address the students' learning needs and their interactions with each other to produce a robust form of support for students (Tabak, 2004). According to Soderberg and Price (2003), teachers should discuss and challenge students' own ideas as well as highlight the limitations of computer models themselves. The effectiveness of whole class instruction of animations might improve if teachers challenge and question the inconsistencies and contradictions between verbal explanations and the corresponding molecular representations (Ardac & Akaygun, 2005). In addition, connections should be made between students' lives and the subject matter being learned, between principles and practice, as well as between the past and the present.

The role of the teacher is also central in the dissemination of curricular initiatives (Barab & Luehmann, 2003). More specifically, the successful introduction of computer-aided instruction, as a tool for enhancing learning as well as teaching, depends on positive attitudes of the teachers (Dori & Barnea, 1997). Science teachers' beliefs affect their attitudes, and these attitudes affect their intentions to incorporate computer-aided instructional tools into class (Zacharia, 2003). Consequently, while examining the enactment of animations in class, it is important to study the teacher's perspective, namely, the teachers' perceptions, challenges, and recommended pedagogical strategies. In the following part, we describe the pedagogical characteristics of two teachers enacting animations for teaching the biotechnological methods in our study.

Enacting Biotechnological Methods Using Animations: Two Case Studies

In this study (Yarden & Yarden, 2011), we attempted to study two teachers' potential contribution to teaching biotechnological methods using animations in two exemplary case studies. Two biotechnology teachers, Ravit and Dora (pseudonyms), enacted several animations while teaching biotechnological methods in their classes. Our analysis revealed that the two teachers contributed to the enactment of animations in the following three aspects: establishing the hands-on point of view, helping students deal with the cognitive load that accompanies the use of animations, and implementing constructivist aspects of knowledge construction.

Establishing the Hands-On Point of View

Analysis of class observations obviously showed that both Ravit and Dora often discussed with their students about how the biotechnological methods—both the rationale and the practical procedure behind various steps in the biotechnological methods that were introduced and demonstrated in the animations—are actually carried out in practice in the laboratory. They gave their students the hands-on point of view by making them aware of the existence of some steps skipped in the animations but are nevertheless important when performing the relevant biotechnological method in the laboratory.

Guided Watching: Help Dealing with the Cognitive Load

Both Ravit and Dora guided their students while watching the animations. Ravit, who tended to be more teacher centered, did this by leading her students' navigation through the animations. Dora, who tended to employ a more student-centered approach, supported her students on several occasions during the learning activity with the animations whenever they had misunderstandings. Both teachers focused their students' attention on important details in the animations and kept asking them different questions about objects in the animations which they were watching.

Ravit explained in an interview that by guiding students through watching animations, she was making the animations more comprehensible for her students:

> Look, I could sit, read a book, and let them watch the animation alone to the end. I believe that in that way they would lose some important points which they might miss because they did not notice them through all the details and changes in the animation.

In addition to the nature of animations, with their dynamic changes and intrinsic visual and cognitive load, Ravit explained that she was directing the students while they watched the animations because of the nature of the subject matter (the biotechnological methods) which is abstract and complex, and therefore, careful watching is needed, especially in animations on this topic in order to identify, for instance, fundamental differences between the structures of similar molecules.

Dora summarized in an interview the type of support she believed she had given her students:

> Focus is the key word here in order to cope with the visual load while they watch. The students could have looked over and over again at the different kinds of bacteria in the animation, but they really need my help to look for the five different plasmids, to focus on each of the plasmids and on its unique elements.

Implementing Constructivist Aspects of Knowledge Construction

Both Ravit and Dora implemented elements of constructivist teaching while using the animations in class, in keeping with the constructivist perspective of Ausubel (1963), for example, more effort should be made by the teacher to engage students more deeply and thoughtfully in any kind of subject-matter learning. Both teachers considered the animation activity as important in the construction of students' understanding of the biotechnological methods. For example, Ravit clearly established the animation activity on students' prior knowledge in biotechnology in order to make it more relevant and meaningful. Also both teachers made the animation activity more meaningful by connecting it explicitly to other activities in the students' learning sequence, such as laboratory experiences.

In her interview, Ravit stressed why she believed that it is so important to link the animation activity to other learning activities to which students have been exposed:

> It is most important to link the animation activity to the trip, to experiences we have had in the lab. Otherwise the student might say: "this belongs to the lab, this to the animation, there is no connection between them."

With different teaching styles, the two teachers tended to perform differently with regard to supporting students' understanding of biotechnological methods while watching the animations. Ravit, with her teacher-centered approach, supported her students by explaining and expanding on the meaning of concepts she believed are crucial for their understanding, and in her interview, she explained why conceptualization of the process that the students had just watched in the animation is so important: "The students are watching a process in the animation but they must know its name, the concept behind what is being demonstrated in the animation."

Whereas Ravit based her supporting efforts while enacting the animations on her own pedagogical and content knowledge, Dora based her supporting efforts on students' difficulties and misunderstandings to which they were exposed during the enactment of the animations. In response to the student's question, Dora discussed the process of plasmid replication beyond what was shown in the animation in order to make the processes in the animation more understandable for her students:

> Student: Dora, I don't understand. Why do we need the origin of replication in the plasmid?
> Dora: Why is it important that the plasmid replicate? Where does it replicate?
> Student: I don't know.
> Dora: In a test tube? Inside a living cell?
> Student: It can do that inside a cell.
> Dora: Only inside a cell. What is needed in order to replicate DNA?

In her interview, Dora revealed that after examining the animation with her students, she became aware of places in which they needed assistance to gain a meaningful understanding. Her presence at that point of the animation enabled her to support the students whenever they encountered concepts or objects which they found not so comprehensible.

Discussion and Conclusions

Learning from animations is not a simple task, even though it might appear to be so. Although animations can provide learners with explicit dynamic information, the inclusion of a temporal change introduces additional information-processing demands, and the transitory nature may lead to a cognitive load because learners have less control of their cognitive processing (Lewalter, 2003; Lowe, 2003). This chapter attempts to represent the complexity of viewing animations while learning biotechnological methods in terms of the cognitive and pedagogical factors involved.

Our first study (Yarden & Yarden, 2010) enabled us to show that animations do have a unique contribution in promoting biology majors' and biotechnology majors' conceptual understanding of biotechnological methods. Previous studies have already shown significantly higher understanding among students who used animations compared with those who used still images in their learning of molecular motion (e.g., Ardac & Akaygun, 2005; Williamson & Abraham, 1995). Animations can give an accurate and rich picture of the dynamic nature of molecules and molecular interactions which is often very difficult to understand (National Science Foundation, 2001). Our findings were also similar to those of Marbach-Ad, Rotbain, and Stavy (2008) who showed that computer animations are effective for learning molecular genetics, especially about the dynamic processes. We also have shown that prior knowledge is not an essential factor when learning using animation. The explicit, expert-like, dynamic representations of the phenomena in animations might explain why students depended less on their prior knowledge when learning with animations as opposed to when learning with static illustrations (Williamson & Abraham, 1995).

In addition to identifying the advantage that animations have on still images for visualizing biotechnological methods, we also found in our study (Yarden & Yarden, 2010) that the use of the animation gave the students an advantage in understanding the mechanistic aspects of these methods, namely, the ontological function of different molecules and the causal mechanism that invokes them. This advantage was also reflected while analyzing biotechnology students' concept maps, before and after viewing the restriction enzyme's animation (Yarden, 2010). According to Pallant and Tinker (2004), molecular dynamics tools help students develop more scientifically accurate mental models of molecular-scale phenomena. Our findings also implied that for such students' tasks, animations serve as a better alternative than the static visuals.

Due to the cognitive load involved, it was reported that students sometimes miss essential features when they watch animations alone (e.g., Hegarty, 2004; Kelly & Jones, 2007). Consequently, it seems that the teacher's role is important in structuring tasks and questions in ways that prompt students' thinking about underlying concepts and relationships being introduced in animations (e.g., Soderberg & Price, 2003) and guiding and helping them to reformulate their thinking when learning with animations (e.g., Parker, 2004). Indeed, in another study (Yarden & Yarden, 2012), we identified three aspects of the contribution of two exemplary biotechnology teachers—to the enactment of animations in class while learning biotechnological methods—(1) establishing the hands-on point of view, (2) helping students deal with the cognitive load that accompanies the use of animations, and (3) implementing constructivist aspects of knowledge construction.

Both teachers in our study (Yarden & Yarden, 2012) implemented elements of constructivist teaching while they used animations in class, namely, they established clearly the activity with the animation on students' prior knowledge as well as connected it explicitly to other activities on the students' learning sequence such as laboratory experiences. Constructivist teachers tend to explore how their students see any problem or issue they encounter in any learning situation and why their path toward understanding seems promising (Glasersfeld, 1998). Thus, the role of the teacher while enacting animations in class is critical in order to make learning of biotechnological methods meaningful.

The findings of our studies presented in this chapter might be usefully extended—beyond the context of visualizing biotechnological methods—to other diverse topics and biological processes that involve motion and interactions between different key factors. Such processes might include macroscopic interactions, for instance, in ecology, as well as molecular processes, which are not visible in the real world. These findings strengthen our assumption that students and teachers should work together in transforming knowledge while learning with animations.

References

Ainsworth, S., & van Labeke, N. (2004). Multiple forms of dynamic representations. *Learning and Instruction, 14*, 241–255.

Ardac, D., & Akaygun, S. (2005). Using static and dynamic visuals to represent chemical change at molecular level. *International Journal of Science Education, 27*(11), 1269–1298.

Ausubel, D. P. (1963). *The psychology of meaningful verbal learning*. New York/London: Grune and Stratton.

Ausubel, D. P. (1968). *Educational psychology: A cognitive view*. New York: Holt, Rinehart & Winston, Inc.

Baddeley, A. (1997). *Human memory: Theory and practice* (Rev. ed., pp. 9–27). Hove, UK: Psychology Press.

Barab, S. A., & Luehmann, A. L. (2003). Building sustainable science curriculum: Acknowledging and accommodating local adaptation. *Science Education, 87*, 454–467.

Blissett, G., & Atkins, M. (1993). Are they thinking? Are they learning? A study of the use of interactive video. *Computers in Education, 21*, 31–39.

ChanLin, L. J. (2001). Formats and prior knowledge on learning in a computer-based lesson. *Journal of Computer Assisted Learning, 17*, 409–419.

Chi, M. T. H., Feltovich, P. J., & Glaser, R. (1981). Categorization and representations of physics problems by experts and novices. *Cognitive Science, 5*, 121–152.

Cook, M. P. (2006). Visual representations in science education: The influence of prior knowledge and cognitive load theory on instructional design principles. *Science Education, 90*(6), 1073–1091.

de Jong, T., & van Joolingen, W. R. (1998). Scientific discovery learning with computer simulations of conceptual domains. *Review of Educational Research, 68*(2), 179–201.

Dori, Y. J., & Barnea, N. (1997). In-service chemistry teachers' training: The impact of introducing computer technology on teachers' attitudes and classroom implementation. *International Journal of Science Education, 19*(5), 577–592.

Dori, Y. J., Tal, R. T., & Tsaishu, M. (2003). Teaching biotechnology through case studies – Can we improve higher order thinking skills of nonscience majors? *Science Education, 87*, 767–793.

Falk, H., Brill, G., & Yarden, A. (2008). Teaching a biotechnology curriculum based on adapted primary literature. *International Journal of Science Education, 30*(14), 1841–1866.

Falk, H., Piontkevitz, Y., Brill, G., Baram, A., Yarden, H., & Yarden, A. (2003). *Gene tamers: Studying biotechnology through research*. A web site, from http://stwww.weizmann.ac.il/g-bio/biotech/

Hegarty, M. (2004). Dynamic visualizations and learning: Getting to the difficult questions. *Learning and Instruction, 14*, 343–351.

Hennessy, S., Deaney, R., & Ruthven, K. (2006). Situated expertise in integrating use of multimedia simulation into secondary science teaching. *International Journal of Science Education, 28* (7), 701–732.

Hewson, P., & Lemberger, J. (2000). Status as the hallmark of conceptual learning. In R. Millar, J. Leach, & J. Osborne (Eds.), *Improving science education: The contribution of research*. Buckingham, UK: Open University Press.

Hoffler, T. N., & Leutner, D. (2007). Instructional animation versus static pictures: A meta-analysis. *Learning and Instruction, 17*(6), 722–738.

Kelly, R. M., & Jones, L. L. (2007). Exploring how different features of animations of sodium chloride dissolution affect students' explanations. *Journal of Science Education and Technology, 16*, 413–429.

Large, A. (1996). Computer animation in an instructional environment. *Library and Information Science Research, 18*(1), 3–23.

Lewalter, D. (2003). Cognitive strategies for learning from static and dynamic visuals. *Learning and Instruction, 13*, 177–189.

Linn, M. C. (2004). Using ICT to teach and learn science. In R. H. E. Scanlon (Ed.), *Mediating science learning through information and communications technology* (pp. 9–26). London: Routledge Falmer.

Lock, R., Miles, C., & Hughes, S. (1995). The influence of teaching on knowledge and attitudes in biotechnology and genetic engineering contexts: Implications for teaching controversial issues and the public understanding of science. *Secondary Science Review, 76*(276), 47–59.

Lowe, R. (2003). Animation and learning: Selective processing of information in dynamic graphics. *Learning and Instruction, 13*, 157–176.

Malacinski, G. M., & Zell, P. W. (1996). Manipulating the "invisible": Learning molecular biology using inexpensive models. *The American Biology Teacher, 58*(7), 428–432.

Marbach-Ad, G. (2001). Attempting to break the code in student comprehension of genetic concepts. *Journal of Biological Education, 35*(4), 183–189.

Marbach-Ad, G., Rotbain, Y., & Stavy, R. (2008). Using computer animation and illustration activities to improve high school students' achievement in molecular genetics. *Journal of Research in Science Teaching, 45*(3), 273–292.

Mayer, R. E. (1996). Learners as information processors: Legacies and limitations of educational psychology's second metaphor. *Educational Psychologist, 31*, 151–161.

Mayer, R. E., & Moreno, R. (2002). Animation as an aid to multimedia learning. *Educational Psychology Review, 14*(1), 87–99.

Michael, D., & Yarden, A. (2007). *Genetic engineering: From principles and methods to research and applications* (2nd ed., 2008 ed.). Rehovot, Israel: The Amos de-Shalit Center for Science Teaching.

National Science Foundation. (2001). *Molecular visualization in science education*. Report from the molecular visualization in science education workshop. Arlington, VA: National Science Foundation.

Olsher, G., & Dreyfus, A. (1999). Biotechnologies as a context for enhancing junior high-school students' ability to ask meaningful questions about abstract biological processes. *International Journal of Science Education, 21*(2), 137–153.

Paivio, A. (1986). *Mental representations: A dual coding approach*. New York: Oxford University Press.

Pallant, A., & Tinker, R. F. (2004). Reasoning with atomic-scale molecular dynamic models. *Journal of Science Education and Technology, 13*, 51–66.

Parker, J. (2004). The synthesis of subject and pedagogy for effective learning and teaching in primary science education. *British Educational Research Journal, 30*(6), 819–839.
Seufert, T. (2003). Supporting coherence formation in learning from multiple representations. *Learning and Instruction, 13*(2), 227–237.
Soderberg, P., & Price, F. (2003). An examination of problem-based teaching and learning in population genetics and evolution using evolve, a computer simulation. *International Journal of Science Education, 25*(1), 35–55.
Steele, F., & Aubusson, P. (2004). The challenge in teaching biotechnology. *Research in Science Education, 34*, 365–387.
Stith, B. J. (2004). Use of animations in teaching cell biology. *Cell Biology Education, 3*, 181–188.
Sweller, J. (1994). Cognitive load theory, learning difficulty, and instructional design. *Learning and Instruction, 4*, 295–312.
Tabak, I. (2004). Synergy: A complement to emerging patterns of distributed scaffolding. *The Journal of the Learning Sciences, 13*(3), 305–335.
Tobias, S. (1990). *They're not dumb, they're different: Stalking the second tier*. Tucson, AZ: Research Corporation.
Trowbridge, J. E., & Wandersee, J. H. (1996). How do graphics presented during college biology lessons affect students' learning? *Journal of College Science Teaching, 26*(1), 54–57.
Tsui, C.-Y., & Treagust, D. F. (2007). Understanding genetics: Analysis of secondary students' conceptual status. *Journal of Research in Science Teaching, 44*(2), 205–235.
Tversky, B., & Morrison, J. B. (2002). Animation: Can it facilitate? *International Journal of Human Computer Studies, 57*, 247–262.
von Glasersfeld, E. (1998). Cognition, construction of knowledge and teaching. In M. R. Matthews (Ed.), *Constructivism in science education: A philosophical examination* (pp. 11–30). Dordrecht, The Netherlands: Kluwer Academic Publishers.
Wellington, J. (2004). Multimedia in science teaching. In R. Barton (Ed.), *Teaching secondary science with ict*. Cambridge, UK: Open University Press.
Williamson, V. M., & Abraham, M. R. (1995). The effects of computer animation on the particulate mental models of college chemistry students. *Journal of Research in Science Teaching, 32*, 521–534.
Wittrock, M. C. (1974). Learning as a generative activity. *Educational Psychologist, 11*, 87–95.
Yang, E. M., Andre, T., Greenbowe, T. J., & Tibell, L. (2003). Spatial ability and the impact of visualization/animation on learning electrochemistry. *International Journal of Science Education, 25*(3), 329–349.
Yarden, H. (2010). *Learning and teaching biotechnological methods using interactive animations*. Ph.D., thesis Weizmann Institute of Science, Rehovot, Israel.
Yarden, H., Marbach-Ad, G., & Gershony, J. M. (2004). Using the concept map technique in teaching introductory cell biology to college freshmen. *Bioscene-Journal of College Biology Education, 30*(1), 3–13.
Yarden, H., & Yarden, A. (2007). *Genetic engineering: From principles and methods to research and applications*. A website, from http://stwww.weizmann.ac.il/g-bio/geneengine/animations.html
Yarden, H., & Yarden, A. (2010). Learning using dynamic and static visualizations: Students' comprehension, prior knowledge and conceptual status of a biotechnological method. *Research in Science Education, 40*, 375–402.
Yarden, H., & Yarden, A. (2011). Studying biotechnological methods using animations: The teacher's role. *Journal of Science Education and Technology, 20*(6), 689–702. doi:10.1007/s10956-010-9262-3.
Zacharia, Z. (2003). Beliefs, attitudes, and intentions of science teachers regarding the educational use of computer simulations and inquiry-based experiments in physics. *Journal of Research in Science Teaching, 40*(8), 792–823.

Chapter 7
Experts' Views on Translation Across Multiple External Representations in Acquiring Biological Knowledge About Ecology, Genetics, and Evolution

Konrad J. Schönborn and Susanne Bögeholz

Introduction

Students' successful communication as biologists is closely related to their competence in interpreting multiple external representations (MERs). Acquiring knowledge in the domains of ecology, genetics, and evolution involves *translating* across and between MERs that depict concepts and principles at different levels of biological organization and in varying modes of representation. Promoting translation processes in learners is pivotal to the development of biological understanding. This study is a follow-up from the research reported in Schönborn and Bögeholz (2009). A Delphi approach was adopted to collect a second round of data from the same expert panel that was interviewed 3 years ago. Specifically, the purpose of the study was (1) to investigate the validity of four types of biological knowledge identified in the first expert data collection, (2) to elucidate experts' views on the challenges facing learners upon engaging translation processes in constructing biological knowledge, and (3) to reveal experts' opinions of what overarching requirements are necessary for effective translation in the development of biological knowledge. The content focus of the present study was directed to the domains of ecology, genetics, and evolution.

K.J. Schönborn (✉)
Division of Media and Information Technology, Department of Science and Technology (ITN), Linköping University, Campus Norrköping, SE-60174 Norrköping Sweden
e-mail: konrad.schonborn@itn.liu.se; konsc@itn.liu.se

S. Bögeholz
Biologische Fakultät, Georg-August-Universität Göttingen,
Waldweg 26, 37073 Göttingen, Germany
e-mail: sboegeh@gwdg.de

D.F. Treagust and C.-Y. Tsui (eds.), *Multiple Representations in Biological Education*, Models and Modeling in Science Education 7, DOI 10.1007/978-94-007-4192-8_7,

© Springer Science+Business Media B.V. 2013

Theoretical Background

Types of Biological Knowledge

Analysis of the German national standards for biology education (Kultusministerkonferenz, 2005) and the core biology curriculum for the federal state of Lower Saxony (Niedersächsisches Kultusministerium, 2007) identified four hierarchical types of biological knowledge that learners are expected to acquire at the secondary level (see Fig. 7.1) (Schönborn & Bögeholz, 2009). Use of *types* of knowledge refers to "static knowledge about facts, concepts, and principles that apply within a certain domain" (de Jong & Ferguson-Hessler, 1996, p. 107).

The four types of knowledge (see Fig. 7.1) are defined as follows. *Type 1* knowledge (*biological terms*) constitutes the building-block *elements* of biological knowledge and could include *predator*, *prey*, *DNA*, and *genotype*. When the semantic relationship between two or more biological terms conveys biological meaning (e.g., a biological process), then this relationship exists as a biological

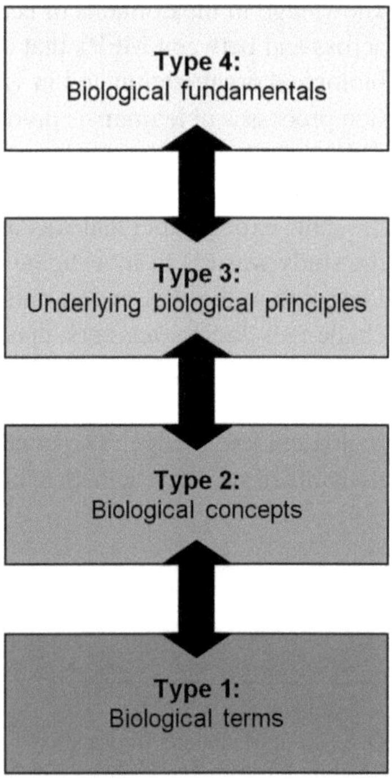

Fig. 7.1 Four types of biological knowledge that learners are required to develop at the secondary school level in Germany

concept (*type 2* knowledge: *biological concepts in* Fig. 7.1). At the school level, each biological concept exists on a continuum ranging from *broad* to *narrow* depending on the degree of the biological meaning that is communicated (e.g., *protein synthesis* vs. *DNA-methylation*). When a collection of biological concepts mutually communicates an underlying biological meaning, then this relationship exists as an underlying biological principle (*type 3* knowledge: *underlying biological principles*) (cf. Niedersächsisches Kultusministerium, 2007). Examples of biological principles could include the *principle of recapitulation* and the *competitive exclusion principle*. Lastly, when an underlying biological principle shares meaning with others, then together, they constitute components of *type 4* knowledge: *biological fundamentals*. For example, three fundamentals operationalized in the Kultusministerkonferenz (2005) document consist of *system, structure and function*, and *development*.

In this chapter, the knowledge types (see Fig. 7.1) are applied to the domains of *ecology*, *genetics*, and *evolution*. These three domains provide a concrete platform from which to consider learners' construction of biological knowledge. For example, Kinchin (2010) described evolution as a *disciplinary threshold* and guiding principle in biological understanding, Tsui and Treagust (2007) highlighted the centrality of genetics in modern biology education, and Kuechle (1995) described ecological knowledge as a principal field for an integrated biology education.

Translation Processes and Communication Competencies

Contemporary curricula stress the development of core *competencies* for promoting biological understanding (e.g., Labov, Reid, & Yamamoto, 2010). For example, in Germany, such an orientation (Kultusministerkonferenz, 2005; Niedersächsisches Kultusministerium, 2007) includes the competence area of *communication*, which also contains the ability to use MERs, such as photographs, micrographs, diagrams, drawings, graphs, and physical models in biology learning.

In biology, MERs communicate knowledge at different *levels* of biological organization that include the *subcellular*, *cellular*, *organ*, *organism*, and *population* levels and in different *modes of external representation (ER)* (e.g., *realistic* vs. *abstract* ERs) (Schönborn & Anderson, 2009). Kozma and Russell (1997) referred to the skills associated with interpreting different ERs as *representational competence*. A central cognitive component of engaging MERs in learning biology is the process of *translation*, which concerns the processing, mapping between, and moving across ERs (Ainsworth, 1999). Translation requires comprehending relationships between MERs and linking different ERs to the idea that is represented (Ainsworth, 2006). Engaging translation processes is necessary for successful biology learning (e.g., Tsui & Treagust, 2003).

Translation Across MERs in the Acquisition of Biological Knowledge

Schönborn and Bögeholz (2009) postulated the role of translation across ERs in the acquisition of different types of biological knowledge (see Fig. 7.1). To construct knowledge about a biological concept (type 2), learners may need to interpret and link the MERs that all depict the concept in the same or in varying modes of representation. Doing so may require applying knowledge about biological terms (type 1) to the necessary biological concept that is being represented and vice versa (bidirectional arrow in Fig. 7.1). For instance, examples A_1 and A_2 (biological concept) provided in an online Appendix I[1] require *translating horizontally* from one ER to another at the same level of biological organization. Examples B_1 and B_2 (biological concept) in Appendix I require *translating vertically* between ERs at different levels of biological organization.

To construct knowledge about an underlying biological principle (type 3), learners may need to interpret and link the MERs that each represent different biological concepts but collectively, depict one underlying principle. The MERs could depict the biological principle in the same or in varying modes of representation. Acquiring type 3 knowledge may require applying knowledge about biological concepts (type 2) to the underlying biological principle that is being represented and vice versa (see bidirectional arrows in Fig. 7.1). For instance, examples A_1 and A_2 (biological principle) in Appendix I require *translating horizontally* across ERs at the same level of biological organization. Examples B_1 and B_2 (biological principle) in Appendix I require *translating vertically* between ERs at different levels of biological organization. We hypothesize here that performing *horizontal* and *vertical* translation across MERs constitutes essential processes in students' acquisition of biological concepts and principles.

Delphi Approach for Obtaining Experts' Views

Delphi studies have the overall goal of attaining agreement or stability in an expert panel's opinions and judgments about a particular *problem* (e.g., Linstone & Turoff, 2002). Two main features of the Delphi technique are anonymity among participants and multiple *rounds* of data collection (e.g., Murry & Hammons, 1995).

The *first round* of a typical Delphi study is an open-ended collection of experts' opinions, often through open-ended questions or interviews. Following this, the researchers qualitatively summarize the responses, which inform the design of more focused questions. Together with communicating a summary of results

---

[1] Appendices I and II are permanently available at http://www.ep.liu.se/PublicationData/diva-85510/Appendix_I.pdf and http://www.ep.liu.se/PublicationData/diva-85510/Appendix_II.pdf, respectively.

from the first round to the panel, more focused questions constitute data collection in the second and subsequent rounds. The Delphi approach is considered complete once consensus or stability is reached.

Examples of Delphi studies in science education research include those by Osborne, Collins, Ratcliffe, Millar, and Duschl (2003) on experts' views of what key ideas should comprise school science curricula and by Häussler and Hoffmann (2000) on experts' views in developing a curricular framework for physics education. An assumed strength of the Delphi technique is that soliciting a group of experts' views increases the likelihood of *honing in* on the identified problem with greater validity (cf. Osborne et al., 2003).

Research Questions

In pursuit of further investigating experts' views on translation across MERs in acquiring biological knowledge in the domains of ecology, genetics, and evolution, the following three research questions were formulated:

1. To what extent do experts agree that the biological knowledge framework (see Fig. 7.1) can be applied to each of the knowledge domains?
2. What do experts view as the challenges associated with horizontal and vertical translation in the construction of knowledge in each domain?
3. What are experts' overarching requirements for students' effective translation in developing biological knowledge about each domain?

Methods

As part of the *second round* of a Delphi approach, this study elicited and analyzed an expert panel's responses to a written questionnaire.

Expert Sample

The ten experts from *Round I* (Schönborn & Bögeholz, 2009) were invited to participate in *Round II* 3 years later in July 2010. A questionnaire (see online Appendix II[2]) was electronically mailed to them together with a summary of experts' views obtained from Round I (see Appendix I). Seven experts responded

[2] Expert responses are presented verbatim. Words between square brackets were inserted to improve readability. An ellipsis denotes the exclusion of four words or less of response text. An ellipsis between square brackets designates the exclusion of five or more words. Each expert was assigned an anonymous identification (*E.a* through *E.e*). The expert and respective question item (see Appendix II) associated with a response follow each datum.

to the questionnaire. Of these, one expert's responses were incomplete, and another stated that s/he was uncertain of how to interpret certain items. Thus, the expert panel for Round II consisted of five experts. Threats to internal validity (for $n = 5$) were minimized in light of considering other Delphi approaches in the literature. For example, Bourrée, Michel, and Salmi (2008) demonstrated that groups of four experts can render valid Delphi results, whereas Yousuf (2007) asserted that a Delphi study is only as good as the quality of the expert participants. The five experts in our study were leading biology education specialists all with a deep understanding of competency-based curriculum reform. Expert validity was reinforced by the following self-ratings. First, the average rating of experts' biological content knowledge was 76% for ecology, 84% for genetics, and 80% for evolution. Second, experts rated their knowledge about different ER types as 87% on average and their knowledge of the communication competence as defined in the *Bildungsstandards* as 84%. Lastly, experts rated their expertise in each of *horizontal* and *vertical* translation as 90% on average, respectively.

Design and Implementation of the Expert Questionnaire

A questionnaire focused on the nature of biological knowledge and translation across MERs served as the data-collection instrument for Round II. A preliminary version was piloted with six biology education colleagues to validate item syntax and clarity. The final questionnaire sent to the expert panel consisted of an electronic form (see Appendix II) and corresponding information booklet (see Appendix I). The questionnaire was divided into a self-rating section (Section 0) and three main sections, namely, *Framework of Biological Knowledge* (Section 1), *Translation Processes and Challenges* (Section 2), and *Designing Translation Situations for Acquiring Knowledge* (Section 3). Sections 1–3 comprised four five-point Likert items ranging from "I completely disagree" to "I completely agree" and 21 open-ended items. The information booklet contained a summary of the results from *Round I* (Schönborn & Bögeholz, 2009). The experts responded in English.

Analysis of Expert Questionnaire Responses

Data were treated with a mixed deductive-inductive analysis (e.g., D'Amour, Goulet, Labadie, San Martín-Rodriguez, & Pineault, 2008). First, the authors used a deductive analysis to code expert responses to the Likert items and sought representative datum examples of expert responses corresponding to each of the domains. In this deductive stage, the authors intended to establish the following: (1) whether the experts agreed that the types of biological knowledge (see Fig. 7.1) could be applied to the domains of ecology, genetics, and evolution; (2) examples of such application in each domain; (3) whether the experts agreed that the nature of the knowledge needed for horizontal versus vertical translation was different; and (4) ways in which the nature of the knowledge required for translation could be different in each domain.

Second, any themes in the data were iteratively developed (e.g., Björnsdóttir, Almarsdóttir, & Traulsen, 2009) during an inductive analysis. This inductive stage intended to uncover experts' views on challenges facing learners in the engagement of (1) horizontal and (2) vertical translation processes and (3) over-arching requirements for effective translation in students' acquisition of biological knowledge.

Results

The findings of this study are structured in response to the three research questions posed.

To What Extent Do Experts Agree That the Biological Knowledge Framework (See Fig. 7.1) Can Be Applied to Each of the Knowledge Domains?

The first result section presents experts' application of the types of biological knowledge framework (see Fig. 7.1) to the domains of ecology, genetics, and evolution.

Ecology Domain

All five participants agreed (3/5 completely and 2/5 partially) that the structure and components of the framework (see Fig. 7.1) could be applied to ecology. Consider the following response[2] obtained from one of the two partially agreeing participants:

> From a pedagogical point of view you have to regard ecology as an applied science. As a consequence you have to consider ethical principles like sustainability, common wealth, utility. So, the 4 types of knowledge are necessary but not enough [. . .] (E.a., 1.1.2.).

The response above suggests that ecological understanding also requires incorporation of other knowledge forms (e.g., Kuechle, 1995). In conjunction with the revealed agreement, all five experts (5/5) demonstrated application of the framework in identifying examples of ecological knowledge corresponding to each knowledge type (see Fig. 7.1), as represented by the example below:

> The biological terms *predator* and *prey* together form the biological concept of *predator-prey relationship*. This concept, together with *competition* (e.g., for food) and *symbiosis*, conveys the principle of *interaction of organisms*. Furthermore, to understand *system* as a biological fundamental in the context of ecology, students need to have knowledge of some more examples for ecological principles [. . .] (E.d., 1.1.3.).

The expert's opinion quoted above also clearly elucidates potential interrelationships between different knowledge components of the framework (see Fig. 7.1).

Genetics Domain

Agreement on application of the framework (see Fig. 7.1) to a genetics domain was reached among four (4/5) experts (two completely and two partially agreeing), whereas one expert was undecided. The response from one of the partially agreeing experts was as follows:

> In addition to these principles [mentioned in response to 1.1.2.] you have to regard the principle of dignity (e.g., genetic fingerprinting, prenatal diagnosis, newborn screening. . .). (E.a., 1.2.2.)

The response expresses the need to include other ideas into the notion of genetics knowledge (e.g., France, 2007). Coupled to the observed agreement in the panel as a whole, the following expert's formulation of examples was related to types of genetics knowledge (see Fig. 7.1):

> Example: sickle cell anemia on the level of molecules
> Type 1/biological term[s]: DNA triplet. . . characteristics of amino acids, amino acid sequences. . .
> Type 2/biological concept: point mutation and molecular structure of proteins (primary to quaternary structure)
> Type 3/underlying biological principle: genetic code determines the molecular structure of proteins
> Type 4/biological fundamental: structure and function (E.b., 1.2.3.)

Evolution Domain

Of all participants (5/5) showing consensus, two experts (2/5) completely agreed, whereas three (3/5) partially agreed that the framework (see Fig. 7.1) can be applied to evolution. A response that represented partial agreement was as follows:

> . . .there are subjects/issues to be regarded in education which are not included in the four types [of knowledge]: [e.g.] epistemology in connection with the dispute on evolution/creation; cultural evolution. (E.a., 1.3.1.).

This same expert mirrored his/her response to the previous domains by suggesting that certain epistemological ideas need to be considered in evolution knowledge. All five experts provided application of the framework in an evolution context, as represented by the following two examples (*1* and *2*) obtained from one expert:

> Fundamental: development (of populations and species)
> Principle *1*: variability and adaptation
> Concepts *1*: mutation
> Terms *1*: DNA, gene, genotype
> Principle *2*: reproduction
> Concepts *2*: selection
> Terms *2*: phenotype, offspring (E.d., 1.3.2.)

In addition to mapping evolutionary knowledge onto the four framework components (see Fig. 7.1), this response provides an example of how different principles can mutually contribute to the same biological fundamental.

What Do Experts View as the Challenges Associated with Horizontal and Vertical Translation in the Construction of Knowledge in Each Domain?

Experts' views on translation processes and challenges facing students' construction of knowledge in ecology, genetics, and evolution are structured in three subsections: (1) the *nature* of the knowledge engaged in *horizontal and vertical translation*, (2) the challenges inherent in *horizontal* translation processes, and (3) the challenges inherent in *vertical* translation processes.

The Nature of the Knowledge in Horizontal Versus Vertical Translation Processes

A split in experts' agreement was revealed as to whether the *nature* of the biological knowledge—which students needed to access in horizontal versus vertical translation across MERs—is fundamentally different. One (1/5) expert completely disagreed, two (2/5) partially disagreed, whereas the remaining two (2/5) completely and partially agreed, respectively. With respect to ecology, the response from the expert who partially agreed was as follows:

> *Horizontal translation* means just [being able] to *apply* a concept, principle, or fundamental to *different examples* (e.g., predator-prey relationship to different species). The idea (model) remains the same, the context changes. With regard to MERs, this means [being able] to recognize the core idea in different ERs. *Vertical transfer [translation]* requires knowledge of new characteristics, that is, there is a *new quality* or a new idea (model), if you go 'level-up'. . ., for example, the relationship between predator and prey could not be predicted from the characteristics of a predator and of prey alone. [. . .] (E.d., 2.1.2.a.)

Regarding genetics, the opinion from the expert who completely agreed that the nature of knowledge is fundamentally different in horizontal and vertical translation was as follows:

> I (as a student) acquire factual knowledge about the terms homo- and heterozygosity by analyzing monohybrid crosses of peas. Thus, I acquire knowledge at the organismic level (e.g., by comparing attributes of pea seeds). For horizontal transfer [translation] to other crosses [. . .] I do not need any new knowledge. I just have to identify the known attributes of those terms [. . .]. However, for explaining the phenotypic differences between the pea seeds, I need additional knowledge, because I have to change to other levels, for instance, the cellular level (comparing homologous chromosome pairs and its [their] distribution during meiosis) or the molecular level (comparing DNA molecules and its [their] distribution during meiosis). Thus, vertical translation again requires that [. . .] a student has to connect those knowledge items [. . .]. (E.e., 2.1.2.b.)

The opinions above drawn from the ecology and genetics domains demonstrate that horizontal translation does not involve *any new knowledge* during linking knowledge to the new context, but vertical translation requires *additional* and a *new quality* of knowledge when changing levels of biological organization, as well as bridging knowledge between the levels. Lastly, with respect to evolution, the following response is from the expert who completely disagreed that the nature of biological knowledge accessed in horizontal versus vertical translation is different:

> [. . .] if we change e.g., [for example] from homologies on [at] the organ level to molecular homologies, we change the level of organization but not the nature of the knowledge. Furthermore, in a phylogenetic tree you make [perform] a vertical transfer [translation] in quite [an]other sense than explained in [Appendix I]. (E.a., 2.1.2.c.)

The view above suggests that the *nature* of biological knowledge can sometimes remain constant—even when the level of organization changes in that *vertical* translation *within* a phylogenetic tree ER, as interpreted by the expert in this context—does not necessarily entail switching *levels* of biological organization.

Challenges Inherent in Horizontal Translation Processes

In responding to a request—to apply their examples of knowledge in each domain for considering the core challenges that learners face in engaging horizontal translation in building such knowledge—two experts had the following responses regarding the domain of ecology:

> The most important challenge in ecology is the fact that ecological systems are constructs (models) and not reality itself, that is, to distinguish between objects (reality) and systems (constructs) [. . .]. (E.c., 2.2.1.a.)

> Biological phenomena. . . in the domain of ecology in biology classes are represented by visualizations that are often very concrete, i.e. they are vivid and taken from the macro world (e.g., prey, predator). To get the idea behind the phenomena (What is prey? What is a predator?) learners have to think on a more abstract level. (E.d., 2.2.1.a.)

In view of the above, one challenge that learners may face in engaging horizontal translation in building ecological knowledge is to discriminate between ecological systems represented in external models and the ecological *reality* itself (e.g., Westra, Boersma, Waarlo, & Savelsbergh, 2007). Another challenge is being able to access the knowledge residing *behind* realistically visualized ecological ideas. With respect to translating horizontally across MERs in the acquisition of genetics knowledge, one expert view was as follows:

> For building up an internal representation of the term DNA [type 1 knowledge], students have to use external representations of different modes. For instance, learners [may] have acquired knowledge about DNA structure by analyzing. . . a schematic drawing. For a horizontal transfer [translation] of their knowledge they are [could be] prompted to build a model of the DNA structure (e.g., 2-D or 3-D). (E.e., 2.2.1.b.)

With regard to evolution, the following expert suggested that one main challenge for learners is to horizontally move across depictions of different evolutionary processes in a manner where underlying principles can be clearly interpreted:

> A major challenge for horizontal transfer [translation] in the domain of evolution lies in the very different examples for evolutionary processes. A huge amount of morphological, physiological, and behavioral features can serve as examples for evolutionary processes. (E.d., 2.2.1.c.)

In addition to considering each of the knowledge domains alone, the experts also provided views on the *overall* challenges faced by learners for performing horizontal translation in the construction of biological knowledge:

> The differences between the three biological domains are: ecology is a describing [descriptive] biological area; genetics is more abstract and with a lot of chemical aspects, and evolution is extremely analytical. The way of thinking differs a lot [between these three domains] [. . .]. (E.b., 2.2.2.)

The datum above implies that genetics knowledge is often communicated at the submicroscopic level, which in turn, requires interpreting ERs that are abstract, whereas ecology often necessitates descriptively interpreting (more) realistic ERs. Overall, core challenges which learners face in *horizontal* translation processes are to:

- Access the underlying knowledge, or *biological reality*, that lies embedded across ERs, which are only models of the represented phenomenon (4/5 experts).
- Appropriately apply the necessary knowledge when interpreting a different ER at the same level of organization and/or map the interpretation of one ER to another that represents the same concept or principle being represented at the same level of organization (3/5 experts).
- Realize the different communicative goals associated with ER interpretation in each domain, where the representation mode is often a function of the qualities of that domain (e.g., abstract ERs in genetics vs. realistic ERs in ecology) (2/5 experts).

Challenges Inherent in Vertical Translation Processes

Experts' opinions concerning challenges in engaging vertical translation in the construction of biological knowledge were also divulged. With regard to ecology, the following is an example of an expert's viewpoint:

> In ecology, the learner must be aware that he or she has to [often] go down to another biosystem with its own relations, which are different from ecological relations, for example, physiological relations of [within] the organism. (E.c., 2.3.1.a.)

The aforementioned expert viewed one challenge in vertical translation as the ability to consider the biosystem *relations* specific to a particular level of ecological organization (e.g., Westra et al., 2007). For the acquisition of genetics knowledge, the following expert's opinion can be considered:

> In genetics, the fundamental processes take place on the molecular level. Visualizations of these have to be schematic, compared to photo-realistic pictures. There is a cognitive distance that has to be bridged in order to connect the abstract molecular level with the real world phenomena on the level of organisms and individuals. [It is hard to connect] an illustration of a gene mutation... directly with the phenotype of, for example, albinism. (E.d., 2.3.1.b.)

The datum above highlights the linking between levels and suggests that this often requires bridging across a great *cognitive distance*. The following two responses were examples of vertical translation challenges facing learners in the evolution domain:

> Learners will often mingle the individual and populational level. (E.c., 2.3.1.c.)

> In the domain of evolution, there might be a problem [for students] with [interpreting] the time evolutionary processes typically span [...]; to reason [about] phylogenetic development from single mutations on the organismic level is not easy. Regarding MERs, different hominid species can be depicted very vividly... by photo-realistic illustrations. But the diagrammatic visualization of mutations underlying the phylogenetic development of hominids might appear unsatisfying and insufficient for learners to make a connection between the two levels of biological organization. (E.d., 2.3.1.c.)

The experts' opinions quoted above both point to the challenge of making appropriate vertical connections between biological properties specific to the individual level with those for the population level. The second expert described this difficulty relative to conceptualizing the time involved in evolutionary processes, such as visualizing the concept of phylogenetic development based on ERs describing micro- and macroevolutionary processes (e.g., Catley & Novick, 2008).

Further to their viewpoints about each domain, the expert panel also offered opinions on the *overall* challenges faced by students for executing vertical translation in constructing knowledge. An example of an expert's view about such challenges was as follows:

> Common challenges [across the three domains]: the way of visualizing (pictures, micrographs, tables, diagrams, symbols, and so on) [in] ecology and evolution are [for] visualizing long-time[term]-processes, vertical transfer [translation] seems to be more seldom[ly represented], not a lot of examples [are available] in [at] different levels. Differences [between the three domains]: Genetics has a lot of in-between-levels, more thinking in short processes and needs more linking of facts (E.b., 2.3.2.).

The response above suggests that there is limited MER support for visualizing different levels of biological organization for expressing time-based phenomena to learners. In summary, experts' opinions on the core challenges facing learners (and teachers) in engaging *vertical translation* in the building of knowledge were to:

- Engage the *abstract* thinking necessary for connecting knowledge represented by an ER at one level of biological organization with knowledge represented at a different level (5/5 experts).

- Provide teaching methods that initiate the shifting between levels of biological organization and corresponding MERs in the construction of knowledge (3/5 experts).
- Gain access to ERs that have been purposefully designed around facilitating links between different biological levels, and relative magnitudes of size, scale, and time (3/5 experts).

What Are Experts' Overarching Requirements for Students' Effective Translation in Developing Biological Knowledge About Each Domain?

Upon revisiting their examples of knowledge they had provided for each domain, the experts described examples of MERs they would employ to develop students' biological understanding. These examples ranged from references to ERs in textbooks and to ERs designed by the experts themselves. The following is one expert's authentic example for the genetics domain (cf. Response E.b., 1.2.3. above):

> I take some pipe cleaners and [. . .] different [colored] beads are representative for [of] different amino acids [see Fig. 7.2, left]. . . the primary structure of [a] protein. If I roll [twist] the pipe cleaners around my finger I produce an alpha-helical structure [Fig. 7.2, center]. I can fold two parts of the long structure [in]to a beta-sheet structure, I demonstrate what tertiary structure means with this model and point out the quaternary structure [Fig. 7.2, right] [. . .] In the case of sickle-cell anemia, I can demonstrate. . . what kind of negative effects the point mutation has in [the] beta-sheet structure of hemoglobin [. . .] (E.b., 3.1.2.).

As per this expert's description, teachers (and learners) can manipulate the physical ER (see Fig. 7.2, left) to visualize and communicate aspects of primary and secondary protein structure (see Fig. 7.2, center), as well as model the effects of genetic mutations on tertiary and quaternary protein structures (see Fig. 7.2, right).

The expert panel also provided views of overarching critical requirements for effective translation in developing sound biological knowledge, such as the two views below:

> Learners have to recognize, how an idea visualized on one level of biological organization corresponds to the visualization on another level of biological organization. The referential connections have to be stimulated explicitly. If different modes of representation are used to visualize a concept or a principle. . . on the same level of biological organization or on different levels of biological organization, learners must be able to translate between modes by themselves. Therefore the modes of representation should be chosen carefully and dependent on learners' abilities [. . .] Learners have to understand how the types of biological knowledge are linked together in a hierarchical way. They have to be able to change between these types of biological knowledge and the corresponding MERs. (E.d., 3.2.)

> . . . I think that both the prior knowledge and students' abilities to analyze external representations are required. The latter [abilities] include a competence to communicate scientifically. . . in an appropriate mode [of representation]. I think that teachers have to

Fig. 7.2 Authentic examples of physical ERs provided by an expert (E.b.) for visualizing amino acids (*left*) and initiating students' translation between primary and alpha-helical secondary (*center*), and tertiary and quaternary levels (*right*) of protein structure with respect to point mutations in genetics

> practice these competencies with their students—they do not arise by themselves. Additionally… designing ERs should consider cognitive load effects known since [for] the last two decades (e.g., split attention effect, redundancy effect). (E.e., 3.2.)

Overarching requirements in the first response above suggest that connections be *stimulated explicitly* for learners to effectively translate between different modes of representation and corresponding biological knowledge. The second response echoes the need of a communicative competence that acknowledges ER-related skills (cf. Lachmayer, 2008) and an alignment of ER design with theoretical information-processing principles. Overall, for effective translation in developing biological knowledge, learners require:

- Explicit visual support for changing levels of biological organization during vertical translation across MERs (3/5 experts)
- Practice in developing the specialized competence of interpreting different modes of representation for communicating biological knowledge (3/5 experts)
- To be overtly taught the skills for translating horizontally and vertically across MERs in the construction of biological knowledge (3/5 experts)

Discussion and Implications

This study has revealed an agreement in experts' application of the biological knowledge framework (see Fig. 7.1) to the domains of ecology, genetics, and evolution. Experts' views on the challenges concerning translation across MERs in the building of biological knowledge were reduced to three viewpoints for horizontal and three for vertical translation processes. Experts' opinions on overarching requirements for effective translation across MERs in the development of biological knowledge were exposed as three overall themes.

With respect to research question (i), the results reflected a consensus that the framework (see Fig. 7.1) can be applied to the knowledge components of ecology, genetics, and evolution. It is important that this stability in agreement served to validate experts' subsequent opinions on translation across MERs because expert viewpoints emanated from a common ground. Although consensus was reached, experts suggested that other dimensions also constitute biological knowledge. For instance, sustainability and citizenship were felt closely related to the ecology domain (e.g., Kuechle, 1995), whereas ethics and morals were deemed a *higher-order* component of genetics knowledge (e.g., France, 2007), and facets of belief and cultural evolution intertwined with evolutionary knowledge (e.g., Kinchin, 2010).

In response to research question (ii), experts did not converge in agreement as to whether the *nature* of the knowledge—which learners need to deploy in engaging horizontal versus vertical translation—is fundamentally different. It is interesting that this divergence has been carried over from *Round I* (Schönborn & Bögeholz, 2009). Although consensus was not reached in Round II, the expert panel clearly revealed that while the nature of knowledge remains *constant* in horizontal translation, a *new quality*, *additional*, and *combinatorial* knowledge is certainly involved in *connecting* different biological levels during vertical translation.

In terms of specific challenges inherent in *horizontal translation*, one core obstacle facing learners is to be able to *comprehend* the biological idea embedded *behind* ERs pitched at the same level of organization. In support of this in an ecological context, Westra et al. (2007) indicated the importance of students *getting hold* of underlying ecological ideas represented in ERs such as food webs since models will never contain *all the features of reality*, and different ER types serve different communicative goals. With respect to the fact that ecological concepts are often communicated through graphical ERs (e.g., Bayrhuber, Hauber, & Kull, 2010), Roth, Bowen, and McGinn (1999) found that novices often interpret graphs as *obtrusive tools* and struggle to extract the intended ecological ideas.

Given that gaining biological knowledge inevitably involves translating across different representation modes, experts often associated ecology with a pronounced use of macroscopic realistic ERs, whereas the genetics domain was viewed as often being communicated through abstract representations. This view was confirmed in our own informal analysis of MERs in a prominent upper secondary school textbook (Bayrhuber et al., 2010), which demonstrated ecology to be associated with a high frequency of realistic pictures, whereas genetics regularly incorporated abstract ERs of structure and process at the submicroscopic level.

In terms of specific challenges inherent in *vertical translation*, learners need to engage in the necessary level of *abstractness* for connecting knowledge represented at different levels. For example, with respect to evolution, experts felt that a major challenge is for learners to make appropriate vertical connections between biological properties specific to the individual with those of the population. This challenge is emulated in the study of Catley and Novick (2008) who indicated that ERs of evolution must support learners' discrimination between macroevolution processes and changes within populations. By the same token, constructing genetics knowledge regularly requires students to *bridge* the submicroscopic and the macroscopic (Bayrhuber et al., 2010), a process which experts often view as a demanding *cognitive distance*, and this is somewhat synonymous with *high-road* transfer, which requires learners' *mindful abstraction* of the possible connections and bridges between knowledge areas (Salomon & Perkins, 1989).

In order to shorten the *transfer distance*, teaching must actively initiate students' shifting between biological levels. In terms of evolution, a further demand placed on students in vertical translation is conceptualizing the relative time periods of evolutionary change, as well as visualizing how changes at the organism level can be *mapped* onto phylogenetic development. In this regard, Catley and Novick (2008) stated the importance of visualizing a *true sense* of time in evolutionary ERs. The expert data divulged that vertical translation could be facilitated by purposeful ER design that centers on a meaningful visualization of relative scale and time magnitudes.

In light of responding to research question (iii), opportunities for effective translation lie in providing students with explicit *visual support*. For example, deployment of ER forms such as those depicted in Fig. 7.2 could actively stimulate learners' connections between levels of biological organization. Such visual communication is paralleled in Halverson's (2010) visualization of phylogenetic tree knowledge in evolution. Some experts also felt it necessary to consider the nature of visual support in view of contemporary cognitive theory (e.g., Ainsworth, 2006, 1999). A central expert opinion was that learners require *specialized competencies* for interpreting biological ERs. In backing this view, Roth et al. (1999) suggested that experienced ecologists interpret graphs *transparently* and perceive the intended concepts directly. Hence, *graphing* competencies must be viewed as a fundamental component of biological communication and teaching (e.g., Lachmayer, 2008). In a similar direction, Halverson (2011) identified core representational competence skills for reading and constructing phylogenetic trees in the evolution domain.

Overall, learners need to be *taught* the skills for horizontally and vertically translating across MERs in the construction of biological knowledge. On this aspect, Westra et al. (2007) state that ecological literacy must involve teaching specific skills associated with moving between individual, population, and ecosystem levels. Verhoeff, Waarlo, and Boersma (2008) also demonstrated that teaching specific modeling skills can promote students' acquisition of knowledge through the horizontal and vertical *interrelation* of concepts at different levels of organization.

In conclusion, this study has yielded experts' views on the challenges and requirements for effective translation across MERs in acquiring biological knowledge. The results substantiate the assertion that students' construction of knowledge in biology is closely related to an ability to translate across and between MERs represented at various levels of organization. Promoting skill-based translation practices for advancing our students' biological understanding should be viewed as a key enterprise of modern biology teaching.

Acknowledgments We are grateful to the experts, and we also thank Shu-Nu Chang, Helge Gresch, Gunnar Höst, Frauke Ostermeyer, Silvia Schönfelder, and Stephan Teschner for piloting the questionnaire.

References

Ainsworth, S. (1999). The functions of multiple representations. *Computers in Education, 33*(2–3), 131–152.

Ainsworth, S. (2006). DeFT: A conceptual framework for considering learning with multiple representations. *Learning and Instruction, 16*(3), 183–198.

Bayrhuber, H., Hauber, W., & Kull, U. (Eds.). (2010). *Linder Biologie–Gesamtband. Lehrbuch für die Oberstufe. 23., neu bearbeitete Auflage* [Linder biology: Textbook for the upper secondary level]. Braunschweig, Germany: Schroedel.

Björnsdóttir, I., Almarsdóttir, A. B., & Traulsen, J. M. (2009). The lay public's explicit and implicit definitions of drugs. *Research in Social and Administrative Pharmacy, 5*(1), 40–50.

Bourrée, F., Michel, P., & Salmi, L. R. (2008). Consensus methods: Review of original methods and their main alternatives used in public health. *Revue d'Épidémiologie et de Santé Publique, 56*(6), 13–21.

Catley, K. M., & Novick, L. R. (2008). Seeing the wood for the trees: An analysis of evolutionary diagrams in biology textbooks. *BioScience, 58*(10), 976–987.

D'Amour, D., Goulet, L., Labadie, J.-F., San Martín-Rodriguez, L., & Pineault, R. (2008). A model and typology of collaboration between professionals in healthcare organizations. *BMC Health Services Research, 8*, article 188. Available at: http://www.biomedcentral.com/1472-6963/8/188

de Jong, T., & Ferguson-Hessler, M. G. M. (1996). Types and qualities of knowledge. *Educational Psychologist, 31*(2), 105–113.

France, B. (2007). Location, location, location: Positioning biotechnology education for the 21st century. *Studies in Science Education, 43*(1), 88–122.

Halverson, K. L. (2010). Using pipe cleaners to bring the tree of life to life. *The American Biology Teacher, 72*(4), 223–224.

Halverson, K. L. (2011). Improving tree-thinking one learnable skill at a time. *Evolution: Education and Outreach, 4*(1), 95–106.

Häussler, P., & Hoffmann, L. (2000). A curricular frame for physics education: Development, comparison with students' interests, and impact on students' achievement end self-concept. *Science Education, 84*(6), 689–705.

Kinchin, I. M. (2010). Solving Cordelia's dilemma: Threshold concepts within a punctuated model of learning. *Journal of Biological Education, 44*(2), 53–57.

Kozma, R. B., & Russell, J. (1997). Multimedia and understanding: Expert and novice responses to different representations of chemical phenomena. *Journal of Research in Science Teaching, 34* (9), 949–968.

Kuechle, J. (1995). Ecology: The last word in biology textbooks. *The American Biology Teacher, 57*(4), 208–210.

Kultusministerkonferenz. (2005). *Bildungsstandards im Fach Biologie für den Mittleren Schulabschluss* [National education standards for biology]. Munich, Germany: Wolters Kluwer.

Labov, J. B., Reid, A. H., & Yamamoto, K. R. (2010). Integrated biology and undergraduate science education: A new biology education for the 21st century? *CBE Life Sciences Education, 9*(1), 10–16.

Lachmayer, S. (2008). *Entwicklung und Überprüfung eines Strukturmodells der Diagrammkompetenz für den Biologieunterricht* [Development and examination of a structural model of graph competence for biology education]. Dissertation an der Universität Kiel. Kiel, Germany: Universitätsbibliothek.

Linstone, H. A., & Turoff, M. (Eds.). (2002). *The Delphi method: Techniques and applications.* Reading, MA: Addison-Wesley.

Murry, J. W., & Hammons, J. O. (1995). Delphi: A versatile methodology for conducting qualitative research. *The Review of Higher Education, 18*(4), 423–436.

Niedersächsisches Kultusministerium. (2007). *Kerncurriculum für das Gymnasium Schuljahrgänge 5-10: Naturwissenschaften* [Science school curriculum for years 5-10 in Lower Saxony], Biology. Hannover, Germany: Unidruck.

Osborne, J., Collins, S., Ratcliffe, M., Millar, R., & Duschl, R. (2003). What "ideas-about-science" should be taught in school science? A Delphi study of the expert community. *Journal of Research in Science Teaching, 40*(7), 692–720.

Roth, W.-M., Bowen, G. M., & McGinn, M. K. (1999). Differences in graph-related practices between high school biology textbooks and scientific ecology journals. *Journal of Research in Science Teaching, 36*(9), 977–1019.

Salomon, G., & Perkins, D. N. (1989). Rocky roads to transfer: Rethinking mechanisms of a neglected phenomenon. *Educational Psychologist, 24*(2), 113–142.

Schönborn, K. J., & Anderson, T. R. (2009). A model of factors determining students' ability to interpret external representations in biochemistry. *International Journal of Science Education, 31*(2), 193–232.

Schönborn, K. J., & Bögeholz, S. (2009). Knowledge transfer in biology and translation across external representations: Experts' views and challenges for learning. *International Journal of Science and Mathematics Education, 7*(5), 931–955.

Tsui, C.-Y., & Treagust, D. F. (2003). Genetics reasoning with multiple external representations. *Research in Science Education, 33*(1), 111–135.

Tsui, C.-Y., & Treagust, D. F. (2007). Understanding genetics: Analysis of secondary students' conceptual status. *Journal of Research in Science Teaching, 44*(2), 205–235.

Verhoeff, R. P., Waarlo, A. J., & Boersma, K. T. (2008). Systems modelling and the development of coherent understanding of cell biology. *International Journal of Science Education, 30*(4), 543–568.

Westra, R., Boersma, K., Waarlo, A. J., & Savelsbergh, E. (2007). Learning and teaching about ecosystems: Systems thinking and modelling in an authentic practice. In R. Pintó & D. Couso (Eds.), *Contributions from science education research* (pp. 361–374). Dordrecht, The Netherlands: Springer.

Yousuf, M. I. (2007). Using experts' opinions through Delphi technique. *Practical Assessment, Research & Evaluation, 12*(4), 1–8.

Chapter 8
Evolution Is a Model, Why Not Teach It That Way?

Paul Horwitz

Introduction: The Model

Evolution is not an intuitive idea. At first blush, in fact, it seems ridiculous to suppose that the wondrous interdependency and exquisite adaptations of living creatures could have evolved by natural causes without conscious planning. Not only is the concept counterintuitive, but the evidence for it is mostly indirect and cannot be appreciated without prior knowledge of seemingly unrelated sciences. And of course in some circles, particularly in the United States, the theory of evolution is in conflict with firmly held religious convictions (Scott, 2004; Sinatra & Nadelson, 2010; Verhey, 2005). No wonder evolution is so hard to teach!

On the bright side, the process of evolution by natural selection is ideally suited to teaching via computer simulations, which can transcend space and time constraints to model processes that take place on scales from molecules to ecosystems and over times ranging from milliseconds to billions of years (Horwitz, 2010; Ottino-Loffler, Rand, & Wilensky, 2007; Rosca, O'Dwyer, Lord, & Horwitz, 2010; Wilenski & Novak, 2010). A very simple model, in fact, can demonstrate how evolution occurs. Here's an example—imagine a highly simplified model of a plant that needs only one thing to grow: light. But it is not enough to have any old amount of light; our plant is very picky. In too much or too little light, it will wither and die, but if it gets just the right amount, it will flower and produce seeds. When winter comes, our plant will die, but if it has made seeds they will germinate, and come spring they will produce other plants, which will produce more seeds, and so on. So if the conditions are just right, even though the original plant has died, our model will support a *population* of plants that can live forever as long as the light level doesn't change.

P. Horwitz (✉)
Modeling Center, Concord Consortium, 25 Love Lane, Concord, MA 01742, USA
e-mail: phorwitz@concord.org

D.F. Treagust and C.-Y. Tsui (eds.), *Multiple Representations in Biological Education*, Models and Modeling in Science Education 7, DOI 10.1007/978-94-007-4192-8_8,

© Springer Science+Business Media B.V. 2013

In fact, if we're not careful, this model will *blow up*! If each plant produces multiple seeds that in turn grow into viable plants that produce their own seeds, there will be more plants in each succeeding generation, and the number of plants will grow without limit. If, on average, each plant produces fewer than one viable seed, the opposite will happen: The plants will all die off. The model is *unstable*: Unless we can somehow contrive to make the birthrate exactly equal the death rate, our model plants will either grow without limit or go extinct. If we do succeed in exactly balancing the two rates, the population will remain exactly the same. Boring—and not at all what you would expect in nature! What's wrong?

The problem is that our model is too simple; in mathematicians' terms, it is *linear*, meaning that there is nothing in it that sets the scale for how many plants can be supported at once. We need to add to our model the concept of the finite *carrying capacity* of the environment. We can accomplish this by adding the feature that the plants they compete for scarce resources and when they are overcrowded they become sickly and produce fewer seeds. This kind of thing is called a *negative feedback loop*, and it's very common in nature, so we're well within our rights to add it to our model.

With this addition, we've got a model that is stable, in the sense that there will be different numbers of plants each year, but they will never exceed a certain number, nor will they go extinct. So far, so good, but what does this have to do with evolution? Evolution depends on three things: inheritance, variation, and fitness. Our model incorporates the first of these; it's time to add the next two.

Imagine that our model plants come in different varieties, distinguished by the size of their leaves. Some plants have big, bushy leaves with lots of surface area for photosynthesis, so they need very little light. Other plants have small, skinny leaves, and they need a lot of light to survive. Still other plants are in between these two extremes: They have medium-sized leaves and are adapted to moderate amounts of light. For simplicity, let's label these different varieties of plant numerically according to size of their leaves: Level 1 plants have very small leaves so they need a lot of light, level 10 plants have big leaves and need very little light, and the other levels are in between. Figure 8.1 shows what these plants might look like.

Now here comes the tricky part: In our model, all the plants depicted in Fig. 8.1 are *different varieties of the same species of plant*. What exactly does that mean? We know that offspring don't always look exactly like their parents; even the littermates of purebred dogs show some variation. So let's add this important feature of the real world into our model by setting up a rule that says that when a plant produces seeds, *the offspring sometimes are shifted in level by one unit*. A level 5 plant, for instance, will mostly produce level 5 offspring, but every once in a while, *by accident*, it will make a level 4 or a level 6 plant. In the presence of a uniform environment suited to level 5 plants, most of the offspring of our plant will do just fine; they will germinate, grow into adult plants, and produce seeds of their own. The occasional level 4 and level 6 seeds, however, will be at a disadvantage. They will grow up withered, and they won't produce a flower, so they will produce no seeds and have no offspring. So after a while, each generation will

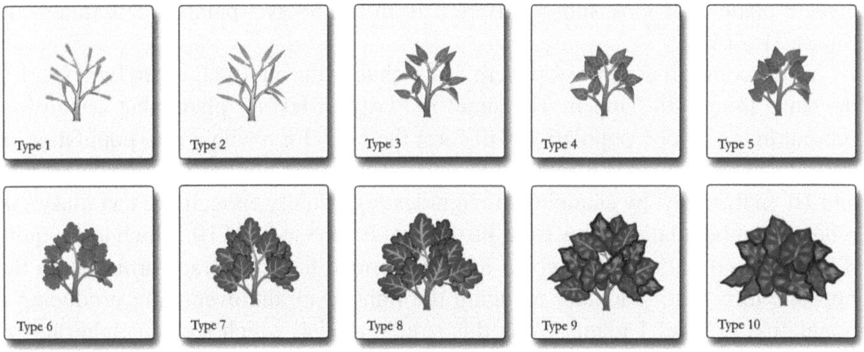

Fig. 8.1 Ten types of plant in the model. Plants with thin leaves are adapted to abundant light; plants with bushier leaves are adapted to shade. All plants are varieties of the same species

consist mostly of level 5 plants, with the occasional, infertile, level 4 or 6 plants randomly mixed in.

Now our model has inheritance, variation, and fitness: the three things that are essential for natural selection to take place. To get evolution, we need to add one little thing: change. Environments are not eternal. A grassy plain that gets plenty of sunshine may over time become a forest of tall trees. When that happens in nature, the original plants that were adapted to lots of sunshine will give way to other kinds more adapted to shady conditions. Our model will do the same thing, *if the change is gradual enough.*

Here's a mental exercise for you. Imagine that we add to our computer model a slider with a range of 1–10 that controls the amount of light available to our plants. You can think of it as a simple way to control the growth of those trees, except that you can do it in seconds instead of having to wait for decades, and we can make the trees grow shorter as well as taller. If we set the slider to 5, which corresponds to a forest of medium-tall trees, we will reproduce the situation described above: a hardy, healthy population of level 5 plants with a few 4s and 6s appearing in each generation. What do you think will happen if we abruptly move the slider to 10, suddenly increasing the light level (the real-world equivalent might be clear cutting that forest)?

The answer, of course, is that all the plants will die because none of them is adapted to live in such a high-light environment. In evolutionary terms, our plants will go *extinct*. Is there any way to avoid this dire fate? What if we were to move the slider just a bit so that the light level changes from 5 to 6? Now the level 5 plants, which are the vast majority of the plant population, can no longer survive; neither can the small minority of level 4 plants. But the level 6 plants, and remember there will be a few in every generation, will thrive in the new environment—all the more so since there will be no other plants around to hog those scarce resources! So even though there may be very few level 6 plants at first, each one will flower and drop seeds, and in just a few generations, their numbers will grow to reach the carrying capacity of the environment, and we will be right back where we started, but with level 6 plants that look subtly different from the level 5 plants we started with. Simple, isn't it?

And of course it doesn't stop there. If we change the environment to light level 7, the same thing will happen: The small minority of level 7 plants that are always present in the level 6 population will form the basis for a whole new population of level 7 plants, adapted to the new environment. And this goes on through levels 8, 9, and 10. In this way, by changing the light level gradually enough, we can make our model plant population grow from its original level 5 to level 10, which looks quite different. And of course we could have performed the same transformation in the opposite direction, gradually reducing the light level and eventually producing a population of level 1 plants. With this simple model, which includes inheritance, variation, and fitness, our different varieties of plants are capable of keeping pace with changes in the environment, evolving into one another (and back again!) as long as the changes are gradual enough to allow the variant plants to take hold and prosper each time the environment changes.

Note that our ability to create and run such a model says nothing at all about whether evolution actually happens in nature! After all, we created all those different levels of plants, specifically designed to be able to live and reproduce in different light conditions, before we even ran the model! So the level 5 plants evolved, yes, but they evolved into something that was in the model to begin with. The model I have described doesn't *prove* evolution by natural selection—no model could do that!—it simply *illustrates and explains* it. And that, with support from the National Science Foundation, is what we set out to do in a recent project called.

Evolution Readiness[2]

The goal of the project is to introduce students in the fourth grade—10-year-olds—in the United States to the concept of evolution by natural selection. Working with school systems in three states, Massachusetts, Missouri, and Texas, we have been presenting students with computer-based learning activities that incorporate models of plants and animals similar to the one described above. The activities present themselves to students in the form of educational video games, in the sense that they have a definite goal and provide context-sensitive scaffolding in the form of helpful hints and congratulatory messages when the goal state is attained. Many of the activities offer a *back story* in the form of real-world examples associated with the students' explorations of the model. All these activities keep track of everything the students do, including their answers to embedded questions, and report back to the teachers and to the research team. In addition, the teachers were requested to fill out a brief survey at the end of each lesson, with comments on their students' reaction to the activities. Some of these comments are included in the descriptions below.

Description of the Learning Activities

Plant Activities

The Virtual Greenhouse

The goal of this activity is to teach the students that plants with different types of leaves are adapted to different amounts of light. The students are given three different types of seeds and are challenged to determine by experiment in which of five virtual flower boxes—differing in the amount of light they receive—each of three types of seeds grows best. They may keep track of their data by taking snapshots of each experiment and saving them in an online laboratory notebook that is incorporated into the program. The activity also introduces a bar graph that shows how many plants of each type have produced flowers, indicating that they are healthy and their environment is optimal for them. This activity is depicted in Fig. 8.2.

The Virtual Field

In this activity, students plant seeds in a field with a gradient of illumination. Plants at the top of the field receive less light than those at the bottom. (Note that the direction of the gradient is reversed from that in the flower box arrangement of the virtual greenhouse activity above, so that students do not confuse location with the critical environmental factor: light.) As in the flower box environment, plants with big leaves can only live where the light is least, whereas those with the smallest leaves must be planted in the part of the field that receives the most light if they are to survive, produce a flower, and drop seeds. The students discover this by experimenting with the same three seeds as before. If they plant their seeds in the wrong place, the plants will wither or die and fail to produce seeds. This activity also introduces the plant life cycle. *Winter* arrives at regular intervals, and all the plants in the field die and disappear. Their seeds, if any, survive the winter and grow into plants the following spring. This feature of the model is pedagogically important because it reinforces the point that the evolutionary changes the students observe take place over many generations and affect the population of plants rather than individuals. Initially, all the *offspring* plants are identical to the *parent* plant—no new types appear, and after many generations, the field is populated by three distinct rows of plants, corresponding to the three types of seeds the student was able to plant. This situation is depicted in Fig. 8.3. The activity ends with a *zoomed-in* simulation of a single plant that produces exactly six seeds—two of which grow into plants that are slightly different from those of the parent plant. These *mutant* plants wilt and do not produce seeds in the environment into which they were born, but the student can pick them up and move them to a slightly different environment where they will do well.

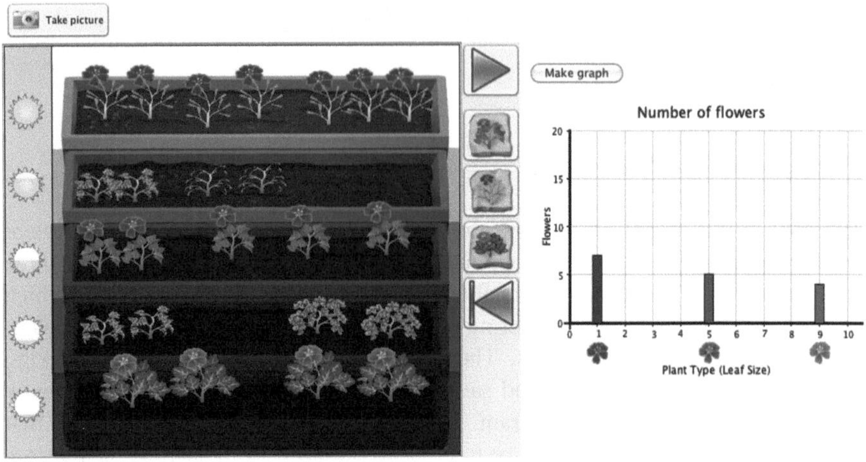

Fig. 8.2 The virtual greenhouse. The bars are color coded to match the colors of the flowers

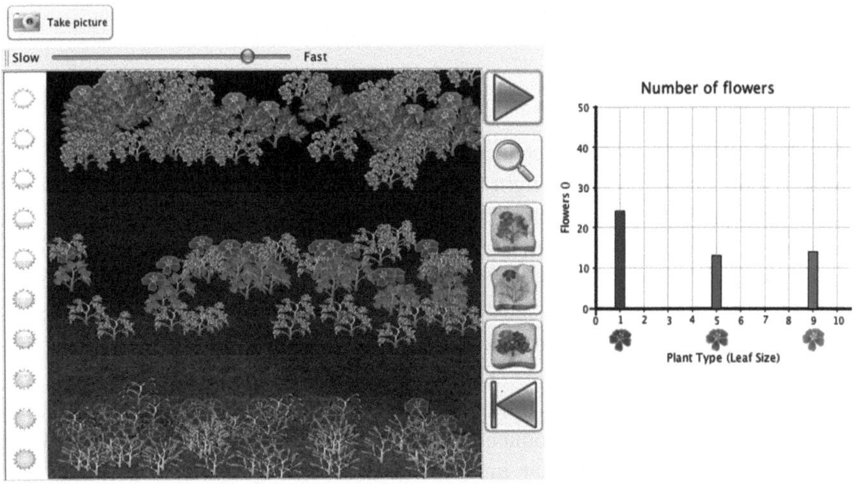

Fig. 8.3 The virtual field. Note that without variation, the three types of plants occupy distinct regions in the field, due to the gradient of light across it. The bar graph shows only those three types

Mystery Plant Adaptation

The third activity revisits the zoomed-in scenario of inheritance with variation which ended the previous activity. It then returns to the same field as before, with the ambient light level varying smoothly from top to bottom. The students are given only a single type of seed to plant: the type that grows best in the center of the field. But this time the model has been altered to include a critically important feature: *variation*. A small

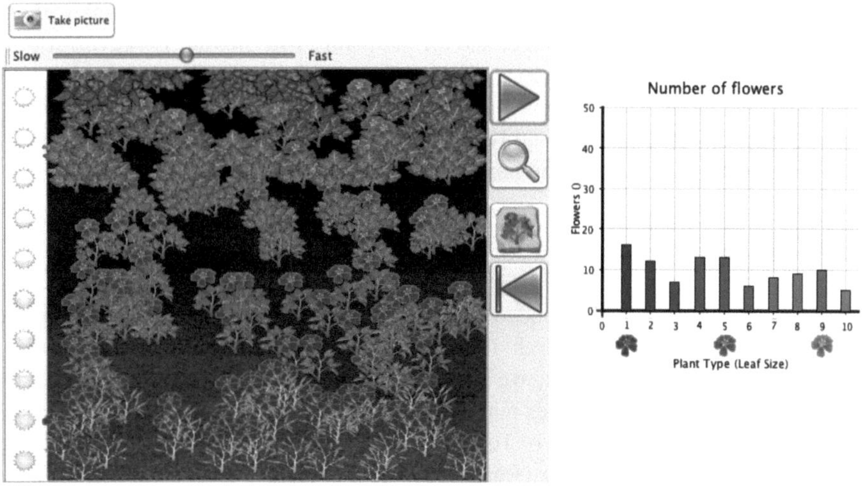

Fig. 8.4 Mystery plant adaptation. Note the bar graph which indicates that every type of plant is present in the population

fraction of the seeds from a plant will grow into new plants that differ slightly from the parent and thrive in the row just above the parent plant's row or just beneath it. Since each plant scatters seeds randomly, it happens occasionally that some of these different seeds fall in a location where the light level is just right for it. When this happens, the seed will grow into a healthy plant that will produce seeds of its own. In this way, the single type of plant planted by the student, which could only live in a particular horizontal slice of the field, eventually evolves into the full spectrum of different varieties that we observed in Fig. 8.1. When this happens, the population of plants is capable of living and reproducing in every area of the virtual field. In this activity, a small fraction of the planted seeds from one type of plant will actually grow into a neighboring type. In the presence of this source of variation, a single type of plant—capable of growing only in one region—can evolve to cover the entire field. The effect is quite dramatic, as shown in Fig. 8.4.

Changes in the Environment

The fourth activity places the control of the environment under the control of the student. The field starts off with a uniform light level midway between the maximum and the minimum, thus capable of growing plants with medium-sized leaves. Students can alter the environment, however, by "growing" a chain of mountains of variable height right down the middle of it. In the presence of these mountains, depending on their height, the light level increases by 1–4 units on one side of them and decreases by the same amount on the other side. Students are challenged to grow the mountains to their maximum height (corresponding to the maximum change in light level) while maintaining a viable population of plants on each side. If the students make the changes too abruptly, their plant populations will not have time to adjust to the change, and all the plants will die out. However, if they change the environment one step at a time, being careful to wait before making each change until there are sufficient numbers of *mutant* plants on each side, then the *normal* plants will die, but roughly half of the mutants will survive, and it is they who constitute the basis for the next generation.

Mystery Plant Mystery

The final plant activity is intended to assess what the students have learned in the first four. In previous research (Horwitz & Christie, 2000; Horwitz, Gobert, Buckley, & O'Dwyer, 2009), we have found quite often that students who are taught with game-like activities may get proficient at the game but fail to learn the science concepts that underlie it. To test whether this was happening, we introduced a new environmental variable (water level) and added 10 new varieties of plants with different root types, ranging continuously from deep to shallow, adapted to different water levels. (Plants with long "taproots" are adapted to dry conditions; those with shallow, wide-spreading roots need lots of water.) Using these plants, we constructed an activity to use as a transfer exercise and a test of whether or not a student has really understood the target concepts. The new activity involves the same concepts of reproduction with variation, natural selection, and adaptation but uses the water level to root type mapping, rather than the light level to leaf size mapping. This is a significant change, particularly since the roots of the plants are not normally visible: They can only be seen if the student *uproots* the plant with a special *hand* tool or observes it closely with a *magnifying glass* tool.

The activity starts with five flower boxes, as in the virtual greenhouse, and three types of seed. The flower boxes differ in the amount of water they receive, and the challenge, as before, is to discover which seeds thrive in which environment. This time, though, the plants all look the same above the ground (they all have medium-sized leaves and pink flowers), so it is not obvious that they are different. Beneath the surface, however, their roots are different. Once the students have discovered this, using the hand tool or the magnifying glass to examine the roots, they are presented with a field where the water level varies continuously from left to right, from one end of the field to the other. They are provided with a packet of seeds, all of which grow the same type of plants. The seeds cost virtual money, and the challenge to the students is to spend as little as possible on seeds but still produce a bumper crop of plants that can grow everywhere in the field. To do this, the students must notice and take advantage of the small variation in root type from one generation to the next.

Animal Activities

For pedagogical purposes, the main difference between plants and animals is that plants, in our model at least, depend only on abiotic (nonliving) factors, such as light and water, while animals consume other living things—plants and other animals.

So by bringing in animals, we are able to introduce the concept of a food chain, with its related notion of competition for scarce resources. Moreover, the interdependence of species at each level of the food chain means that the environment of each species comprises, in part, all the other species with which it interacts. Thus, evolutionary changes in one species will affect others, and vice versa, resulting in a sort of *adaptation arms race* qualitatively different from the one-way response of the plant population to external changes in a nonliving environment. In this, the third year of the project, the *Evolution Readiness*, students in all three school districts are exploring these related concepts through a sequence of five animal activities, which we describe below.

The Virtual Ecosystem

With this activity, we introduce students to the idea that all living organisms must compete for food with other living organisms. We do this interactively by having the students take on the role of a rabbit in a field with edible plants. The students can control the movements of their rabbit using the arrow keys on the keyboard, and in this way they move the rabbit from one plant to another. When the rabbit moves onto a plant, it *eats it*, the plant's icon disappears, and the rabbit's hunger level is decreased. At first the students' rabbit is alone in the field, but then other, computer-controlled, rabbits appear, one by one. With all this competition, it becomes harder and harder for the students to keep their rabbit alive.[3] Even if their particular rabbit starves, however, the population of rabbits survives, and from the evolutionary point of view, that's all that matters. Accordingly, an important goal of this activity is to encourage students to think globally: shifting from a focus on individual organisms to a concern for the well-being of the population as a whole.

Variations and Adaptations

This activity introduces three varieties of plant: tall, medium, and short; students experiment to determine how climate can affect ecosystems. First, they investigate the effect of rainfall on the plants and discover that the larger plants can live in near-drought conditions, while the smaller ones perish. Next, we introduce variation in the rabbit population and challenge the students to figure out which variety of rabbit eats which kind of plant. The students are encouraged to make the connection between rainfall amount and the rabbit population's ability to survive by thinking first about rainfall and plants, then about plants and rabbits, to infer that when certain plants cannot grow and reproduce, the rabbits that eat those plants will not have enough food to survive. In this way, students are introduced to the concept of interdependence in an ecosystem and its effect on the evolution of populations.

Natural Selection

In the third activity of the animal sequence, students explore how changes in the environment affect both the plants and animals in a simple ecosystem with just two species living in it: grass and rabbits. They build a dam in the middle of the field, dividing the ecosystem in half. The area below the dam gradually dries out, which affects both the grass and the rabbit populations in that region. As the smaller plants die out, the rabbits that eat them soon follow suit. Once the students have observed this progression and entered data into their virtual laboratory notebooks, they remove the dam and observe as the ecosystem slowly returns to its original state.

Predators and Prey

This activity uses a model of the virtual ecosystem with three species in it—grass, rabbits, and hawks—enabling the students to explore the effect of predation on the prey population. At first, they *become* a hawk and try to catch and eat brown and white rabbits on a snowy field. The latter blend into the background and are harder to see, so they have a selective advantage. Having discovered through personal experience the reason for this selective advantage, the students proceed to explore an environment that changes over time starting out white and turning brown as the snow melts. A line graph shows plainly the shifting of the relative proportions of white and brown rabbits in response to this environmental change.

Experiment with Ecosystems

This is the most open-ended of all the *Evolution Readiness* activities and perhaps the most challenging for students. The goal is to give the students the opportunity to *think like a scientist*, making hypotheses, doing experiments, observing what happens, and analyzing and thinking about data. Students are encouraged to construct and conduct their own experiments with ecosystems comprising grass, rabbits, and up to two predator species: hawks and foxes. First, they are prompted to come up with a hypothesis for a particular question—for example, *What will happen to the hawk population if the grass is removed from the field?* Then, they are challenged to experiment with the model ecosystem in a way that allows them to test their hypothesis.

Off-Line Activities and Teacher Support

We supplemented the computer-based activities described above with off-line activities involving manipulable objects of various kinds. These activities were borrowed or adapted from existing curricula. Any required physical materials were supplied by the project to all the participating teachers. These materials included

- Several books about evolution written for children
- An 18-ft-long vinyl timeline with graphics and text depicting the evolution of life over the past 600 million years
- A set of fast plants[4] together with a simple lighting and watering system, designed by the project, to facilitate their maintenance
- A game called the Lego Tree of Life designed to illustrate phylogenetic trees; materials included sets of large Lego pieces and special-purpose plastic laminated cards
- Another game called Clip Birds that illustrates selective pressure by challenging students to pick up three different sizes of *seeds* using three different kinds of clips
- An activity that introduces the complex interdependence of species in an ecosystem by having students literally construct a *food web* by passing a ball of yarn between them to illustrate interactions between different trophic levels

The subject matter of the *Evolution Readiness* project is challenging for teachers as well as students. Accordingly, we offered extensive support for teachers through a variety of channels: face-to-face workshops, an online course, and a comprehensive teacher guide that introduces each of the activities and covers both content and pedagogical content knowledge. Teachers were compensated for the time they spent on professional development, as well as any other time devoted to activities outside their normal duties (e.g., administering tests).

Results from Second-Year Implementation

In the second year of the project, that is, Year 2, we evaluated the plant activities and the first four of the off-line activities in all three participating school districts. In what follows, we refer to this treatment as the *trial curriculum*. At this writing, halfway through Year 3, implementation of the full curriculum, which includes the animal activities and food web off-line activity, has begun with an implementation in the Massachusetts school district. Results from the full curriculum are not yet available, so we report only on the trial curriculum here.

We compared the learning gains of students exposed to the trial curriculum in Year 2 to a baseline cohort consisting of students taught by the same teachers using a traditional curriculum in Year 1. The comparison is meaningful because the topics covered by the *Evolution Readiness* materials, designated by us as *Big Ideas*, as shown in Table 8.1 are all contained within the science standards of each of the three states we worked in, Massachusetts,[5] Missouri,[6] and Texas,[7] and were therefore covered by the traditional curriculum, but without the assistance of the online and off-line activities, and lacking the integrative, evolution-based explanatory approach adopted by the *Evolution Readiness* project.

Table 8.1 Big ideas of evolution readiness

Big ideas and standards	Learning goals
1. Basic needs of organisms	Both plants and animals need air and water; plants also need light and nutrients; animals also need food and shelter
	Different species have different preferred conditions for survival
2. Life cycle—birth and death cycle	Organisms are born, live, and die
	A species can survive even though every individual in a given generation eventually dies
	All organisms have a finite lifetime, and populations will survive only if their constituent organisms have enough offspring over time to compensate for the number of deaths
3. Organisms and their environment	Organisms thrive in environments that match their specific needs
4. Classification of organisms	Plants and animals are classified into species and other groups based on shared characteristics
5. Interspecific differences	There are differences between species
6. Interactions between species	Organisms with similar needs compete with one another for resources
	Animals obtain energy and resources by eating other animals and plants. Plants produce their own food
	The presence of other plants and animals, as well as environmental factors, can affect the survival of plants and animals
7. Intraspecific differences	Individuals of the same species may differ. Not all offspring from the same parents look alike, even with respect to inherited traits
	Purposeful selection of certain traits over many generations can result in substantial changes in the physical characteristics of organisms in a population
8. Adaptation and evolution	Species are adapted to their environments. If the environment changes, only certain species survive
	Organisms carrying traits that are better suited for a particular environment will have more offspring on average
	Selection pressure can lead to a change in the characteristics of a population
9. Heritability of traits	Offspring inherit some, but not all, of their traits from their parents
10. Reproduction	Organisms have offspring, and without reproduction, the species cannot continue. Only members of the same species can have viable fertile offspring
11. Descent with modification	Species evolve from common ancestors. Different species can arise from one species if different groups have different selection pressures

Development of the Assessment Instrument

In Year 1 of the project, we developed a Concept Inventory for Evolution Readiness (CIER)[8] that covers the projects learning goals (see Table 8.1) and is aimed at uncovering students' preconceptions. Designed to be administered in two sessions, the CIER includes 32 multiple-choice, 5 short-answer, and 24 open-response questions and measures students' understanding of the fundamental concepts related to the theory of evolution.

We conducted Rasch analyses before we used the CIER and measured high item and person reliability (0.88 for person reliability and 0.97 for item reliability). The Wright map from Rasch measurement and person-item separation indices indicated that the CIER was a valid measure and its results matched expected typical fourth grade students' ability. We include the Wright maps from the baseline and Year 2 cohorts as shown in Figs. 8.5 and 8.6, respectively.

In the northern spring of Year 1 of the project, we used the CIER to collect baseline data from 132 students (Cohort 1) taught using the traditional curriculum in each state. In Year 2, we used the same instrument to collect data from 186 students (Cohort 2) in the same schools taught by the same teachers but using the *Evolution Readiness* trial curriculum (all the plant activities and four out of five off-line activities). To avoid unintentional bias, the tests from both cohorts were combined and scored by trained scorers who did not know which student belonged to which cohort. Estimates of students' knowledge of the concepts were computed using both classical test theory and item response theory.

The test results indicated that the students in the post-implementation cohort had a deeper understanding of the concepts underlying the theory of evolution than did the pre-implementation cohort and that this difference was sharpest for the more advanced topics. For instance, none of the students in Cohort 1 achieved a maximum score on the open-ended response questions relating to adaptation and evolution, indicating that the pre-implementation cohort did not have a deep understanding of these core concepts. In contrast, several students in Cohort 2 did achieve the maximum score on these questions. The Cohort 2 also outperformed Cohort 1 on questions relating to descent with modification, indicating that they understood that new species could arise from a single species if different subgroups were subjected to different selection pressures for a long time.

Overall, the mean for the pre-implementation Cohort 1 was 530.87 (SD = 67.78), and the mean for the post-implementation Cohort 2 was 555.71 (SD = 78.97). An independent means t-test showed that the students in Cohort 2 performed significantly higher on the CIER than did students in Cohort 1, with an effect size difference of 0.35 standard deviations. It should be noted that the test instrument was identical for the two different cohorts.

What Did We Leave Out and Why?

According to national polls conducted in the United States,[9] approximately half of the US adult population does not *believe in* evolution (the exact number depends on how the question is asked), and a substantial majority believe that the various creationist theories should be given *equal time* in precollege science courses.[10] Should we be concerned about that? If it's a problem, is it one that a model-based pedagogy can address?

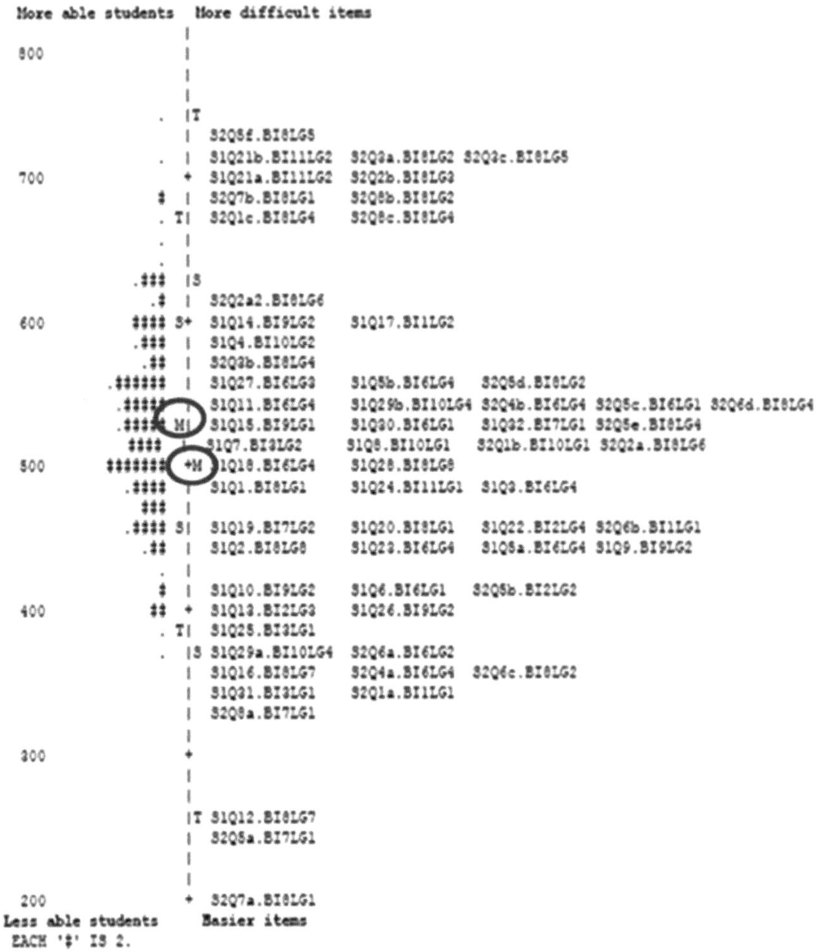

Fig. 8.5 Wright map of baseline data from Cohort 1, collected in Year 1 prior to treatment

I have commented elsewhere, for example, in an interview by Sparks (2010), that the goal of the *Evolution Readiness* project is not to try to persuade students to believe in evolution but rather to help them to understand it as an explanatory model that ties together diverse findings from a wide variety of fields. I would generalize that statement: I don't think the primary goal of any course in science should be to induce the students to believe in the science being taught—in fact, the whole idea of *believing in* science strikes me as somewhat bizarre.

Students in high school are taught the Pythagorean theorem, but we do not therefore infer that the primary purpose of their geometry course is to induce those students to believe that the square of the hypotenuse of a right triangle is equal to the sum of the squares of the other two sides. We recognize that the

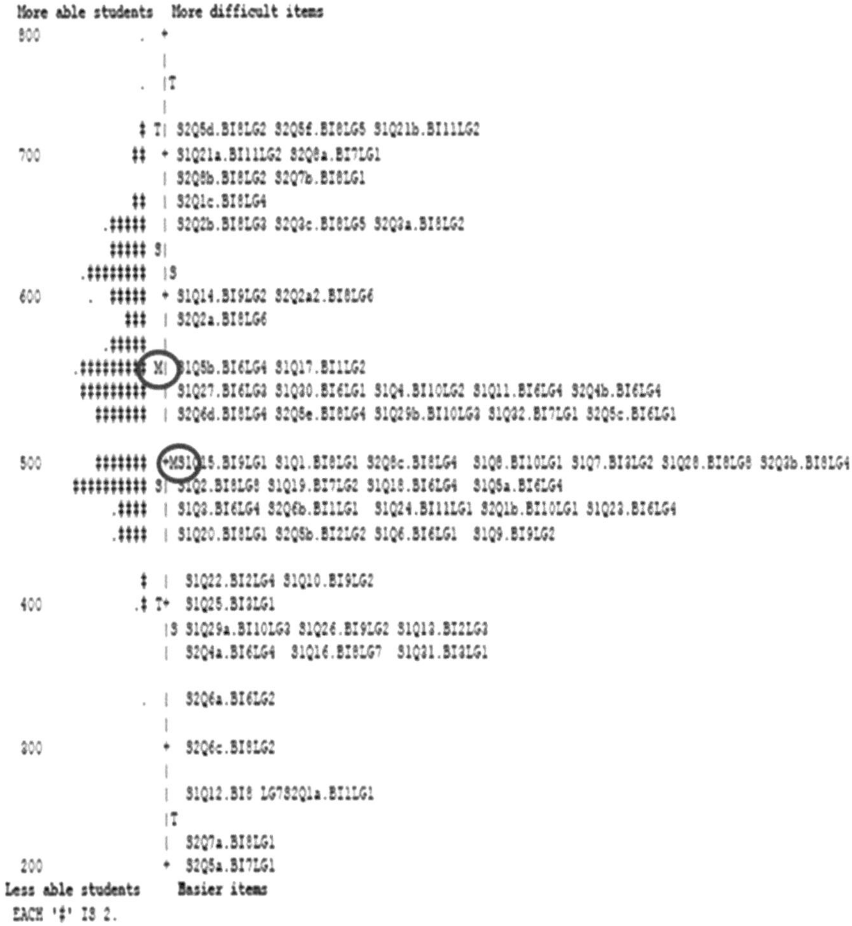

Fig. 8.6 Wright map of data from Cohort 2 collected in Year 2, after treatment

important aspects of this celebrated result of Euclidean geometry lie in the various ways that it can be proved, as well as in the multitude and variety of its applications to mathematics and other disciplines. So it is with evolution: Whether or not students come away with a firm belief that every living thing on Earth evolved is less important than that they understand the model of evolution driven by natural selection[11] and appreciate how such a model is supported by evidence. We believe that our project is accomplishing the first of these goals; the second we have largely ignored.

At the start of the *Evolution Readiness* project, we were faced with the task of identifying which aspects of the evolutionary model we were going to try to teach to fourth graders. After much discussion, we decided to leave out those aspects of the model that take place on time and space scales that are unfamiliar and largely inaccessible to the young children who were our audience. Accordingly, we left out phenomena and processes that are either very small or very slow: We do not introduce the molecular basis for inheritance, for instance, nor do we emphasize, with the exception of the timeline, the nature and interpretation of the fossil record. This intentional pruning of the curriculum has the somewhat unfortunate consequence that we have had to skip over much of the supporting evidence for the evolutionary model; we have instead resorted to presenting that model in a manipulable form and guiding students to explore and come to understand it by experimentation, in response to specific prompts. For 10-year-olds, we feel, this is challenge enough; we look forward to developing similar interactive curricula, based on more complex challenges and models, for use with older children.

Acknowledgments The research described in this chapter was supported by grant number 0822213 from the United States National Science Foundation. I am deeply grateful to his colleagues, Carolyn Staudt, Cynthia McIntyre, and Trudi Lord, of Concord Consortium, and Laura O'Dwyer and Camelia Rosca, of Boston College, for their contributions to the project itself and for numerous insightful conversations on this and many other topics.

Notes

1. This project is supported by the US National Science Foundation under grant # 0822213. For more information, visit http://er.concord.org
2. http://concord.org/projects/evolution-readiness
3. It turns out, in fact, that the only way to stay alive for the required 100 s is not to eat if you are not hungry, thereby conserving resources that you're going to need later on when more and more rabbits arrive—a useful lesson even without evolution!
4. See examples at http://www.fastplants.org
5. See http://www.doe.mass.edu/frameworks/current.html for a detailed description of the standards for this state.
6. http://dese.mo.gov/divimprove/curriculum/frameworks/science.html and ancillary documents available for download from this site.
7. http://ritter.tea.state.tx.us/rules/tac/chapter112/ch112a.html#112.15 gives an overview of the Texas standards for 4th grade life science.
8. This work was done primarily by the research team at Boston College.
9. A CBS poll conducted in 2006 reported that 55 % of those questioned believed that "God created humans in their present form," 27 % believed that "humans evolved but God guided the process," and only 13 % believed that "humans evolved and God did not guide the process." A 2007 Gallup poll found that when asked "Do you personally believe in evolution?" 49 % of the respondents answered "yes," and 48 % answered "no"—a statistical tie. (2 % had no opinion.) Both polls were restricted to adult citizens of the United States. Evidently, the explicit mention of humans had a dramatic effect on the result.
10. In 2005, a poll conducted by the Pew Forum on Religion and Public Life and the Pew Research Center for the People and the Press found that nearly two-thirds of Americans say that creationism should be taught alongside evolution in public schools. Sixty-four percent of the respondents said they were open to the idea of teaching creationism in addition to evolution, while 38 % favored replacing evolution with creationism.
11. In fact, several forces drive evolution, but natural selection is foremost among them and was the focus of the Evolution Readiness project, as we have seen.

References

Horwitz, P. (2010). *Teaching and assessing "evolution readiness" to fourth graders using games.* Paper presented at the American Educational Research Association annual meeting, Denver, CO.

Horwitz, P., & Christie, M. A. (2000). Computer-based manipulatives for teaching scientific reasoning: An example. In M. Jacobson (Ed.), *Innovations in science and mathematics education: Advanced designs for technologies of learning*. Mahwah, NJ: Lawrence Erlbaum Assoc.

Horwitz, P., Gobert, J., Buckley, B., & O'Dwyer, L. (2009). Learning genetics with dragons: From computer-based manipulatives to hypermodels. In M. J. Jacobson & P. Reimann (Eds.), *Designs for learning environments of the future: International perspectives from the learning sciences*. New York: Springer.

Ottino-Loffler, J., Rand, W., & Wilensky, U. (2007). *Co-evolution of predators and prey in a spatial model*. Paper presented at the genetic and evolutionary computation conference, London.

Rosca, C., O'Dwyer, L., Lord, T., & Horwitz, P. (2010). Ready, set, go, evolution! *@ Concord, 14*(2), 4–6.

Scott, E. C. (2004). *Evolution vs. creationism: An introduction.* Berkeley, CA: University of California Press.

Sinatra, G. M., & Nadelson, L. (2010). Science and religion: Ontologically different epistemologies. In R. S. Taylor & M. Ferrari (Eds.), *Epistemology and science education.* New York/London: Routledge.

Sparks, S. (2010, November 17). Efforts to improve evolution teaching bearing fruit. *Education Week*, pp. 16–17.

Verhey, S. D. (2005). The effect of engaging prior learning on student attitudes toward creationism and evolution. *Bioscience, 55*(11), 996–1003.

Wilenski, U., & Novak, M. (2010). Teaching and learning evolution as an emergent process: The beagle project. In R. Taylor & M. Ferrari (Eds.), *Epistemology and science education.* New York/London: Routledge.

Chapter 9
Multiple Representations of Human Genetics in Biology Textbooks

Pierre Clément and Jérémy Castéra

Introduction: Different Representations of the Genetic Determinism

For many years, the teaching of genetics has only been centered on the determinism of the phenotype by the genotype as shown in Fig. 9.1.

This linear and deterministic representation is important in explaining what genetics is but is limited and even dangerous, at least in human biology and more generally. The human phenotype includes not only the anatomy and physiology of any person but also his/her appearance built during his/her life; his/her illnesses and health; and his/her behavior, emotions, intelligence, skills, and any other learned competence. Most of these features are sociocultural and cannot be reduced to genetic determinism. For the learner, these ideas must be introduced in a more systemic manner. This does not mean that the genome is not important in explaining some phenotypic aspects, but to reduce all the phenotypes to a genetic influence is more ideological than scientific, expressing innatist ideas, namely, hereditarianism.

Several works illustrate the danger of this deterministic reductive representation. For instance, to justify sexism, a sociological analysis showed the imprinting of innatist values (Nelkins & Lindee, 1995). These authors suggested a parallel between DNA and the soul:

> Today, these are the genes that allow [us] to talk about personality traits, the nature of immortality, and the sacred meaning of life, in a way that resembles that of religious narratives [. . .]. DNA took in mass culture, the aspect of an entity similar to the soul. (p. 67)

P. Clément (✉)
The Interdisciplinary Laboratory of Didactical Research and History of Science and Techniques, University Lyon 1, 38 bd Niels Bohr, 69622 Villeurbanne Cedex, France
e-mail: Pierre.Clement@univ-lyon1.fr

J. Castéra
Science Education Centre, University of Tartu, Estonia
e-mail: jcastera@ut.ee

D.F. Treagust and C.-Y. Tsui (eds.), *Multiple Representations in Biological Education*, Models and Modeling in Science Education 7, DOI 10.1007/978-94-007-4192-8_9,
© Springer Science+Business Media B.V. 2013

Fig. 9.1 The most common representation of genetic determinism

Other authors, in biology or in epistemology of biology, developed the idea that genes take the place of God to explain the determinism of human behavior and performances and that the *genetic program* is a kind of predestination by God: Everything happening was written in advance (Clément & Castéra, 2007; Keller, 2003; Kupiec & Sonigo, 2000).

Innatist ideas were very strong during the first half of the twentieth century, with a large extension of the Nazis' ideology. During this period, the research in genetics was growing and being structured in France into institutes of *genetics and eugenics*. More recent works—such as those of the psychologists Keller (2005) in Germany, or Dambrun, Kamiejski, Haddadi, and Duarte (2009) in France—showed that, even today, differences among genres or ethnic groups are often justified by this deterministic representation of genetics, which is therefore linked to sexism and racism.

The idea of environmental influence on human phenotypes and performances became nevertheless more and more accepted during the twentieth century. Consequently, the traditional debate *nature VERSUS nurture* was progressively replaced by a new representation *nature AND nurture*. Percentages of contributions from both genotype and environment were proposed, for instance, to explain intelligence from *research on twins* (see the famous fraud of Burt reported by Lewontin, Rose, and Kamin (1984)). This additional representation (genes + environment, see Fig. 9.2) is still very popular in students' conceptions (e.g., Lewis, 2004; Lewis, Leach, & Wood-Robinson, 2000) and even in school textbooks and teachers' conceptions (Clément & Forissier, 2001; Forissier & Clément, 2003).

Nevertheless, this representation is scientifically incorrect because the genes and their environment cannot be added; rather they interact, as has been demonstrated by researchers in genetics (e.g., Jacquard, 1972). What is the part played by each in the development of the heart, the brain, or the liver in our body? When the genotype and the environment interact, it is impossible to evaluate their importance by a percentage. All biologists agree with an interactive representation shown in Fig. 9.3 (e.g., Atlan, 1999; Jacquard & Kahn, 2001; Lewontin, 2000).

However, the interaction can be more complex than that in the representation in Fig. 9.3, as, for example, Lewontin's (2000) description in *Triple Helix: Genes, Organism and Environment*. Similarly, Forissier and Clément (2003) described three levels of interaction (see Fig. 9.4):

1. Between the genes and their environment (epigenetics) (see Fig. 9.5)
2. Between the phenotype and its environment, for example, when one has an accident resulting in amputation of one's leg
3. Between (1) and (2), for example, genetic manipulation or the use of a diet without phenylalanine to correct the effects of the gene mutation that causes phenylketonuria (Jacquard, 1972)

Since the late twentieth century, the reductionist representation of genetic determinism—all by the genes—has declined (Atlan, 1999), giving more and

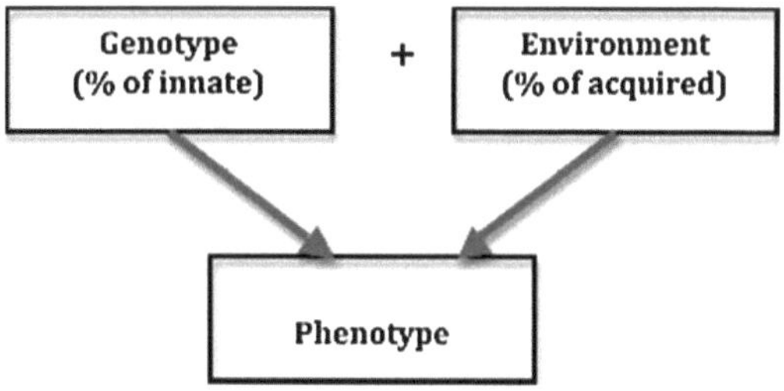

Fig. 9.2 The additive representation of the genetic determinism (genotype + environment)

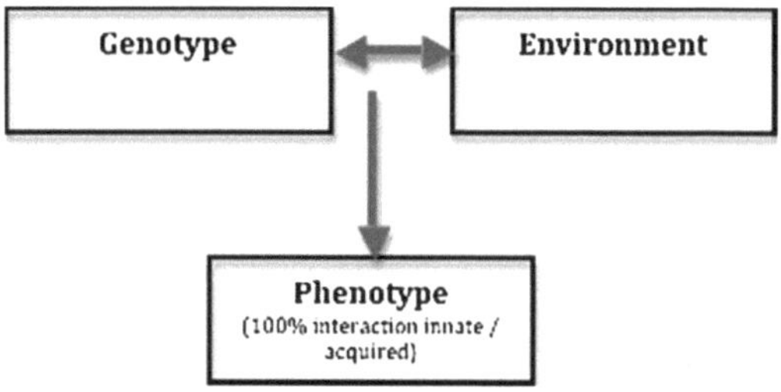

Fig. 9.3 The interactive representation of the genetic determinism (genotype in interaction with environment)

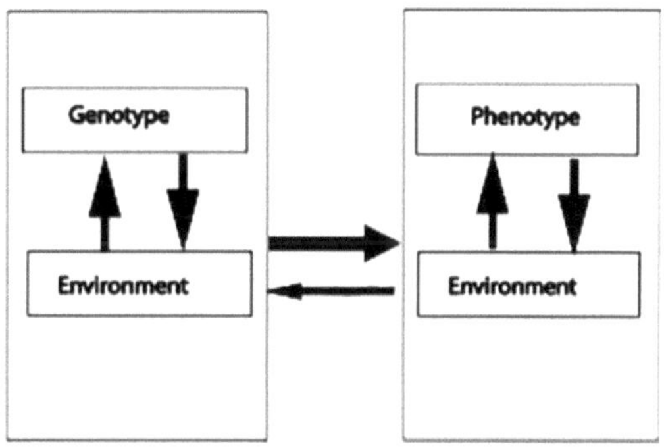

Fig. 9.4 Interactive representation with three levels of interaction (Adapted from Forissier & Clément, 2003)

more importance to the theories of complexity (Atlan, 1979; Morin, 1990) with the new paradigms of *emergence* (Stengers, 1997) and *epigenetics* (Morange, 2005a, 2005b; Wu & Morris, 2001). The interaction between the genes and their environment (see Figs. 9.3 and 9.4) is now accepted by the scientific community opening a *postgenomic* period that follows the limited information from human DNA sequencing. The main results of the Human Genome Project were first published in the special issues of *Nature* and *Science* in February 2001. Most authors have insisted on the importance of interactions between several levels of biological organization, with DNA being one of these levels (Képès, 2005; Lewontin, 2000; see Fig. 9.5). Nevertheless, there are two kinds of definition and representation of epigenetics (see Fig. 9.5).

The strict scientific definition of *epigenetics* (Pouteau, 2007, p. 155) is only concerned with the control of the activity of genes by chemical modifications of the DNA itself (e.g., by methylation) or of proteins of the chromatin around the DNA (e.g., histone acetylation). A broader definition of epigenetics is concerned with all

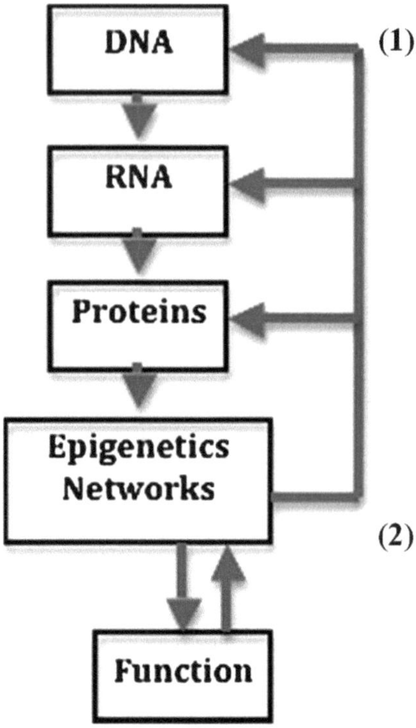

Fig. 9.5 Epigenetic feedback loops. Epigenetics *sensu stricto* (1) is the feedback involving DNA (e.g., methylation, histone acetylation). Epigenetics *sensu largo* (2) involves all the levels of feedback. At each level, there is an interaction between a biological organization and its environment (Modified from Atlan, 1999)

the nongenetic processes which, in interaction with the genes, are acting to build a phenotype. This last definition does not differ from the notion of epigenesis used during the seventeenth century against the *preformationist* ideas and was used again in the 1940s by Waddington for the nongenetic processes by which the phenotype is emerging (for Waddington, "epigenesis + genetics = epigenetics") (Van Speybroeck, 2002). The human *cerebral epigenesis* is a possible illustration of this kind of emergence, by interaction between the human genes (as the basis of the *cerebral ontogenesis*) and other processes such as selection of neural networks by individual activity—*natural selection of synapses* (Changeux, 1983) and *neural Darwinism* (Edelman, 1987).

More generally, the concepts of *determinism* or *instructions* are much debated today in biology. The alternative representation, from the consensual schema of the Darwinian theory of evolution (see Fig. 9.6), shows that several different structures, coming from processes of differentiation, are selected by their interaction with the environment and only the most adapted ones survive. This schema is documented for the cerebral epigenesis, with a first step of having redundant innervations, some of which are then selected for differentiation when they become functional depending on the activity of the organism (Changeux, 1983; Edelman, 1987). More recently, this same schema exists for embryology and cellular differentiation, namely, cellular or molecular Darwinism (Gayon, 2009; Kupiec, 2008, 2009; Kupiec, Gandrillon, Morange, & Silberstein, 2009; Pàldi & Coisne, 2009).

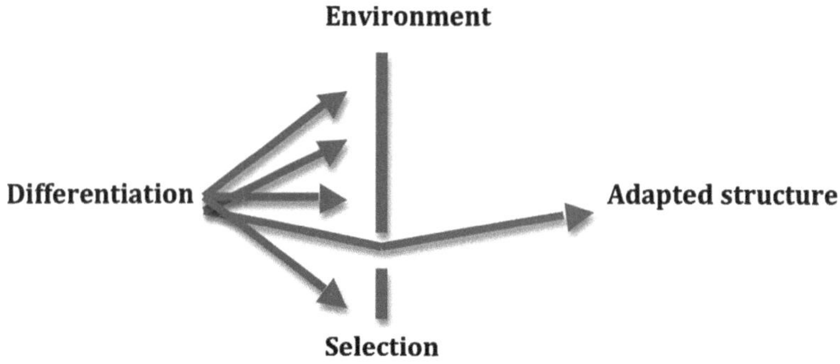

Fig. 9.6 Darwinian selection as an alternative to the biological determinism

In all these cases, stochastic processes are very important for the differentiation, even if they are in interaction with the preexisting biological structures. For instance, a mutation can be stochastic but dependent on the preexisting gene. Similarly, the first redundant synapses can be stochastic, but the functional synapse depends on the interaction between cells in a precise location and so on. The same stochastic processes exist in molecular and cellular biology for understanding cell differentiation, embryology, and other biological processes. Cell differentiation results from the general schema, shown in Fig. 9.6, of an interaction between stochastic processes and biological preexisting structures.

In consequence, any determinism seems to be a special case of a probabilistic phenomenon though the regularity of such emergence is not in contradiction with the stochastic processes. For instance, when tossing a coin, each event (head or tail) is totally stochastic but is regularly 50%/50% after a thousand tosses for a fair coin.

To summarize this part, Figs. 9.1, 9.2, 9.3, 9.4, 9.5 and 9.6 illustrate possible multiple representations of genetic determinism. The additive representation in Fig. 9.2 is out of date because it is in contradiction with the process of interaction. Nevertheless, the other representations are never totally false. For instance, causal determinism of phenotype by genotype in Fig. 9.1 is often used by researchers who try to identify the genetic determinism of a precise feature; however, this procedure is dangerous and scientifically wrong when it reduces any feature and particularly any sociocultural human feature to genetic determinism. Moreover, any research in biology today focuses on the analysis of greater complexity more with a systemic approach including interaction between biological structures, between them and their environment, and also with stochastic processes.

The social challenges of these scientific debates about multiple representations of human genetics are very important for improving health as well as for arguing against fatalism and exclusion of citizenship based on sexism or racism. The dangers of innatist ideas have been stated by researchers in several countries—for instance, in the USA, by Lewontin et al. (1984), Lewontin (2000), Beckwith (1993), and Gould (1997); in the UK, by Rose et al. (1977); and in France, by Jacquard (1972), Clément, Blaes, and Luciani (1980), Stewart (1993), Atlan (1999), Kupiec and Sonigo (2000), Clément and Forissier (2001), Jacquard and Kahn (2001), Séralini (2003), and others.

In the second part of this chapter, we analyze the multiple representations of genetic determinism and discuss the related issues from two sets of data: (1) an analysis of genetic diseases in French textbooks and (2) an analysis of two indicators of genetic determinism (photos of human twins and the metaphor *genetic program*) in biology textbooks of 16 countries.

Analysis of Genetic Diseases in French Textbooks

Human genetic diseases are frequently used as examples in chapters on genetics in French secondary school textbooks. But the choice of a particular example over another can significantly impact the message to the learner. For this reason, we chose to list genetic disease examples found in chapters on genetics in school textbooks and to explore the way these examples are presented. Do they illustrate genetic determinism in a strict reductive sense (see Fig. 9.1) or rather do they introduce more complexity and systemic approach—showing interactions between the genome and its environment (see Figs. 9.3 and 9.4)—and possibly notions of epigenetics (see Fig. 9.5) or debates on genetic determinism (see Fig. 9.6)?

Every human genetic disease is the result of interactions between the genotype and its environment (Chakavarti & Little, 2003) even when they appear illustrative of a simple, linear model of genetic determination (see Fig. 9.1). For example, in the case of phenylketonuria (a monogenic disease), a special diet can completely prevent the occurrence of mental retardation. Furthermore, a genetic disease represents a malfunctioning of or mutation in one or multiple genes but is not always hereditary (Séralini, 2003). In our analyses, we separate genetic diseases into two large categories—monogenic diseases, caused by a mutation in a single gene, and polygenic diseases, where multiple genes play a role in symptom development (Swynghedauw, 2000). Our research addresses both monogenic and polygenic diseases as well as chromosomal anomalies.

We analyzed 18 biology textbooks, published by four different French publishers, containing chapters dealing with human genetics (Castéra, Bruguière, & Clément, 2008). As genetics is only taught in the last 4 years of secondary education in France (i.e., students aged 15–18 years), only these school levels were included in the study. The main results are now summarized in the following section.

The first result indicates that the examples of genetic diseases or anomalies present in the textbooks are not reflecting their prevalence around the world. Monogenic diseases, which are rare, are the most represented (between 51 and 91% of examples, depending on the school level; see Table 9.1). For instance, cystic fibrosis affects 1 in 13,000 births in France[1]. Even the chromosomal

[1] Based on the online data from France's Centre d'Epidémiologie sur les causes médicales de décès (CéPiDc) at http://www.invs.sante.fr/surveillance/maladies_rares/mortalite_mucoviscidose.htm (data retrieved in August 2011).

Table 9.1 Occurrences of genetic diseases by both school level and type of genetic determinism in the school textbooks analyzed

Genetic determinism / School level	Monogenic	Polygenic	Chromosomal anomaly	Total (%)	Number of occurrences per school level
3ème (14–15 year olds)	14 (64%)	0 (0%)	8 (36%)	100	22
2nde (15–16 year olds)	10 (91%)	1 (9%)	0 (0%)	100	11
1ère (16–17 year olds)	27 (51%)	25 (47%)	1 (<1%)	100	53
Terminale S (17–18 year olds)	12 (57%)	3 (14%)	6 (27%)	100	21
Terminale S Spé Bio (17–18 year olds)	10 (71%)	0 (0%)	4 (29%)	100	14

anomalies are relatively rare (1 in 800 births in France has Down syndrome[2]). On the other hand, polygenic diseases are much more frequent: One American in two and one European in three will develop a cancer during the course of his or her lifetime (Séralini, 2003). Also there were 177 million diabetic patients in the world in the year 2000, and, according to Shaw, Sicree, and Zimmet (2010), this figure was predicted to rise to 300 million by the year 2025.

The prevalence of examples of monogenetic diseases probably corresponds to textbook authors' wish to present conceptual ideas as simply as possible for the youngest students (14–15 years old). They start with the visual images of the chromosomes of the Down syndrome, the definition of a gene as a portion of chromosome, and then as a portion of DNA—while the definition of a gene is currently under debate (e.g., Abrougui & Clément, 2005; Chevassus-Au-Louis, 2001; Keller, 2003)—and finally they introduce a clear example of a monogenetic disease such as Duchenne muscular dystrophy. Nevertheless, the danger of the choice of these examples is to deeply anchor the deterministic representation of Fig. 9.1 in the students' minds.

More complexity is introduced to students of 15–16 years old, with nearly half of the examples dealing with polygenic diseases, and even sometimes with modulator genes. The main examples are diabetes (sometimes presented as monogenic in the previous school levels, but no more here), and half of the examples are cancers.

Environmental influence is rarely addressed in the school level *3ème* and is included in no more than one-third of genetic disease examples presented in the textbooks for the school levels *2nde*, *1ère*, and *terminale*. These results demonstrate the predominance of simplistic, causal deterministic mechanisms of genetic diseases (see the representation in Fig. 9.1) in the French secondary school textbooks studied (Table 9.2).

However, diseases significantly influenced by environmental factors (e.g., diabetes, cancers) are relatively frequent in textbooks for students of 15–16 and 16–17 years old. The presence of such examples contributes to a less simplistic representation of genetic determinism. As shown in Fig. 9.7, the environment is mentioned

[2] Depending the age of the mother, from 1/1,500 (20 years old) to 1/100 (40 years old) (Herman et al. 2002)

Table 9.2 Environmental influence on genetic disease examples

School level	3ème (14–15 year olds)	2nde (15–16 year olds)	1ère S (16–17 year olds)	Terminale *S* (17–18 year olds)	Terminale spécialité (17–18 year olds)	Total
1. Number of occurrences of genetic disease examples	22	20	144	24	25	235
2. Number of occurrences of genetic diseases mentioning environmental influence	3	6	49	6	4	68
"(2) Number of . . . environmental influence" as percentage of "(1) Number of . . . genetic diseases examples"	14%	30%	34%	25%	16%	28%

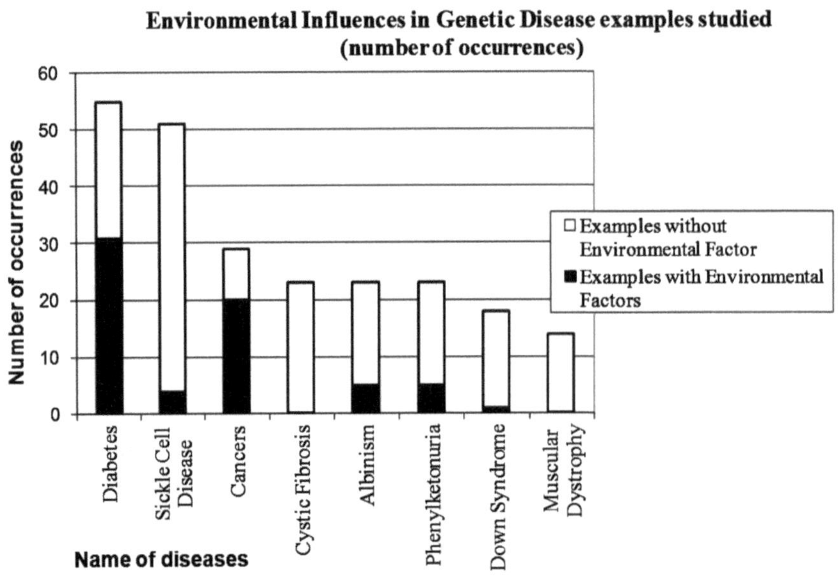

Fig. 9.7 Environmental influence in ten genetic disease examples in the analyzed French textbooks

in the discussion of cancers and diabetes in more than 50% of cases studied. In case of monogenic diseases or chromosomal anomalies, the environment can influence a disease by rendering the disease either partially reversible (e.g., by means of an adapted education in the case of Down syndrome) or completely reversible (e.g., in the case of phenylketonuria); however, environmental influence is rarely mentioned in French textbooks. Such representation is only justified for cystic fibrosis and Duchenne muscular dystrophy, both of which are strongly subject to monogenic determinism. With regard to Down syndrome, only one textbook (Bordas publisher, level *Terminale scientifique*) (Castéra, Bruguière, et al., 2008) makes reference to environmental factors capable of influencing the disease, by clearly indicating that an education program may allow better integration of Down syndrome patients in society.

This underrepresentation of the environmental influence in the examples of genetic diseases is consistent with the biomedical model of health predominating in French biology textbooks. In contrast, textbooks from other countries are more focused on the model of health promotion (Carvalho et al., 2008).

A final point is that not even one of the analyzed textbooks contained the word *epigenetics* nor any one explained its content. Furthermore, there was no debate on the notion of *genetic determinism* and no mention of the possible alternatives as shown in Fig. 9.6. These new concepts and representations of the human genetics were too recent (1999–2005) to be introduced in the analyzed textbooks (published between 2001 and 2004). The Didactic Transposition Delay (DTD)—the time between the scientific publications and their introduction in syllabuses and textbooks (Quessada & Clément, 2007)—is longer, and we hope these new representations will be present in the more recently published textbooks.

Nevertheless, important advancements have been made since publication of the previous French curricula and biology textbooks in 1995 toward a less simplistic presentation of the causes of genetic diseases—no longer only limiting these diseases to a single gene (that was totally dominant in the previous textbooks; see Abrougui, 1997). In the new syllabuses, multiple genes, as well as the interaction between the genome and its environment, are presented. Complex determinism models for diseases (such as cancers or diabetes) help to prevent students' minds from being ingrained with the single, oversimplified conception of genetics, which nevertheless continues to predominate in biology textbooks.

Images of Twins and the Metaphor Genetic Program in Textbooks of 16 Countries

The results presented here come from research in the context of the project BIOHEAD-Citizen (Biology, Health and Environmental Education for Better Citizenship) (BIOHEAD-Citizen, 2004–2008). This research project included a comparison of the biology textbooks dealing with six selected topics, one of which is human genetics.

We analyzed 50 textbooks in 16 countries (with number of textbooks analyzed given in parentheses): Cyprus (2), Estonia (2), Finland (2), France (11), Germany (3), Hungary (3), Italy (7), Lebanon (4), Lithuania (2), Malta (2), Morocco (2), Poland (1), Portugal (4), Romania (1), Senegal (1), and Tunisia (3). The number of textbooks studied in each country differed because (i) in some countries, human genetics was taught at only one school level, whereas in other countries, it was taught at two or more school levels; and (ii) in some countries, there was only one official national publisher for school textbooks, whereas in others, there were several private publishers. In the latter case, the most significant publishers were chosen. For each analyzed textbook, a long grid was completed. We present here only some results, dealing with two indicators.

First Indicator: Photos of Human Twins

Photographs are considered as scientific images because they convey a scientific message (Clément, 1996). In this specific case, the message is the morphological similitude of identical twins, which corresponds to the identity of their genotypes. In contrast, some images are intended to show morphological differences between fraternal twins. Nevertheless, the images of monozygotic twins—the same clothes, the same hairstyle, the same behavior, and so on—can also have an implicit ideological message when they strongly suggest that features other than morphological ones can be genetically determined. Consequently, for each image of twins in the textbooks, we examined whether the twins were presented as having the same clothes, style and behavior or, on the contrary, if the images showed differences illustrating the *paradox of the twins*—psychologists such as Zazzo (1984) showed that identical twins tend to differ in their psychological characteristics and sociocultural appearances. Identical twins are also a good illustration of possible epigenetic differences. For instance, Fraga et al. (2005) showed that in 35% of the monozygotic twins studied, there were differences in the methylation of their DNA and histone acetylation; these epigenetic differences are more important in the older twins and in twins with different lifestyle or medical history.

The results of the analysis are spectacular (Castéra & Clément, 2007; Castéra et al., 2008; Clément & Castéra, 2007). In all the images of identical twins in the textbooks, twins had exactly the same clothes, hairs, and so on (except one case where the color of the jacket was different, as well as the length of the hair), whereas the images clearly differ for the fraternal twins. Consequently, for all these images, the representation of genetic determinism corresponded to Fig. 9.1 ("genotype → phenotype") with, moreover, implicit innatist ideas that are not scientifically correct, suggesting that the sociocultural features (e.g., clothes, hairstyle) would be determined by the genes.

Second Indicator: Occurrence of the Metaphor Genetic Program

This hereditarianist expression *genetic program* now appears questionable. The deterministic representation of genetics (see Fig. 9.1) suggesting that everything of our life is written in our DNA program—as a predestination written in the plan of God for some religions—is no longer accepted by biologists (Abrougui & Clément, 1997; Atlan, 1999; Clément, 2007; Kupiec & Sonigo, 2000; Morel & Miquel, 2001; Noble, 2007). Consequently, Atlan and other authors proposed to replace *genetic program* by *genetic information*. Is the Didactic Transposition Delay (DTD) sufficient for observing a diminution or suppression of this expression *genetic program* in the analyzed textbooks? The results (see Table 9.3) showed important differences in using this expression among the textbooks of 16 countries. In the textbooks of some countries, the expression *genetic program* is very commonly used:

Table 9.3 Total occurrences of "genetic program" and some expressions with the same implicit meaning

Textbooks for / Country	11–12 years old	12–13 years old	13–14 years old	14–15 years old	15–16 years old	16–17 years old	17–18 years old	18–19 years old
Cyprus					0		0	
Estonia				0		0		
Finland			17			53		
France				66-7[a]	18-49[a]	1-2[a]	0-0[a]	
Germany				0			0-0	
Hungary						1	0-0[a]	
Italy	0-0-0[a]			0-2-0-0[a]				
Lebanon			27		3-0[b]	1		
Lithuania					0		1	
Malta	No textbook				10-1[a]			
Morocco			23				28	
Poland							0	
Portugal				2		2	0-1[c]	
Romania							5	
Senegal								1
Tunisia				0			0	0

[a]Occurrences in two or more textbooks from different publishers (figures separated by a hyphen)
[b]Occurrences in two textbooks from the same publisher but from science section or humanity section (figures separated by a hyphen)
[c]Occurrences in one biology textbook and one psychology textbook in Portugal (figures separated by a hyphen)

(1) *The Maltese textbooks* are in fact British textbooks published in 1986. At that time, the deterministic model was very popular in several countries (Abrougui, 1997). In these Maltese textbooks, the term *genetic program* was not used, but the same message was in the expression: "the DNA molecule forms a code which instructs the cell" (10 occurrences in a textbook and one occurrence in another textbook for 15–16 year olds) (see Table 9.3). The message is that the cells could simply follow these instructions to produce the phenotype, just like a program.

(2) *In Finnish textbooks*, we found a strong presence of the notion *genetic program*: 70 occurrences versus only five occurrences of *genetic information* in recently published textbooks (2004 and 2006). So even though the scientific community no longer uses the expression genetic program, the Finnish textbooks still do. In another work, we found a trend of the Finnish teachers' conceptions clearly correlated with the implicit metaphor *genetic program* (Castéra, Clément, & Kosonen, 2009).

(3) *In Portuguese textbooks*, there is both a persistence of the notion of a *genetic program* with at least one occurrence by level and an absence of a clear evolution between the various levels. A possible explanation in this case is the difficulty in changing the scientific and ideological content of the chapter on human genetics.

(4) *In France*, there are strong differences in the occurrences of this metaphor *genetic program* among the textbooks from three different publishers and also across school levels (the higher the level, the fewer its use), and it completely disappears at the end of the curriculum. Compared to syllabuses and French textbooks published in the 1990s, the occurrence of the metaphor genetic program is very infrequent in today's textbooks, with this being increasingly replaced by the notion *genetic information*. Nevertheless, the disappearance of the term genetic program is not complete, suggesting the influence of other parameters. One parameter is pedagogical. Teachers start using genetic program for teaching the youngest students and progressively use genetic information more frequently. This initial simplification (to start with the message *genotype → phenotype*) is educationally dangerous; Clément, Forissier, and Carvalho (2003) showed that the first concepts taught are those most memorized by students. The second parameter is the different strategies among the publishers. Thirdly, it is also possible that the difficulty of completely suppressing the notion genetic program was because that this was extremely central in the previous syllabuses. Therefore, the textbook authors and publishers still use this notion with a certain difficulty to change their *traditional* way of thinking.

(5) (6) (7) *In Moroccan, Lebanese, and Senegalese textbooks*, we found the same kind of results as in those in France, possibly indicating a French influence, but sometimes with delay. On the contrary, some countries do not use, or else very rarely use, the metaphor *genetic program* in textbooks.

(8) *Tunisian textbooks* use only the concept of *genetic information* throughout the three school levels where human genetics is taught. The total absence of the use of *genetic program* is probably a consequence of the growing influence of Tunisian researchers in didactics of biology (Abrougui, 1997). Nevertheless, that does not yet mean a total disappearance of innatist ideas from the Tunisian textbooks (Clément, Mouehli, & Abrougui, 2006) nor does it mean its disappearance from the Tunisian teachers' conceptions (Kochkar, 2007, 2010).

(9) *In Germany*, the metaphor *genetic program* is totally absent. As in other Western European countries and the USA, genetic research developed in eugenics institutes has the goal of building genealogical trees and tracking patterns of occurrence of diseases and disabilities (Wolf, 2002). According to O'Mahony and Schäfer (2005), the collective memory of the Nazis' eugenics program is an important background for not referring to a genetic program in communicating about human genetics.

(10) *In Cyprus*, the textbooks use exclusively *genetic information*: eight occurrences in the textbooks for 15–16 year olds and 17 occurrences in the textbooks for 17–18 year olds. The explanation is not easy: It can be an effect of the complex history of this country, as well as a desire to avoid using the metaphor of a computer program.

(11) *In the Italian textbooks*, the precise expression, *genetic program*, occurs only twice and just in one textbook. The textbooks generally use terms like

hereditary patrimony and *genetic patrimony* which are more neutral than genetic program, with less implicit meaning. This could show the same desire as in Germany being cautious in dealing with hereditarianist ideology. The common past of Germany and Italy during the Second World War certainly had an influence on the way human genetics was taught, albeit sometimes awkwardly. For example, in one of the textbooks, the metaphor of *the books of life* is used. According to this metaphor of the book of life, it would be enough to know the alphabet and the genetic syntax to reach the essence of the human being. Today, such a conception is scientifically unacceptable and ethically dangerous.

(12) *In the Lithuanian textbooks*, there is just one occurrence that could be considered as a notion similar to genetic program: The reproduction of cells is *programmed in genes*, but there is never the precise expression: *genetic program*. In Lithuania, the explanation for this absence seems to be deeply rooted in the past: The notion was traditionally absent in the previous syllabuses and textbooks. We have verified this for textbooks published since 1979.

(13) (14) (15) *Estonia, Poland, and Hungary* are three other countries included in or influenced by the former Soviet Union, and their textbooks show the same trend as in Lithuania: no one mention of *genetic program*, but several mentions of *genetic information*. Only one exception was found in a Hungarian textbook with the term *programmed by a gene* (exactly the same expression quoted above in a Lithuanian textbook). Our hypothesis is that in these countries, there was one official line to teach biology based on the work of Lysenko (Лысéнко) and Michurin (Мичурин) in the former Soviet Union, with a negation of the idea of a genetic program, even if at the end of the 1960s, the pseudoscientific, neo-Lamarckism ideas of Lysenko and Michurin were rejected. The role of DNA in hereditary information was then accepted and presented in all textbooks. However, without the idea of a genetic program, and the differences between human individuals were also always explained by environmental or social conditions.

(16) *Romania* also was formerly influenced by the Soviet Union; we found five occurrences of *genetic program* and 19 occurrences of *genetic information* in the unique Romanian textbook, dealing with human genetics (with only 13 pages devoted to this topic).

In conclusion, there were contrasting results from the biology books used in these 16 countries. In each country's textbooks, the representations of identical twins were linked not only to a scientific message (morphological resemblance correlated with the same genome) but also to an implicit ideological message (a suggested genetic determinism of sociocultural features such as clothes or hairstyle). This kind of social representation of identical twins, deeply anchored in nonscientific ideologies, is in contradiction with the renewal of scientific knowledge in human genetics, such as the importance of epigenetic processes in explaining differences even between monozygotic twins.

On the contrary, there were interesting differences among the 16 countries in terms of the use in their textbooks of the metaphor *genetic program*, which was sometimes replaced by another deterministic metaphor the books of life. Language is never neutral and the expression genetic information is less ideological than genetic program, less deterministic and more open to interactions with environmental and epigenetic processes from a systemic perspective. We suggest several hypotheses to explain the occurrence (or absence) of the metaphor genetic program in each of these 16 countries. They illustrate interactions between science (the taught science) and society (its history and other characteristics), as well as ethical, cultural, and social implications of this use, and interactions between the taught scientific knowledge (K) with implicit values (V) and social practices (P)—the KVP model (Clément, 2004, 2006). As a general conclusion, the multiple representations of human genetic determinism in school textbooks not only correspond to the renewal of the scientific knowledge in human genetics but are also correlated with sociocultural parameters, values, and, social practices, which differentiate the way by which human genetics is taught in different countries.

Acknowledgments This research was financed by the BIOHEAD-Citizen project (2004–2008): *Biology, Health and Environmental Education for Better Citizenship* (E.C., STREP n° CIT2-CT2004-506015, FP6, Priority 7). We would like to thank our colleagues from the 16 countries who, in the context of this project, analyzed the textbooks of their country and provided us with the results for writing this chapter.

References

Abrougui, M. (1997). *La génétique humaine dans l'enseignement secondaire en France et en Tunisie.* [Human genetics in secondary schools, in France and in Tunisia]. Thèse Doctorat Université Lyon 1, Lyon, France.

Abrougui, M., & Clément, P. (1997). Enseigner la génétique humaine: Citoyenneté, ou fatalisme? [Teaching human genetics: Citizenship or fatalism?]. In A. Giordan, J.-L. Martinand, & D. Raichvarg (Eds.), *Actes JIES 19e Journées internationales sur l'éducation scientifique* [International days on science education] (pp. 255–260). Chamonix, France: Université Paris Sud.

Abrougui, M., & Clément, P. (2005). Two strategies to introduce genetics in Tunisian and French textbooks. In R. Pinto, & D. Couso (Eds.), *Proceedings of the fifth international conference on contributions of research to enhancing students' interests in learning science* (pp. 1245–1249). Barcelona, Spain.

Atlan, H. (1979). *Entre le cristal et la fumée: Essai sur l'organisme vivant* [Between crystal and smoke: An essay on the living organism]. Paris: Seuil.

Atlan, H. (1999). *La fin du "tout génétique"* [The end of "everything is genetic"]. Paris: INRA.

Beckwith, J. (1993). A historical view of social responsibility in genetics. *BioScience, 43*(5), 327–333.

BIOHEAD-Citizen. (2004–2008). *Biology, health and environmental education for better citizenship.* In G. Carvalho, P. Clément, & F. Bogner (Eds.), *European research project* (Specific

Targeted Research n° CIT2-CT2004-506015, FP6, Priority 7). http://projectos.iec.uminho.pt/projeuropa/

Carvalho, G. S., Dantas, C., Rauma, A.-L., Luzi, D., Bogner, F., Caussidier, C., & Clément, P. (2008). Comparing health education approaches in textbooks of sixteen countries. *Science Education International, 19*(2), 133–146.

Castéra, J., Bruguière, C., & Clément, P. (2008). Genetic diseases and genetic determinism models in French secondary school biology textbooks. *Journal of Biological Education, 42*(2), 53–59.

Castéra, J., & Clément, P. (2007). Hereditarianism and reductionism in biology school textbooks, through images of twins. A comparative study between 15 countries. In *Proceedings of the ESERA (European Science Education Research Association)* (4 p.). Malmö, Sweden (CD-ROM).

Castéra, J., Clément, P., Abrougui, M., Sarapuu, T., Turcinaviciene, J., Agoram, B., et al. (2008). Genetic Determinism in school textbooks: A comparative study conducted among 16 countries. *Science Education International, 19*(2), 163–184.

Castéra, J., Clément, P., & Kosonen, A-L. (2009, August). *The genetic determinism of human performances: A comparison between teachers' conceptions in Finland and France.* Paper Presented at the conference of the European Science Education Research Association (ESERA), Istanbul, Turkey.

Chakavarti, A., & Little, P. (2003). Nature, nurture and human disease. *Nature, 421*, 412–414.

Changeux, J. P. (1983). *L'homme neuronal* [The neuronal man]. Paris: Fayard.

Chevassus-Au-Louis, N. (2001). Dix-huit facettes d'un même concept [Eighteen facets of the same concept]. *La Recherche, 348*, 51–56.

Clément, P. (1996). L'imagerie biomédicale: Définition d'une typologie et proposition d'activités pédagogiques [Biomedical imagery: Definition of a typology and proposition of pedagogical activities]. *Aster, 22*, 87–126.

Clément, P. (2004). Science et idéologie: Exemples en didactique et épistémologie de la biologie [Science and ideology: Examples in didactics and epistemology of biology]. *Actes du colloque "Sciences, médias et société"* [Proceedings of the Conference "Sciences, media and society"], Lyon, ENS LSh (Laboratoire Communication, Culture et Société) (pp. 53–69). Available: http://sciences-medias.ens-lsh.fr/IMG/pdf/actes.pdf

Clément, P. (2006). Didactic transposition and the KVP model: Conceptions as interactions between scientific knowledge, values and social practices. In: *Proceedings summer school ESERA* (pp. 9–18). Minho, Portugal: IEC, University of Minho.

Clément, P. (2007). Introducing the cell concept by both animal and plant cells: A historical and didactic approach. *Science Education, 16*, 423–440.

Clément, P., Blaes, N., & Luciani, A. (1980). Le mythe tenace du "chromosome du crime", encore appelé "chromosome de l'agressivité" [Persistent myth of the "chromosome of crime", also called "chromosome of aggressiveness"]. *Raison Présente, 54*, 109–127.

Clément, P., & Castéra, J. (2007). L'Actualité des images de vrais jumeaux dans les magazines de vulgarisation et dans les manuels scolaires [Pictures of identical twins in popular magazines and in school textbooks. In A. Giordan, J. L. Martinand, & D. Raichvarg (eds.), *Actes JIES (Journées Internationales sur la Communication, l'Éducation et la Culture Scientifiques et Industrielle)* [International days on science education] (6 p.), Chamonix, France (CD-ROM).

Clément, P., & Forissier, T. (2001). L'identité biologique n'est pas que génétique: un défi pour un enseignement citoyen [The biological identity is not only genetic: A challenge for a citizenship teaching]. In *Communication au symposium BioEd 2000: The challenge of the next century* (pp. 15–18). Paris. Available: www.iubs.org/cbe/pdf/clement.pdf

Clément, P., Forissier, T., & Carvalho, G. S. (2003). The structuring influence of first images on the construction of science concepts. An historical and didactical approach. In *Acts of the 4th ESERA conference*. Noordwijkerhout, The Netherlands (CD-ROM).

Clément, P., Mouehli, L., & Abrougui, M. (2006). Héréditarisme, béhaviorisme, constructivisme: le système nerveux dans les manuels scolaires français et tunisiens [Hereditarianim, behaviorism, constructivism: Nervous system in French and Tunisian school textbooks]. *Aster, 42*, 187–222.

Dambrun, M., Kamiejski, R., Haddadi, N., & Duarte, S. (2009). Why does social dominance orientation decrease with university exposure to the social sciences? The impact of institutional socialization and the mediating role of "geneticism". *European Journal of Social Psychology, 39*(1), 88–100.

Edelman, G. (1987). *Neural Darwinism: The theory of neuronal group selection*. New York: Basic Books.

Forissier, T., & Clément, P. (2003). Teaching "biological identity" as genome/environmental interactions. *Journal of Biological Education, 37*(2), 85–91.

Fraga, M. F., Ballestar, E., Paz, M. F., Ropero, S., Setien, F., Ballestar, M. L., et al. (2005). Epigenetic differences arise during the lifetime of monozygotic twins. *Proceedings of the National Academy of Sciences of the USA, 102*, 10604–10609.

Gayon, J. (2009). Déterminisme génétique, déterminisme bernardien, déterminisme laplacien [Genetic determinism, Bernard's determinism, Laplace's determinism]. In J.-J. Kupiec O., Gandrillon M., Morange & M. Silberstein (eds.), *Le hasard au cœur de la cellule, probabilité, déterminisme, génétique* [Chance at the heart of the cell, probability, determinism, genetics] (pp. 79–91). Paris: Syllepse.

Gould, S. J. (1997). *Darwin et les grandes énigmes de la vie* [Darwin and the great mysteries of life] (éd. originale: 1977). Paris: Seuil.

Herman, A., Dreazen, E., Herman, A. M., Batukan, C. E., Holzgreve, W., & Tercanli, S. (2002). Bedside estimation of Down syndrome risk during first trimester ultrasound screening. *Ultrasound in Obstetrics & Gynecology, 20*(5), 468–475.

Jacquard, A. (1972). *Eloge de la différence: La génétique et les hommes* [In praise of difference: Genetics and human affairs]. Paris: Seuil.

Jacquard, A., & Kahn, A. (2001). *L'avenir n'est pas écrit* [The future is not written]. Paris: Bayard.

Keller, E. F. (2003). *Le siècle du gène* [The century of the gene]. Paris: Gallimard.

Keller, J. (2005). In genes we trust: The biological component of psychological essentialism and its relationships to mechanisms of motivated social cognition. *Journal of Personality and Social Psychology, 88*, 686–702.

Képès, F. (2005). L'épigénétique comme aspect de la post génomique [Epigenetics as an aspect of post-genomics]. *Medecine/Science, 21*, 371–376.

Kochkar, M. (2007). *Les déterminismes biologiques. Analyse des conceptions et des changements conceptuels consécutifs à un enseignement sur l'épigenèse cérébrale chez des enseignants et des apprenants tunisiens* [Biological determinism: Analysis of Tunisian teachers' and learners' conceptions and conceptual changes resulting from education on cerebral epigenesis]. Thèse Doctorat Université Lyon 1 & ISEFC – Université de Tunis.

Kochkar, M. (2010) *Enseigner des valeurs ou des connaissances ? L'épigenèse cérébrale ou le "tout génétique"* [Teaching values or knowledge? Cerebral epigenesis or "all genetics"]. Saarbrücken, Germany: Editions Universitaires Européennes.

Kupiec, J.-J. (2008). *L'origine des individus* [The origin of individuals]. Paris: Fayard.

Kupiec, J.-J. (2009). Le hasard au cœur de la vie. Les gènes jouent-ils aux dés ? [Chance at the heart of life: Genes do they play at dice?]. *La Recherche, 434*, 51–53.

Kupiec, J. J., Gandrillon O., Morange, M., & Silberstein, M. (2009). *Le hasard au cœur de la Cellule. Probabilité, déterminisme, génétique* [Chance at the heart of the cell. Probability, determinism, genetics]. Paris: Editions Syllepse.

Kupiec, J.-J., & Sonigo P. (2000). *Ni Dieu, ni gène* [Neither God, nor gene]. Paris: Seuil.

Lewis, J. (2004). Traits, genes, particles and information: Re-visiting students' understandings of genetics. *International Journal of Science Education, 26*, 195–206.

Lewis, J., Leach, J., & Wood-Robinson, C. (2000). All in the genes? Young people's understanding of the nature of genes. *Journal of Biological Education, 34*, 74–79.

Lewontin, R. (2000). *The triple helix: Gene, organism, environment*. Cambridge, MA: Harvard University Press.

Lewontin, R. C., Rose, S., & Kamin, L. J. (1984). *Not in our genes: Biology, ideology and human nature*. New York: Pantheon.
Morange, M. (2005a). L'épigénétique: un domaine aux multiples facettes [Epigenetics: A multifaceted field]. *Medecine/Science, 21*, 339.
Morange, M. (2005b). Quelle place pour l'épigénétique ? [What place for epigenetics?]. *Medecine/Science, 21*, 367–369.
Morel, M. C., & Miquel, P. A., (2001). *Programme génétique: Concept biologique ou métaphore ?* [Genetic program: biological concept or metaphor?]. Paris: éditions Kimé.
Morin, E. (1990). *Introduction à la pensée complexe* [Introduction to complex thought]. Paris: ESF.
Nelkins, D., & Lindee, M. S. (1995). *The DNA mystique: The gene as a cultural icon*. New York: Freeman.
Noble, D. (2007). *La musique de la vie. La biologie au-delà du génome* [The music of life: Biology beyond the genome]. Paris: Seuil.
O'Mahony, P., & Schäfer, M. S. (2005). Comparing German and Irish media discourse on human genome research. *Journal of Social Studies of Science, 35*(1), 99–130.
Pàldi, A., & Coisne, S. (2009). La fin du déterminisme en biologie [The end of determinism in biology]. *La Recherche, 434*, 40–44.
Pouteau, S. (2007). *Génétiquement indéterminé – Le vivant auto-organisé* [Genetically undetermined: The auto-organization of life]. Versailles, France: Quae.
Quessada, M. P., & Clément, P. (2007). An epistemological approach to French syllabi on human origins during the 19th & 20th centuries. *Science Education, 16*(9–10), 991–1006.
Rose, H., Rose, S., Hanmer, J. Enzensberger, H. M., Franck, R., Lévy-Leblond, J. M., & Stéhelin L. (1977). *L'idéologie de/dans la science* [Ideology of/in science]. Paris: Le Seuil.
Séralini, G. E. (2003). *Génétiquement incorrect* [Genetically incorrect]. Paris: Flammarion.
Shaw, J. E., Sicree, R. A., & Zimmet, P. Z. (2010). Global estimates of the prevalence of diabetes for 2010 and 2030. *Diabetes Research and Clinical Practice, 87*, 4–14.
Stengers, I. (1997). *La Vie et l'Artifice: les visages de l'émergence* [Life and Artifice: Faces of the emergence]. Cosmopolitiques. VI Paris: La Découverte.
Stewart, J. (1993). Au delà de l'inné et de l'acquis [Beyond the innate and the acquired]. *Intellectica, 16*, 151–174.
Swynghedauw, B. (2000). *Biologie et génétique moléculaire* [Biology and molecular genetics]. Paris: Dunod
Van Speybroeck, L. (2002). From epigenesis to epigenetics: The case of C. H. Waddington. *Annals of the New York Academy of Sciences, 981*, 61–81.
Wolf, N. (2002). Krankheitsursachen Gene. Neue Genetik und Public Health [Genes and causes of disease: New genetics and public health]. In *Publications series of the research unit Public Health Policy Wissenschaftszentrum Berlin für Sozialforschung*. Berlin, Germany. http://bibliothek.wzb.eu/pdf/2002/p02-202.pdf
Wu, C. T., & Morris, J. (2001). Genes, genetics, and epigenetics: A correspondence. *Science, 293*, 1103–1105.
Zazzo, R. (1984). *Le Paradoxe des jumeaux* [The twins paradox]. Paris: Stock.

Chapter 10
Deconstructing and Decoding Complex Process Diagrams in University Biology

Phyllis B. Griffard

Introduction

Students at all levels learn about biology via numerous communication modes. Direct experience, oral, text, and representations are but a few. These might include any combination of text, animations, verbal explanations, 3-D models, gestures, and printed images. This chapter explores one particular type of printed image, the complex process diagram. These are diagrams that represent complex biological processes that occur in multiple levels of organization over time. Although complex process diagrams are single static images, they are composites of pictorial, symbolic, and text elements related by devices such as telescoping and arrows. Therefore, they can be considered multiple external representations (MERs), and any findings about how learners interact with MERs may be relevant to this specific representation mode.

Let me first begin with a sketch (Fig. 10.1a) created in my office by a scientist offering to have my first-year university students visit his research laboratory. As he was explaining his research, he spontaneously generated this representation on the whiteboard when words alone seemed inefficient. As an impromptu creation for negotiating shared meaning, it can be considered an inscription. It was not designed to be a self-explanatory, stand-alone representation. Rather it evokes a sense that you had to be there and that you need significant background knowledge to understand it. Judging by the common observation of such diagrams in laboratory areas and faculty offices, such inscriptions seem to be an essential communication tool of biologists and biology educators. The adjacent diagram (Fig. 10.1b) from a first-year biology university textbook represents a closely related phenomenon—intracellular calcium homeostasis. Unlike the whiteboard sketch, this diagram was designed to be used without an expert to explain it. The designer of this diagram

P.B. Griffard (✉)
Department of Natural Sciences, University of Houston – Downtown,
One Main Street, Houston TX 77002, USA
e-mail: griffardp@uhd.edu

D.F. Treagust and C.-Y. Tsui (eds.), *Multiple Representations in Biological Education*,
Models and Modeling in Science Education 7, DOI 10.1007/978-94-007-4192-8_10,
© Springer Science+Business Media B.V. 2013

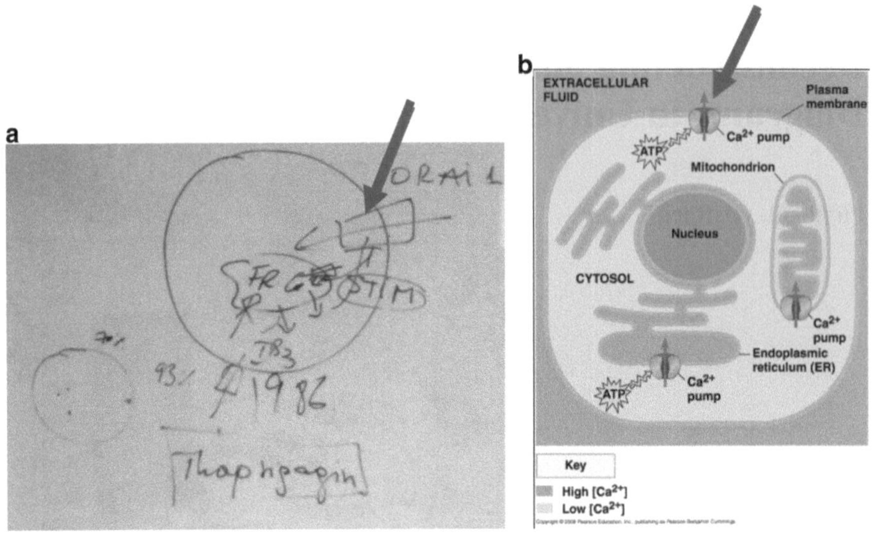

Fig. 10.1 (**a**) A scientist's sketch representing regulation of cytoplasmic calcium. (**b**) Textbook diagram summarizing intracellular calcium homeostasis (Campbell et al., 2008, p. 217). *Red arrows* indicate membrane channel icons (Reprinted with permission)

would have to make assumptions about the audience's prior knowledge of the represented concepts and the meaning of graphic conventions for representing them—such as icons for membrane channels (red arrows in Fig. 10.1)—as well as about how much detail to include and how much to simplify without compromising fidelity to the accepted scientific model or inviting misconceptions.

How such process diagrams are designed and how students make sense of them during learning is the focus of this chapter. First, semiotics and visual cognition are considered with respect to complex process diagrams. Second, recent research on how students use complex process diagrams is summarized. This chapter concludes with a discussion of the pedagogical implications of the research findings.

Diagrams in Biology

Images are ubiquitous in biology instruction and can take many forms. On a continuum of increasing abstraction, they include realistic images such as photographs, micrographs, and naturalistic art; representational images such as process diagrams, molecular structures, classic experiments, biochemical cycles, and cladograms; and symbolic images such as equations, chemical formulae, graphs, gels, and arrays (Pozzer & Roth, 2003). Content analysis of recent editions of a few representative university science textbooks used in North America showed that approximately one-third of page space is occupied by images. Of the textbooks analyzed, representational and realistic images were most frequently encountered in the biology textbooks, whereas symbolic images such as equations, formulae, and graphs were the prominent representations in introductory physics and chemistry textbooks (Griffard, 2010a). The ubiquity of complex process diagrams in biology supports the suggestion that biology has a nature and structure distinct from other sciences (Mayr, 1982) and thus may present unique pedagogical challenges for biology educators.

Diagrams are one type of representational image frequently encountered in biology textbooks. For the purpose of this chapter, a diagram is defined as any graphic art that is designed to depict or explain how something is organized or how it works. This is more general than some definitions that emphasize geometric or schematic features in which pictorial elements are largely absent. On the contrary, the diagrams encountered in biology often contain pictorial elements that are iconic or semi-realistic, many of which have become domain-specific conventions. For example, rectangular or cylindrical shapes representing membrane transport channels in Fig. 10.1a and b are readily recognizable by biologists. It is interesting and relevant to consider how novices to biology come to understand the meanings of such icons and elements over time.

Complex Process Diagrams as MERs

Rich visual narratives that depict complex biological processes can be considered a type of visual confection because they are "visual events, selected ... then brought together and juxtaposed on the still flatland of paper" (Tufte, 1997, p. 121). Unlike some graphics designed for other purposes, textbook diagrams have few or no decorative elements (eye candy) or chartjunk; in other words, they have a parsimonious data/ink ratio. Interaction designer Brad Paley recommended that more research be done on how people extract information from various representation modes (Paley, 2008).

An image is considered a complex process diagram here if it meets these criteria:

- Shapes are used to represent biological entities such as organisms, cells, communities, molecules, and membranes; these can be pictorial, realistic, or metaphorical icons.
- Three dimensions are represented, for example, by shading, layering, or parallax.
- Time or sequence is represented with arrows, placement in reading order, or numbered steps.
- Multiple levels of organization are evident by telescoping multiples or exaggeration of scale.

According to these criteria, the MERs in Fig. 10.2 can be considered complex process diagrams. Each is an association of small multiples connected by arrows with different meanings. In the diagram of water uptake in roots (see Fig. 10.2a), gray arrows represent zooming between levels of organization, whereas red and blue arrows represent direction of movement of water through the tissues. In the diagram of blood clotting (see Fig. 10.2b), the arrows signify changes in the blood vessel cross

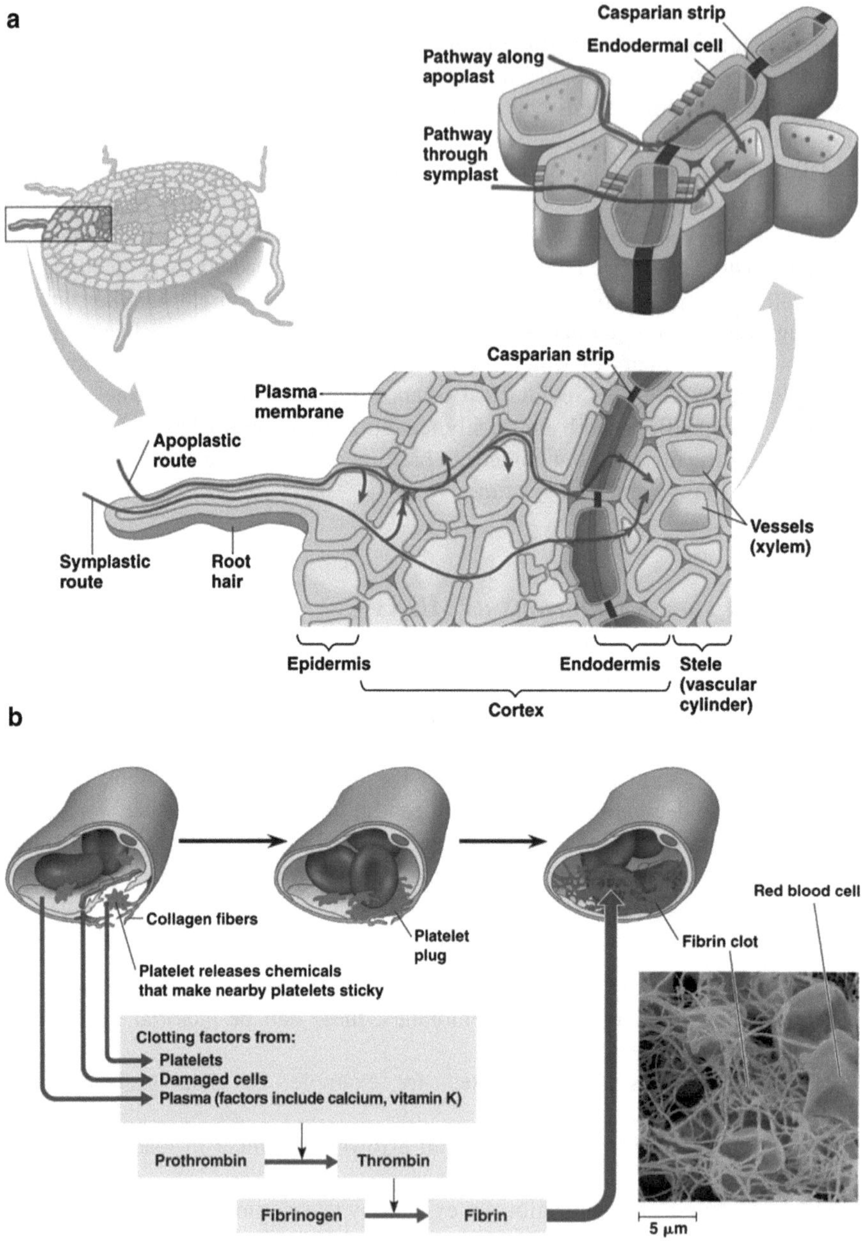

Fig. 10.2 Examples of complex process diagrams illustrating (**a**) water transport into xylem (Campbell et al., 2008, p. 773) and (**b**) blood clotting (Campbell et al., p. 913) (Reprinted with permission)

sections over time in the clotting cascade depicted below them. Complex process diagrams also employ graphic elements such as color, shape, position, and labels to enhance their explanatory power. Some arrow colors are meaningful (e.g., blue for aqueous interior of xylem, red for blood cells), whereas others are arbitrary (magenta for extracellular route, blue for intracellular route). Realistic cell colors, shapes, and layers convey three dimensions, as does the imbedded photomicrograph of a clot.

MERs serve several functions to support learning: they complement, constrain, and construct (Ainsworth, 1999). Static, two-dimensional complex process diagrams can provide these benefits, as can animated, narrated MERs. For example, zooming from macro to micro (roots) and juxtaposing rendered art and real electron micrographs (a clot) provide complementary information about context and ultrastructure, forcing implicit comparison or engagement of more than one cognitive process. Diagrams constrain possible interpretations by focusing the learner's attention to one possible scenario. The images are presented in a reading order (left to right, top to bottom), which suggests a stepwise path by which the learner can construct a linear narrative, complemented by text and scale cues. Therefore, knowing how students use complex process diagrams can contribute to our growing understanding of how MERs function (Ainsworth, 2008; Scheiter, Wiebe, & Holsanova, 2008).

Complex Process Diagrams as Signs

Diagrams can be analyzed from a semiotic perspective, which focuses on the diagram as a sign designed to communicate ideas. Semiotics is the study of signs, which are any images, gestures, sounds, text, models, or textures that communicate information and thus have meaning (Crow, 2003). A sign's meaning as intended by the producer and as interpreted by the user is also considered in semiotic analysis. Iconic shapes and devices such as color coding or layering have to be meaningful. In cell biology diagrams, blobs regularly represent proteins, dots represent ions, cylinders represent channels, and shading represents hollow compartments (Tversky, Zacks, Lee, & Heiser, 2000). Colors take on meaning as arbitrary codes or nonarbitrary metaphors. For example, a popular US university biology textbook (Campbell et al., 2008) uses color as codes: proteins are purple, lipids are yellow, nucleic acids are red, and aqueous compartments are blue. Process diagrams also rely heavily on arrows to represent a great many aspects of molecular processes (Fantini, 2006), including sequences, gradients, pathways, movement, polarity, increases, and decreases. Furthermore, graphic devices—such as cutaways, zooming frames, and shading—convey depth, scale, and three dimensions.

Because there are common patterns of use and interpretation of the codes that compose signs, diagrams have a visual grammar (Kress & van Leeuwen, 1996). Like the grammar of linguistics, visual grammar is not universal but is culturally influenced and changes with invention and adoption of new codes. This is

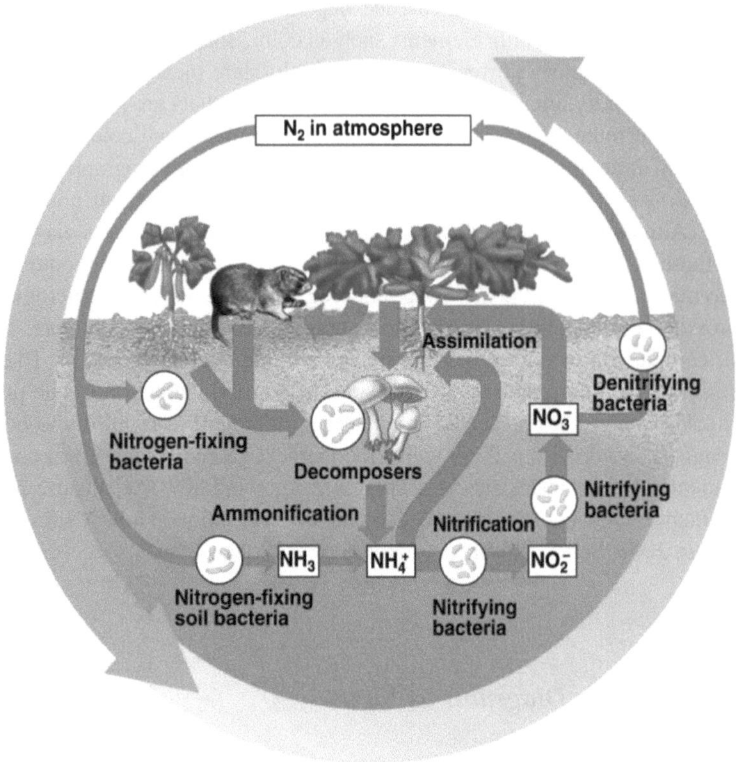

Fig. 10.3 A complex process diagram depicting the cycling of nitrogen through an ecosystem (Campbell et al., 2008, p. 1233) (Reprinted with permission)

especially true in biology, where the enormous expansion of the knowledge base has led to the invention of new icons and devices to represent new phenomena, models, and data, such as those for genomics (Takayama, 2005). There remains a great cognitive distance between abstract external representations generated by these means and the complex process diagrams designed for the general audience. More research is needed to understand how novices to a discipline, such as university biology majors, come to understand these increasingly abstract and domain-specific visual models.

Deconstructing Complex Process Diagrams

The set of marks that compose a printed external representation is arranged in specific positions using ink on a page of paper. One core strategic method in semiotics is deconstructing the marks to interpret underlying meanings (Noble & Bestley, 2005). This representation of the nitrogen cycle (see Fig. 10.3) is an

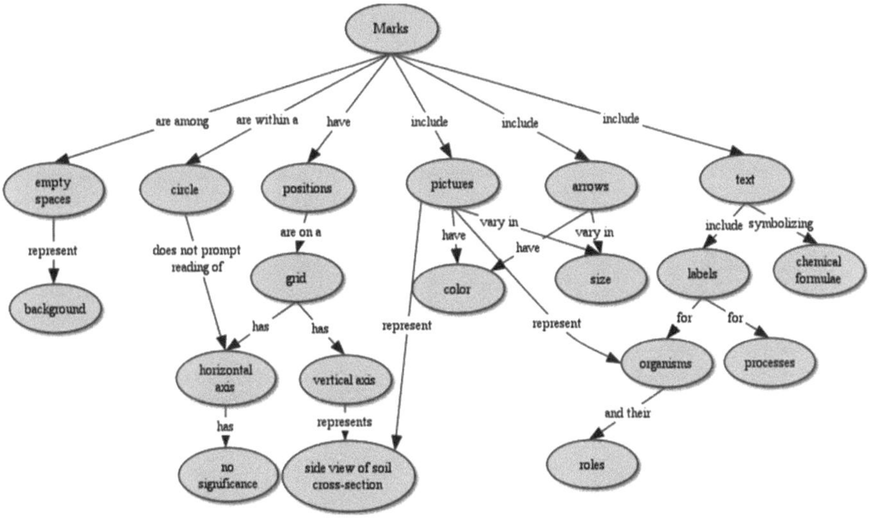

Fig. 10.4 Taxonomy of the properties of the marks composing the process diagram in Fig. 10.3

example of a complex process diagram that meets the aforementioned criteria: iconic irregular shapes represent microbes; three dimensions are evident in shading in the plants, animal, and mushrooms; sequence is represented with arrows to show steps in the nitrogen conversion; and multiple levels of organization are represented by exaggeration of scale of microbes alongside the larger organisms.

First, the marks in a complex process diagram can be categorized as pictures, arrows, or text, each of which has color, size, and position on the background of white space (see Fig. 10.4). In the nitrogen cycle diagram, the semi-realistic pictures represent organisms: an animal (rodent), two plants (different legume species, recognizable by their leaves and pods but with distinct root structures), mushrooms (recognizable by their morphology and the label decomposers), and six white circles containing irregular shapes to represent microbial species. These microbes are not drawn to scale with the other organisms, allowing speculation that their circular white backgrounds were chosen to resemble what might be seen under a microscope. These small pictures are arranged on a background above or below the soil, recognizable by the uneven surface, grainy texture, roots, and darker shading at greater depth. Large arrows on the periphery represent the cyclical nature of nitrogen movement. The blue color was likely chosen for these arrows because nitrogen is generally represented as blue in molecular model kits, for example. Similarly, over a two-page spread in the same textbook, blue and gray arrows are used in the two adjacent diagrams representing, respectively, the water cycle and carbon cycle, whereas arrows in the phosphorus cycle were colored arbitrarily yellow. The arrows within the diagram are shown in various widths to represent relative contributions of each process to the nitrogen cycle. The positions of the arrows on the grid suggest the processes do not occur in a particular stepwise sequence because the processes are ongoing and simultaneous. This is in contrast

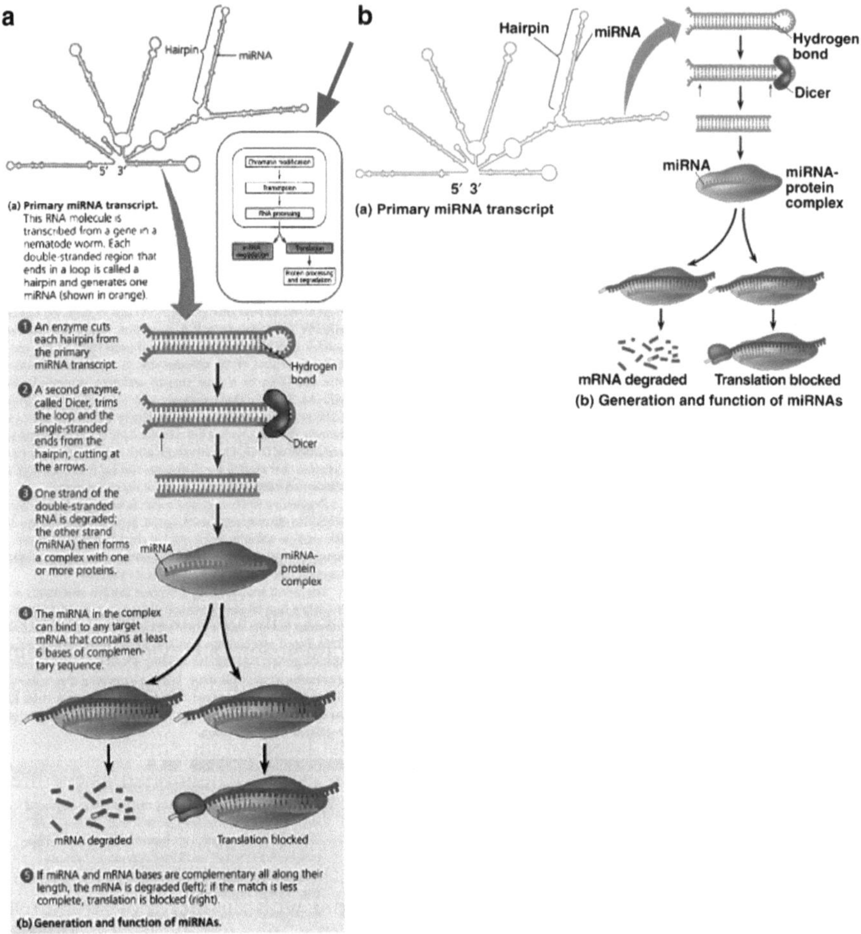

Fig. 10.5 (**a**) Textbook diagram of the processing of RNA to produce microRNAs (Campbell et al., 2008, p. 365). (**b**) Instructor's version of the diagram in (**a**) provided by the publisher with fewer orienting and explanatory cues (Reprinted with permission)

with sequential processes whose steps are often rendered in positions that are read from left to right and top to bottom (see Figs. 10.5 and 10.6).

The text in Fig. 10.3 takes the form of either labels for organisms and processes or symbols for the relevant chemical forms of nitrogen. None is colored or decorated. Most of these, for example, ammonification and NO_2, require prior knowledge for full understanding of their roles in the represented process. Adjacent to the cycle diagram is a caption with headings Biological Importance, Forms Available to Life, and Reservoirs and Key Processes. In addition to the marks themselves, graphic designers also consider the positions, relative sizes, space use and boundaries of the marks, as well as decision about how much white space to

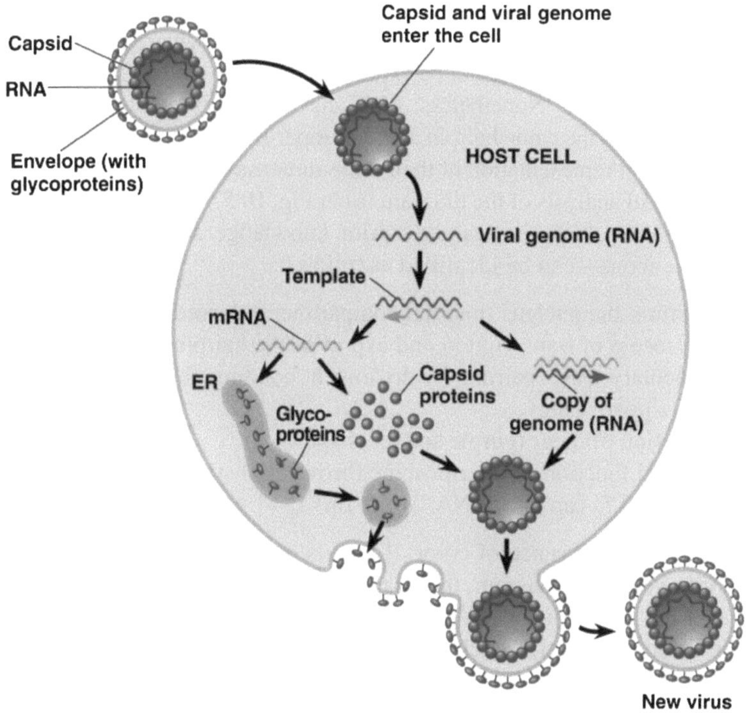

Fig. 10.6 Replication of enveloped viruses (Campbell et al., 2008, p. 388) (Reprinted with permission)

retain. In the diagram in Fig. 10.3, approximately the top one-third of the picture represents atmosphere above ground, presumably to represent the proportion of the nitrogen cycle that occurs in soil.

Graphic designers also practice selective exclusion (Goodsell & Johnson, 2007) to simplify complex phenomenon to its most salient features and "reduce chaos" (D. Mikhael, personal communication, May 30, 2010). This is evident in the nitrogen cycle diagram in that only one representative example of each organism type is shown, and details about the microbial species and their respective biochemical processes were omitted. This nitrogen cycle in Fig. 10.3 is the only one in a set of the textbook's (Campbell et al., 2008) four biogeochemical cycles presented with extensive captions. Therefore, this set of four processes can be considered MERs which complement, constrain, and construct readers' understanding of biogeochemical cycles by virtue of their similar codes, proximity, and juxtaposition with explanatory text.

The nitrogen cycle example is first presented for its relative simplicity and its macroscale elements of everyday experience. However, most process diagrams in university biology textbooks, particularly those about cell and molecular processes, contain more elements and require more prior knowledge to decode, for example, the process diagram for microRNAs in Fig. 10.5 from the chapter of the same textbook about regulation of gene expression (Campbell et al., 2008). Unlike the nitrogen cycle diagram, icons here represent polynucleotides, hairpin RNA structures, proteins, and ribosomes that cannot be experienced directly and do not have referents in everyday experience. Nonetheless, molecular biologists recognize these iconic shapes readily. Even the name hairpin and the zipper-like icon have a basis in analogy rather than a direct representation of their three-dimensional structures.

From a careful analysis of the diagram (a) in Fig. 10.5, several assumptions of its graphic designer—about the learner's prior knowledge and familiarity with the representative icons—can be identified as follows:

- Cell structure: the nucleus (internal compartment) denotes a eukaryotic cell
- Nuclear process of transcription and export of the hairpin (textboxes)
- Complementary base pairing by hydrogen bonding that allows the hairpin structure (zipper shape)
- Enzyme action of dicer (purple scissor shape)
- Structure and function of the ribosome (brown realistic shape)
- Significance of 5′ cap on mRNA (white tip)

Furthermore, significance of color, if any, is often not self-evident. In contrast with the nitrogen cycle diagram, this diagram's vertical orientation is meaningless except as a top-to-bottom reading cue of the sequence. In diagrams for experts, there is significant selective exclusion because of assumptions about the learner's prior knowledge and availability of explanations in adjacent paragraphs.

The schematic map about regulation of gene expression in the form of a cell (labeled by the red arrow in Fig. 10.5a) recurs throughout the chapter. It provides a metacognitive cue to orient the learner to where this process is occurring in the larger context of the cell. Such orienting icons are also offered in the chapters on metabolism (cell and mitochondria) and evolution (cladograms) and in a chemistry textbook (periodic table) (McMurray & Fay, 2008). This schematic map is not present in the instructor's version for professors (see Fig. 10.5b), nor are explicit textual explanations of the process. It seems that the publisher considered this cue as redundant for professors, but it is not known how commonly instructors might verbally cue students to consider the level of regulation at which this step is occurring.

Semiotics of Production of Textbook Diagrams

Where does the kernel of an idea for a diagram come from and how does the idea evolve into a printed figure in a textbook? A medical illustrator said that when he was asked to produce a graphic representation of a process, his first step was to research the topic in order to understand it (M. Marion, personal communication, January 16, 2011). In doing so, authors and illustrators certainly encounter features and devices of similar representations and adapt them for their purposes. This suggests that inscriptions in the public domain become signs when their users find them effective, particularly when elements or devices invented by graphic designers come into common usage and take on a meme-like quality. For example, the complex process diagrams of successive editions of competing textbooks—such as 3-Ds, cutaways, zooming, telescoping, color coding, and recurring multiples—can become de rigueur in a short time.

How does a textbook author's design become a part of a widely distributed printed textbook? It involves an iterative process between an author and a graphic artist assigned to the author by the publisher. First, the author generates hand-drawn sketches based on his experiences as a scientist and educator. A graphic artist then renders the sketches and returns the draft to the author for additional changes. After several iterations and both the author and designer are satisfied, editors with additional marketing or cost considerations may suggest further modifications. The artwork may change again after being reviewed by paid consultants from across the US teaching professoriate. It is not known whether students are involved in reviewing the artwork in textbooks, but the existing process seems dependent on assumptions of professors and graphic designers about how learners use and learn from their artwork. More research is needed to identify and test these assumptions with learners and to inform graphic designers about whether their assumptions work. For example, some cladogram designs (phylogenetic trees), although informationally equivalent, engender misconceptions about speciation (Novick, Shade, & Catley, 2011). Serendipitously, several textbook authors have become aware of this finding and changed their cladograms from ladder to tree formats in their first or successive editions (L. Novick, personal communication, January 18, 2011). It is hoped that communication about such research findings to textbook authors and publishers improves the quality of complex process diagrams in textbooks.

The designers of complex process diagrams must make choices about what to include, what codes (colors, icons, and symbols) to use, and the order and placement of elements. All of these require commitments of ink to paper, and some of these commitments are arbitrary. Biology educators teaching first-year university courses encounter learners with a wide range of requisite prior knowledge that is needed to learn from complex process diagrams. Textbooks developed for these learners include graphical cues to grain size, nestedness, and molecular features that would be unnecessary and distracting in representations designed for experts. An informal vertical comparison of high school textbooks to lower and upper level university textbooks supports this. In the progression of textbooks from more novice to more expert audiences, there is an increase in the number of details and icon use and a decrease in the use of semi-realistic icons or orienting cues such as telescoping and color coding. Even when these cues are offered, many go unnoticed without scaffolding (Ainsworth, 2008; Cromley, Snyder-Hogan, & Luciw-Dubas, 2010).

How Learners Use External Representations in Biology

Knowledge of how students use graphic representations during biology learning has come largely from researchers in science education and educational psychology. These studies have focused on how students make sense of representations of biological structures such as antibodies (Schönborn, Anderson, & Grayson, 2002), chromosomes (Kindfield, 1993), and membrane proteins (Dahmani, Schneeberger, & Kramer, 2009) and processes such as membrane transport (Cook, Carter, & Wiebe, 2008), meiosis (Kindfield), genetics (Tsui & Treagust, 2003), antibody activation of T-cells (Cook et al., 2008; Cromley et al., 2010), and evolution (Catley, Novick, & Shade, 2010; Halverson, Abell, Friedrichsen, & Pires, 2009). Kindfield found that more expert biologists exhibited more flexible use of representations of chromosomes and crossing-over than did less expert participants. She suggested that such graphic use skills and conceptual knowledge coevolve or are mutually reinforcing. Tsui and Treagust used the multimedia learning environment *BioLogica* to assess development of genetics reasoning. They found that this MER was effective in improving easier types of genetics reasoning and only when students were engaged. Using eye-tracking tools, Cook et al. found that domain knowledge affected which fields students noticed in a diagram of membrane transport. Those with high prior knowledge looked at the most thematically relevant parts, whereas those with low prior knowledge focused on surface features. More recently, Cromley et al. used think-aloud interviews to categorize the strategies college biology students used when learning about immune function from a text excerpt and its accompanying diagram. They found that students using a diagram with text used higher-level strategies such as inferencing and summarizing whereas students using text only with no diagram used instead lower level strategies like rereading, paraphrasing, and mnemonics. The findings of these studies are consistent with what is now understood about the general nature of expertise (Chi, Glaser, & Farr, 1988) and collectively contribute to the growing body of knowledge about how learners interact with MERs.

How Students Learn from Complex Process Diagrams

Research is underway to explore how biology students interpret complex process diagrams during learning. My study used in-depth clinical interviews with pre-medical students to reveal the skills, habits, strategies, and prior knowledge these novices use when decoding complex biology diagrams (Griffard, 2010). Diagrams representing viral replication and muscle contraction were used as cognitive probes in these interviews. (In this chapter, only the viral replication example is discussed due to space limitations.) Neither of these topics was taught in the course; however, subordinate concepts needed to understand the topics had been taught. These included cells, membranes, endocytosis/exocytosis, DNA replication, transcription, protein sorting, neurotransmitters, gradients, channels and pumps, depolarization, intracellular compartments, microfilaments, and ATPases.

Qualitative analysis of the think-aloud protocols and subsequent debriefing interviews identified several dimensions of representational competence with complex process diagrams. The purposeful sample began with two pairs of participants: selected with respect to English (Abbie and Bob) and Arabic (Alan and Cathy) language high schooling and success in the author's general biology course. A fifth student (Bill) was added when he volunteered to participate; his language of schooling had been English, and he had been moderately successful in biology. Pseudonyms were assigned to the participants such that the initial letter represents their grade in the introductory biology courses: A (Abbie, Alan), B (Bill, Bob), or C (Cathy) on an A–F grading scale, with a median grade of B + for the entire class. In the interviews, each participant was provided one diagram at a time and instructed to "explain in any amount of detail how you understand it." The following protocols illustrate the contrasting explanations of viral replication (see Fig. 10.6) of a more successful student (Abbie) and a less successful student (Cathy):

Abbie: So we're starting off with a virus I'm guessing, inside a cell, and it's going to enter the cell. It's probably how a virus affects a cell, a host. It shows that when they enter, they lose the coat, so the color is meant to like, yeeah, denote that. And they show different, like how it's going to be changing as it continues to infect and then change over time in the host cell. So you start off with the capsid, then it opens up its coat, then you've got the RNA, the template, the uh the virus comes into the cell, it enters the cell through the membrane, it loses its coat, the viral genome is now replicated due to the, um, the replication that occurs inside the cell. And then you have RNAs used to code for the proteins in the ER [endoplasmic reticulum] as well as the capsid proteins, the new ones that are going to be made. Uh, the ones that are in the ER are expressed on the outer surface of the membrane and then the remaining part of the genome (is still there) [points].

Cathy: This is as written; this is a host cell (reading), ok. And then we have this virus, and this virus is encountered by this cell. And this picture explains the process, like what happens to this virus when it enters this host cell. OK, and I think it's replication of this virus because here you have a virus and here it says new virus, so maybe it's the process, like how it replicates inside the host cell.

These protocols show a trend across all the protocols: More successful students noted many more details in the process and made explicit statements about them, whereas less successful students perceived the task differently and were satisfied with a more general understanding of the process. Given the same instructions to "explain the diagram in any amount of detail," Abbie, Alan, Bill, Bob, and Cathy mentioned, respectively, 10, 11, 10, 8, and 1 of the eleven features in the diagram. Abbie and Alan actively compared, evaluated, and integrated the information gleaned from the diagram into their existing internal representations, whereas Bob and Cathy decoded the diagram at face value by stating propositions that corresponded piecemeal to elements in the diagram. Bill, who had been moderately successful, attended to fewer details than did Abbie and Alan but made comments about this cell in the context of other cells and the process for the organism, extending the represented image beyond the diagram itself.

A semi-structured interview about their think-aloud protocols was conducted in the same session. Participants were asked to elaborate or clarify their meanings and were asked further questions to check for misconceptions. These questions were generally about number, position, color, and orientation of elements in the diagram. For example, participants were asked whether the cell actually sheds a single virus particle as shown or it sheds many particles, represented by a single particle in the diagram. All participants except one correctly assumed one virus particle represented many and that the artist provided only one to keep it simple (selective exclusion and chaos reduction). Bill even chuckled at the notion that the diagram represents replication since production of one particle cannot be considered replication. Only Cathy accepted a face-value interpretation that this single virus particle could be an accurate representation but imagined that a viral infection would be a collective production of single virus particles by many such cells. The participants also were asked whether the position of elements, particularly that of the infecting virus particle, was significant. All responded that the position of virus entry has no top since the cell is a sphere. They understood that the position, as constrained by the ink on paper, was chosen to be at the top to facilitate reading the sequence of events in the process to help them. Each of the participants readily interpreted the significance of color as a code (purple for protein and red for nucleic acid) but overlooked the significance of the yellow membrane surrounding the particles. Taking note of this code would have helped them resolve their question about where the envelope goes when the virus particle enters the cell.

During debriefing, the participants were asked how they used diagrams when studying. Abbie and Alan said that they read the text first so that they could envision the process internally. They then turned to the diagram as a confirmation or check of their internal representation. Bob and Cathy reported going back and forth between the diagrams, as if to use them to clarify the meaning of the text, and vice versa. In this case, their internal representation probably was very similar to the diagram presented. Cathy even reported having somewhat of a photographic memory and could even recall where similar diagrams could be found in her high school textbook. Bill expressed embarrassment that he sometimes took a shortcut when studying by looking first at the textbook diagram before or in lieu of reading the text. In saying so, he seemed to recognize the cognitive value of using both representations actively, as well as the effort required for doing so.

The next phase of interviews was conducted with twelve participants and an additional complex process diagram about the molecular events of seed germination. Preliminary analysis verified that more successful students decoded a complex process diagram in order to understand the germination process rather than to simply read it. In all cases, the participants' attention gravitated first to the familiar features of the diagram, at the expense of attention to contextual cues needed to understand where and why the process was occurring. With adequate wait time, the more successful students noticed the features they overlooked at their first glance and placed the process in a larger context. As in the first phase of the study, more successful students made remarks about familiar features, indicating when they were comparing the external representation with their internal one, again drawing actively upon their prior knowledge. When they were not sure of something, they looked for additional clues in captions and elements of the diagrams they had overlooked previously, but if they recognized something they had learned previously, they did not commit effort to speculation since this would be easy to look up. This was observed less often in less successful students, who were sometimes distracted by these knowledge gaps.

All of the participants, regardless of whether they had been successful in biology, had similar ability to interpret icons and devices in these diagrams. This suggests that the design of these diagrams was effective for this audience or that the students all became familiar with them in the course of using this textbook. However, depth of interpretation corresponded with how well they performed in the course. Where participants had a strong content knowledge, the arrows, shapes, icons, and colors elicited rich explanatory frameworks in their protocols. However, when they lacked the requisite prior knowledge, icons and arrows could not provide the missing information, such as the significance of the branched arrow in expression of viral RNA. This is consistent with the findings elsewhere that prior knowledge strongly affects what someone finds notable or salient to a problem. Additional studies will be needed to ascertain how novices come to understand the meaning of domain-specific representation strategies, icons, and signs and whether instruction can improve the knowledge resources a learner brings to bear on future tasks.

Dimensions of Representational Competence with Complex Process Diagrams

Kozma and Russell (2005) defined representational competence as "a set of skills and practices that allow a person to reflectively use a variety of representations or visualisations, singly and together, to think about, communicate and act on chemical phenomena in terms of underlying, perceptual physical entities and processes" (p. 131). Based on these findings, the following are proposed as dimensions of representational competence exhibited by the successful students in my study when interpreting complex process diagrams:

- They engage with a clear goal of understanding.
- They notice more details and graphic cues.
- They recognize when they can transfer prior knowledge to the task at hand, including the meaning of graphic elements in the diagram.
- They tap prior knowledge to generate, evaluate, and sometimes discard tentative explanations about the process and the signs representing them.
- They identify and hold in memory what information is missing and look for clues among the available information.
- They attend to cues and devices that can provide information about the larger context in which a process is occurring.

Limitations

This research approach has limitations for answering questions about how novices come to learn to decode complex process diagrams. Although the think-aloud approach is a revered standard in cognition research and an improvement over retrospective verbal reports (Ericsson & Simon, 1993), the very act of converting thoughts to verbalizations changes the cognitive process, and thus, think-aloud protocols cannot be considered a faithful record of internal cognitive processing (Schooler, Ohlsson, & Brooks, 1993). Furthermore, the interviewer's act of asking questions about these features calls attention to features that might not be attended in an authentic learning environment. In addition, any interview strategy that uses textbook diagrams in isolation cannot replicate how students learn from a book in which diagrams are imbedded among elaborative text. These methodological constraints prevent the researcher from making assertions about which codes and signs imbedded in complex process diagrams are noticed and correctly decoded during learning. However, identification of habits and skills is a starting point from which further studies can be designed.

Pedagogical Recommendations for Teaching with Complex Process Diagrams

Complex process diagrams are distinct from other MERs in that they represent processes with many small moving parts that interact over time and space under various conditions and at multiple levels of organization. In consideration of this and the research findings summarized here, the following recommendations for teaching with complex process diagrams are proposed:

- *Engage* with a clear goal.
- *Model* complete decoding.
- Identify necessary *prior knowledge.*
- Consider the *production* process.

Engage with a Clear Goal

Educators should make it clear to their students that the goal of learning with a diagram is understanding, not simply encoding or restating the propositions represented. The intent, therefore, should be generation of a memorable internal representation based only loosely on the diagram used. Using multiple sources (e.g., text, animation, diagrams in comparable textbooks) makes this more likely. Such intent can be conveyed by providing explicit learning goals that incorporate but do not correspond exactly to diagrams in a textbook.

Model Complete Decoding

Educators should cue attention to all details, perhaps by deconstructing diagrams interactively and exhaustively. Educators can scaffold this process by having students systematically identify each graphic element in the diagram and providing effective prompts and adequate wait time for them to learn with the diagram. It is possible that students will have allowed their attention to gravitate toward the familiar, and in doing so, they overlooked boundaries, background color, text, or components within larger structures. This is also an opportunity to explicitly identify devices such as color codes, recurring orienting maps, or domain-specific conventions. For example, instructors can ask students to explicitly state the meaning of arrows. Instructors can ask students to suggest where the represented process is occurring at this very moment in time, such as a predator in its ecosystem.

Identify Necessary Prior Knowledge

When teaching a complex process using a diagram, an educator can informally make explicit the concepts represented in the diagram that students have encountered before in a different context. This will cue students' relevant prior knowledge of content as well as graphic conventions and icons. As students progress from novices to experts, they will encounter more and more domain-specific graphic forms and conventions, and their early explicit attention to these graphic devices will facilitate their automaticity and accuracy in decoding in the future.

Consider the Production Process

Educators can cue consideration of the limits of representations by putting the student in the illustrator's shoes. This can be accomplished by asking why the artist drew only one virus or made the arrows in the cycle so large or left out the nucleus. Instructors can cue students to consider when an artist's decisions about color, number, and position were arbitrary (meaningless) or intentional (meaningful). Lastly, educators can remind students to consider the limitations of graphic analogies. For example, some students may wonder if the proteins would be purple in color or the ATP would flash if they could see inside a real cell. Even when students do not make such egregious decoding errors, attention to the production process serves as a reminder that a representation is the map, not the territory.

In spite of the great pedagogical potential of external representations, visual literacy is often overlooked by educators (Mathewson, 1999; Schönborn & Anderson, 2006). Arguments have been made for the inclusion of visual literacy in science pedagogy (Schönborn & Anderson) and for attending to the development of representational competence (Kozma & Russell, 2005). As part of undergraduates' acculturation to the disciplines, particularly biological sciences, novices must learn to recognize and understand the elements that compose complex process diagrams and the represented knowledge.

References

Ainsworth, S. (1999). The functions of multiple representations. *Computers and Education, 33* (2–3), 131–152.

Ainsworth, S. (2008). The educational value of multiple-representations when learning complex scientific concepts. In J. K. Gilbert, M. Reiner, & M. Nakhleh (Eds.), *Visualization: Theory and practice in science education* (pp. 191–208). New York: Springer.

Campbell, N., Reece, J., Urry, L., Cain, M., Wasserman, S., Minorsky, P., et al. (2008). *Biology* (8th ed.). San Francisco: Pearson.

Catley, K. M., Novick, L. R., & Shade, C. K. (2010). Interpreting evolutionary diagrams: When topology and process conflict. *Journal of Research in Science Teaching, 47*, 861–882. doi:10.1002/tea.20384.

Chi, M., Glaser, R., & Farr, M. (1988). *The nature of expertise*. Mahwah, NJ: Lawrence Erlbaum Associates.

Cook, M., Carter, G., & Wiebe, E. N. (2008). The interpretation of cellular transport graphics by students with low and high prior knowledge. *International Journal of Science Education, 30* (2), 239–261. doi:10.1080/09500690601187168.

Cromley, J. G., Snyder-Hogan, L. E., & Luciw-Dubas, U. A. (2010). Cognitive activities in complex science text and diagrams. *Contemporary Educational Psychology, 35*(1), 59–74. doi:10.1016/j.cedpsych.2009.10.002.

Crow, D. (2003). *Visible signs*. Lausanne, Switzerland: AVA.

Dahmani, H. R., Schneeberger, P., & Kramer, I. M. (2009). Analysis of students' aptitude to provide meaning to images that represent cellular components at the molecular level. *CBE Life Sciences Education, 8*(3), 226–238. doi:8/3/226 [pii] 10.1187/cbe.09-03-0023.

Ericsson, K. A., & Simon, H. A. (1993). *Protocol analysis: Verbal reports as data* (2nd ed.). Cambridge, MA: The MIT Press.

Fantini, B. (2006). Of arrows and flows. Causality, determination, and specificity in the central dogma of molecular biology. *History and Philosophy of the Life Sciences, 28*(4), 567–593.

Goodsell, D. S., & Johnson, G. T. (2007). Filling in the gaps: Artistic license in education and outreach. *PLoS Biology, 5*(12), e308. doi:07-PLBI-E-2742 [pii] 10.1371/journal.pbio.0050308.

Griffard, P. B. (2010, April). *Decoding of visual narratives used in university biology*. Paper presented at the National Association for Research in Science Teaching (NARST) Annual Conference, Philadelphia, PA.

Halverson, K. L., Abell, S. K., Friedrichsen, P. M., & Pires, J. C. (2009, April). *Testing a model of representational competence applied to phylogenetic tree thinking*. Paper presented at the National Association of Research in Science Teaching (NARST) Annual Conference, Garden Grove, CA.

Kindfield, A. C. H. (1993). Biology diagrams: Tools to think with. *The Journal of the Learning Sciences, 3*(1), 1.

Kozma, R., & Russell, J. (2005). Students becoming chemists: Developing representational competence. In J. K. Gilbert (Ed.), *Visualization in science education* (pp. 121–146). Dordrecht, the Netherlands: Springer.

Kress, G., & van Leeuwen, T. (1996). *Reading images: The grammar of visual design*. London: Routledge.

Mathewson, J. H. (1999). Visual-spatial thinking: An aspect of science overlooked by educators. *Science Education, 83*(1), 33–54.
Mayr, E. (1982). *The growth of biological thought.* Cambridge: Harvard University Press.
McMurray, J., & Fay, R. (2008). *Chemistry* (5th ed.). San Francisco: Pearson.
Noble, I., & Bestley, R. (2005). *Visual research.* Lausanne, Switzerland: AVA.
Novick, L. R., Shade, C. K., & Catley, K. M. (2011). Linear versus branching depictions of evolutionary history: Implications for diagram design. *Topics in Cognitive Science, 3*(3), 536–559.
Paley, W. (2008). Rich data representation: Sophisticated visual techniques for ease and clarity. In G. Stapleton, J. Howse, & J. Lee (Eds.), *Diagrammatic representation and inference* (Vol. 5223, pp. 2–3). Berlin, Germany/Heidelberg, Germany: Springer.
Pozzer, L., & Roth, W. (2003). Prevalence, function, and structure of photographs in high school biology textbooks. *Journal of Research in Science Teaching, 40*(10), 1089–1114.
Scheiter, K., Wiebe, E., & Holsanova, J. (2008). Theoretical and instructional aspects of learning with visualizations. In R. Zheng (Ed.), *Cognitive effects of multimedia learning* (pp. 67–88). Hershey, PA: IGI Global.
Schönborn, K. J., & Anderson, T. R. (2006). The importance of visual literacy in the education of biochemists. *Biochemistry and Molecular Biology Education, 34*(2), 94–102.
Schönborn, K. J., Anderson, T. R., & Grayson, D. J. (2002). Student difficulties with the interpretation of a textbook diagram of immunoglobulin G (Igg). *Biochemistry and Molecular Biology Education, 30*(2), 93–97.
Schooler, J., Ohlsson, S., & Brooks, K. (1993). Thoughts beyond words: When language overshadows insight. *Journal of Experimental Psychology General, 122*, 166–166.
Takayama, K. (2005). Visualizing the science of genomics. In J. Gilbert (Ed.), *Visualization in science education* (pp. 217–252). Dortrecht, the Netherlands: Springer.
Tsui, C.-Y., & Treagust, D. F. (2003). Genetics reasoning with multiple external representations. *Research in Science Education, 33*(1), 111–135. doi:10.1023/a:1023685706290.
Tufte, E. (1997). *Visual explanations.* Cheshire, CT: Graphics Press.
Tversky, B., Zacks, J., Lee, P., & Heiser, J. (2000). Lines, blobs, crosses and arrows: Diagrammatic communication with schematic figures. *Theory and Application of Diagrams, Proceedings, 1889*, 221–230.

Chapter 11
Learning Tree Thinking: Developing a New Framework of Representational Competence

Kristy L. Halverson and Patricia Friedrichsen

Introduction

A major goal of science education is to develop scientific literacy (National Research Council, 1996; Rutherford & Ahlgren, 1990). One component of scientific literacy is the ability to use common representations of phenomena, such as molecular models (Ferk, Vrtacnik, Blejec, & Gril, 2003), abstract physics diagrams (Chi, Feltovich, & Glaser, 1981), Punnett squares (Cavallo, 1996), and genetic molecules—such as DNA diagrams (Patrick, Carter, & Wiebe, 2005; Takayama, 2005), pedigrees (Hackling & Lawrence, 1988), and phylogenetic trees (Matuk, 2007). Representations affect multiple aspects of learning including the following: reasoning through problems and phenomena, developing deeper understandings of the relationships among phenomena, and improving creativity (Peterson, 1994). We use representations to explain how we make sense of things on a daily basis and are critical for communicating abstract science concepts (Gilbert, 2005a). In science, visual representations are used to display data, organize complex information, and promote a shared understanding of scientific phenomena (Kozma & Russell, 2005; Roth, Bowen, & McGinn, 1999).

Researchers are interested in investigating how visual representations affect content understanding and how students evaluate and interact with visual representations (Ferk et al., 2003). More specifically, visual representations enhance learning from texts, improve problem solving, and facilitate developing connections between new knowledge and prior knowledge (Cook, 2006; Roth et al., 1999).

K.L. Halverson (✉)
Department of Biological Sciences, University of Southern Mississippi, Hattiesburg, MS, USA
e-mail: Kristy.Halverson@usm.edu

P. Friedrichsen
Division of Biological Sciences, Department of Learning, Teaching and Curriculum, University of Missouri, Columbia, MO, USA
e-mail: FriedrichsenP@missouri.edu

D.F. Treagust and C.-Y. Tsui (eds.), *Multiple Representations in Biological Education*, Models and Modeling in Science Education 7, DOI 10.1007/978-94-007-4192-8_11,
© Springer Science+Business Media B.V. 2013

Studies highlight the importance of both content knowledge mastery as well as spatial reasoning (Bodner & Guay, 1997; Halverson, 2010; Maroo & Halverson, 2011). Researchers also agree that students have difficulties understanding and interacting with representations (e.g., Ainsworth, 2008; Anderson & Leinhardt, 2002; Ferk et al., 2003; Reiss & Tunnicliffe, 2001; Tufte, 2001; Zbiek, Heid, Blume, & Dick, 2007). Student difficulties include the following: identifying key structures of representations, interpreting and using representations, transitioning among different modes of representations, and relating abstract representations to content knowledge. However, abilities such as transitioning between different modes of representations (e.g., 2-D and 3-D models) are a vital part of becoming an expert.

Kozma and Russell (2005) proposed, in the context of chemical representations, a set of seven core skills that must be developed in order to develop competence in the use of visual representations:

- Use representations to describe observable phenomena in terms of the underlying entities and processes.
- Generate or select an appropriate representation and explain why it is best suited.
- In the case of nonverbal modes, use words to identify and analyze features of the representation.
- Describe how different representations can illustrate the same idea in different ways and how one representation might illustrate something different or something that cannot be said by another because of differences in limitations.
- Make connections across different representations, to transfer features of one type of representation onto those of another and explain relationships between the features.
- Accept that representations are depictions of phenomena or concepts but are distinct from the actual phenomena.
- Use representations and associated features as evidence to support claims, draw inferences, and make predictions.

However, these skills have not been empirically tested in chemistry education. They assume that once these skills are developed, a learner should be able to effectively use a variety of representations, thereby achieving some level of representational competence varying based upon the problem or representation encountered. The five levels proposed by Kozma and Russell correspond to a progressive developmental gradient moving from the use of surface features to define phenomena to the metaphoric or reflective use of representations (Chi et al., 1981; Kozma & Russell, 1997). Similarly, there is a need to develop a framework for how students gain representational competence in biology education, particularly with phylogenetic trees, in order to maximize the potential of evolution education and improve scientific literacy. In addition, Cook's (2006) study identified a need for exploring the developmental continuum of representational competence as a way to help maximize the potential of visual representations. A previous study of the first author (Halverson, 2011) investigated core skills essential to communicate with and make sense of phylogenetic trees and identified two sets of core skills—one for tree reading and one for tree building. We used this idea that skills lead to levels of competence to guide developing a model of representational competence tested in biology education with tree thinking.

Phylogenetic Tree Representations

Phylogenetic trees play a prominent role in biology textbooks and are often used to present multiple relationships and processes that are difficult to describe. These representations of phylogenetic thinking involve understanding phylogenetic tools, evolutionary mechanisms, inheritance, and genomics. Researchers have argued that understanding phylogenetic trees as representations of evolutionary relatedness is a cognitively complex task, given the numerous misconceptions that students commonly hold (e.g., Baum, Smith, & Donovan, 2005; Gendron, 2000; Gregory, 2008; Halverson, Pires, & Abell, 2011).

Achieving an expert understanding about species relatedness and phylogenetic trees involves multiple biological concepts. Evolutionary biologists recognize relationships among species by using foundational concepts such as inheritance, mechanisms of evolution (mutation, genetic drift, gene flow, and natural selection), and parsimony to develop hypotheses and build phylogenetic trees. Evolutionary biologists interpret phylogenetic trees in accordance with how they illustrate evolutionary histories or inferred evolutionary relationships among a set of organisms. Scientists interpret patterns in phylogenetic representations using an approach that involves mapping descent from common ancestry in order to identify the most recent common ancestor and isolate monophyletic groups, or clades, of species. Scientists also compare phylogenetic representations in search of similar patterns to provide support for hypothesized relationships among taxa. They find similarities by comparing monophyletic groups across representations. Equivalent phylogenetic trees will have identical topologies, illustrating consistent evolutionary histories and common ancestry. Generating a phylogenetic tree involves isolating and interpreting informative data into evidence of evolutionary relationships. Scientifically correct phylogenetic representations share the following features: Relationships are grouped based on evolutionary histories and common ancestry, all organisms are related and are connected within a single representation, taxa are placed at the terminal tips assuming hypothetical ancestors at nodes, and consensus nodes are used when relationships are uncertain (see Fig. 11.1). Being able to correctly interpret, compare, and generate phylogenetic trees is necessary for becoming a highly competent tree thinker (Halverson, 2011).

Many studies (e.g., Baum et al., 2005; Brumby, 1979; Driver, Squires, Rushworth, & Woods-Robinson, 1994; Williams & Tolmie, 2000) have found that students often struggle with accommodating foundational concepts involved in evolution; thus, this lack of scientific knowledge about evolution may hinder students' abilities to interpret evolutionary trees. In particular, Moore et al. (2002) found in their study that many students shared Lamarckian views, believing that acquired traits could be passed down to offspring and these students often based their assumptions on environmental explanations. These students also thought that natural selection occurs with purposeful intent and organisms deliberately selected traits for survival. Brumby also found students believed evolution can alter an individual during its lifetime and that evolution progressively improves organisms

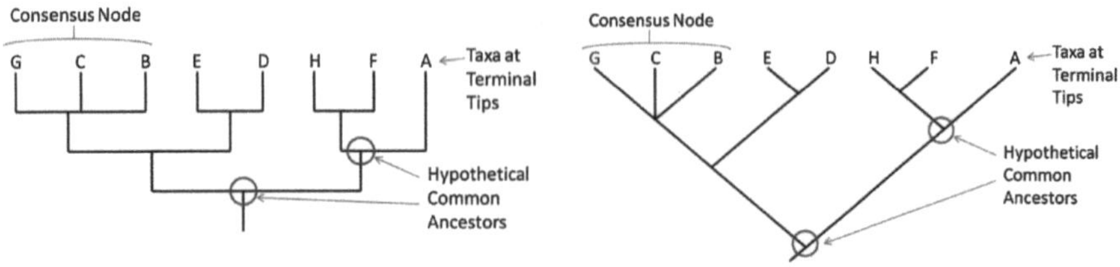

Fig. 11.1 Two examples of phylogenetic trees representing identical relationships

toward perfection. Lord and Marino (1993) reported that many students believed humans directly evolved from monkeys. They also found that some other students believed that species were static; thus, evolution did not occur. All of these alternative views can impact upon student tree-thinking development.

In this chapter, we address how the previously presented representational competence framework for chemistry education (Kozma & Russell, 2005) applies to phylogenetic tree representations as well as providing insights into how undergraduate biology majors develop representational competence in tree thinking. This contributes to the development of a cohesive, empirically based representational competence model that can inform the design of evolutionary biology curriculum.

Method

Research Design

We devised a new theoretical framework for developing representational competence with phylogenetic trees derived from data collected, over 4 years, from 157 students at two American research universities. Participants were all upper level biology students enrolled in courses focusing on a systematics approach, or tree thinking approach, toward evolution. We assessed progress in student learning through regular homework assignments, in-class activities, group discussions, and examinations throughout the semester.

Data Collection

We utilized several methods of data collection to gain an understanding of how undergraduates develop tree thinking over the course of each semester. To elicit students' ideas about phylogenetic representations and challenges they face when developing tree thinking, we used multiple open-ended data sources (Patton, 2002). The primary data sources were students' online reflective journal entries, responses to the pretest/posttest (modified from Baum et al., 2005), and semi-structured interviews with key informants. The secondary data sources for this study included written documents from student coursework and field notes from course observations. By using multiple data sources, we increased the validity of our research by being able to triangulate our findings.

Data Analysis

We utilized all transcripts from the interviews with key informants, field notes, expanded observation notes, and documents in data analysis. Rather than using an approach to collect data with predetermined themes in mind, we used an inductive approach to identify meanings that the students created. We inductively coded the profiles to identify reasoning used by students when interpreting, comparing, and building representations.

We searched for patterns in the data that distinguished levels of representational competence for phylogenetic trees. Once the patterns were identified, we triangulated the findings using secondary data sources to ensure that the research findings represented accurate interpretations of the data.

Interpretations

We identified two major themes after testing Kozma and Russell's (2005) chemistry education theoretical framework with phylogenetic tree representations. First, we found that developing expertise in tree thinking is a cognitively complex task, with students using alternative approaches or ignoring the representation completely when trying to solve phylogenetic problems. Second, we revisited the original model of representational competence and presented a new perspective encompassing milestones associated with developing competence based on evidence we found. Data from this work in biology education support the assumption that developing representational competence is a non-steplike trajectory and students often have varying levels of representational competence dependent upon the nature of the posed problem.

Major Components of Tree Thinking

Over the course of each semester we investigated, we found that not all of the participating students were able to consistently interpret and compare phylogenetic trees as would have been expected of skilled tree readers. The pretest/posttest prompted students to answer multiple choice questions about a phylogenetic tree representation and provide a written rationale for their selection. We found that a correct selection on the multiple choice portion of each question was not necessarily an indicator of scientific thinking when interpreting a phylogenetic tree. For example, one student, Miranda, selected the appropriate option but used a faulty approach to reach her decision. She came to her conclusion by interpreting the relationships represented on the basis of the positions of the organisms on the tree in relation to a main branch. Miranda justified her response by stating, "All are coming off from the same main branch" (see Fig. 11.2).

Emergent trends from the data supported shifts in the rationales that students used when interpreting and comparing phylogenetic trees over the course of a semester. For example, students who interpreted phylogenetic trees based on the proximity of organisms along the terminal tips (see Fig. 11.3) or their knowledge about ecology tended to shift their rationale to rotation-based interpretations by the end of a course. This trend illustrated a shift from students using superficial location of organisms along the tips or ignoring the representation completely when forming conclusions about relationships among the organisms to acknowledging scientific meaning in the representation and recognizing the mobile nature of trees. Some students focused on the number and location of nodes when interpreting relationships represented on a phylogenetic tree (see Fig. 11.4). An emerging trend showed that the students who began the course using a nodal emphasis rationale when interpreting trees shifted their rationale to focus on implied *apomorphies* (derived character states) and common ancestry by the end of the course. While these students still used the nodes to interpret relationships illustrated on the tree, they learned to recognize the symbolism of these intersections to represent common ancestry and divergence events.

A second critical component to developing tree thinking involves being able to correctly build phylogenetic trees. Emergent trends from the data supported shifts in the styles and types of representations that students generated when interpreting phylogenetic scenarios over the course of a semester. For example, the types of representations that students generated were consistent with the approaches they used to interpret and compare phylogenetic trees. For example, students who generated a single progressive tree interpreted relationships represented in phylogenetic trees in the context of a main branch.

Representational Competence Framework for Tree Thinking

We modified the levels of representational competence for phylogenetic trees based on identified milestones and core skills that are used as indicators of students' tree thinking. In order to develop competence with a biological representation such as a phylogenetic tree, a person must achieve the following six milestones:

- Recognize and interpret informative symbolic parts of a representation.
- Compare and contrast multiple representations of similar nature, explaining why one may be more appropriate than another.

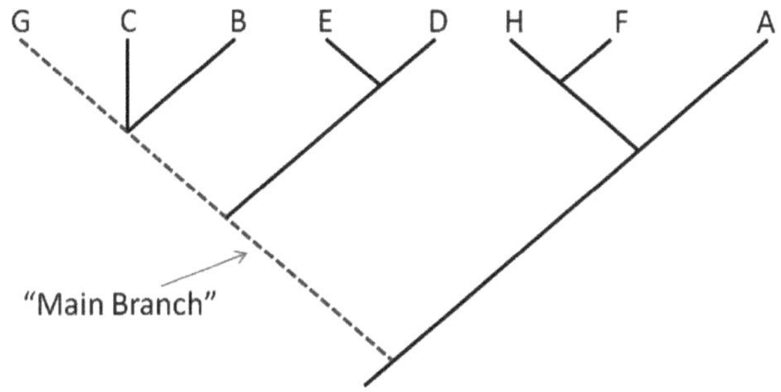

Fig. 11.2 Example illustrating the incorrect idea that a phylogenetic tree has a *main branch*

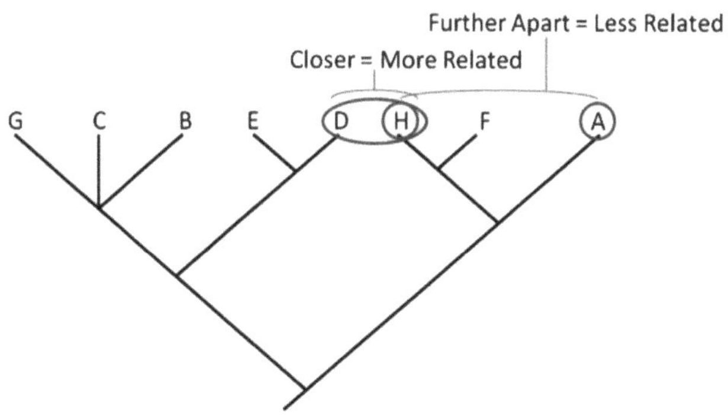

Fig. 11.3 Example representing an incorrect idea that closer in proximity means that the taxa are more closely related

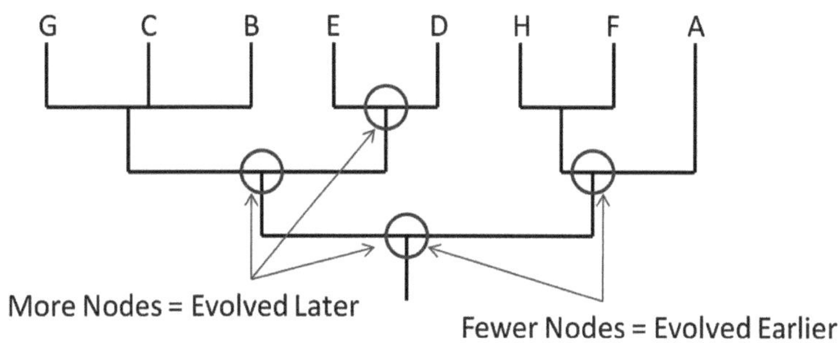

Fig. 11.4 Example representing an incorrect idea that the number of nodes found along a taxon's lineage indicates how long ago it evolved

- Accurately communicate the meaning of a representation to others.
- Make predictions from a representation that supports evidence.
- Test and manipulate a representation given new scenarios or data.
- Generate appropriate and accurate representations to support evidence.

Representational competence is context specific, and these milestones must be achieved for each representation considered. For example, a phylogenetic tree may not be fully understood just because a person has representational competence with atoms, molecules, or a pedigree.

Using these milestones and previously identified core tree-thinking skills (Halverson, 2011), we constructed seven levels of representational competence that a student may hold in tree thinking: (1) no use of representation, (2) superficial use of representation, (3) simplified use of representation, (4) symbolic use of representation, (5) conceptual use of representation, (6) scientific use of representation, and (7) expert use of representation. We separated our descriptions of each of these levels into tree reading and tree building to capture the differences in competency for each task.

Level 1: No Use of Representation

Tree Reading. Prior knowledge about the morphology and ecology of the organisms interferes with students' abilities to recognize information presented in the representation. Thus, these students do not use the representation to make sense of the phylogenetic scenario depicted. Additionally, these students view all phylogenetic representations as unique and cannot make comparisons of similarities across the trees.

Tree Building. Students do not consider or are not able to generate a visual representation as a possible solution to a phylogenetic scenario. At most, these students generate written lists for organizing taxa.

Level 2: Superficial Use of Representation

Tree Reading. Students base interpretations of phylogenetic trees on superficial features of the representation (such as uninformative bends, proximity of the organisms placed along the tips) without connections to the underlying meanings of the phylogenetic relationships illustrated. When comparing phylogenetic trees, students look at the same superficial features and patterns to determine the similarities and differences shown among representations.

Tree Building. Students generate a literal translation of a phylogenetic scenario and create a pictorial image—to represent how they understand organisms existing in the natural world—which are often related to the students' prior knowledge of ecology connected to each organism (see Fig. 11.5).

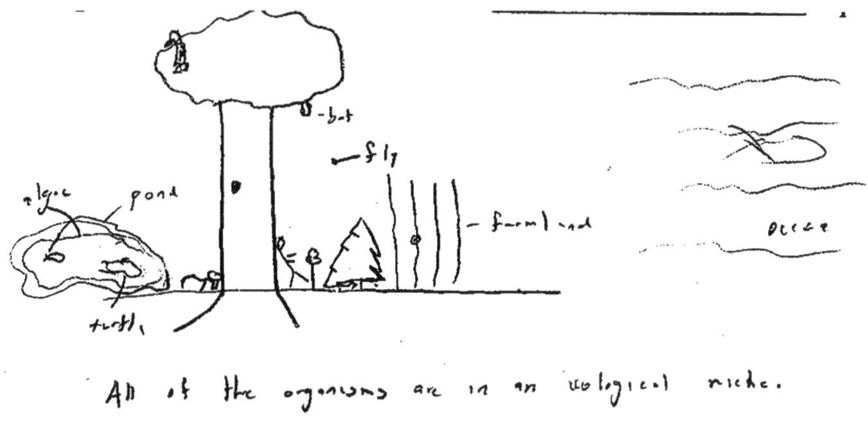

Fig. 11.5 Pictorial image incorrectly representing how organisms are related

Level 3: Simplified Use of Representation

Tree Reading. Interpretations of phylogenetic trees are based on the idea of a main branch, with taxa branching off from a main branch and later branching off from one another. These students compare representations by looking at differences and similarities in branch length stemming from the main branch or last point of divergence. In these instances, branches, or lineages, are viewed as straight and cannot be bent (see Fig. 11.2).

Tree Building. Students at this level recognize that scientists use representations to organize how organisms are related to one another. However, they generate representations based on folk taxonomy, or classification on morphological and/or ecological characteristics rather than evolutionary histories. These students generate dichotomous key visual representations.

Level 4: Symbolic Use of Representation

Tree Reading. Students at this level understand the symbolic elements associated with parts of phylogenetic trees; however, they overly emphasize nodes when interpreting and comparing phylogenetic representations. These students tend to count nodes between taxa and place importance on the location of the nodes to make sense of the phylogenetic scenario represented. In these instances, more nodes are viewed incorrectly as representing more differences between organisms (see Fig. 11.4).

Tree Building. Students rely upon Lamarckian views of evolution (purposeful, progressive evolution with multiple origins of taxa) and generate flow chart representations with taxa evolving into other taxa (see Fig. 11.6). More advanced

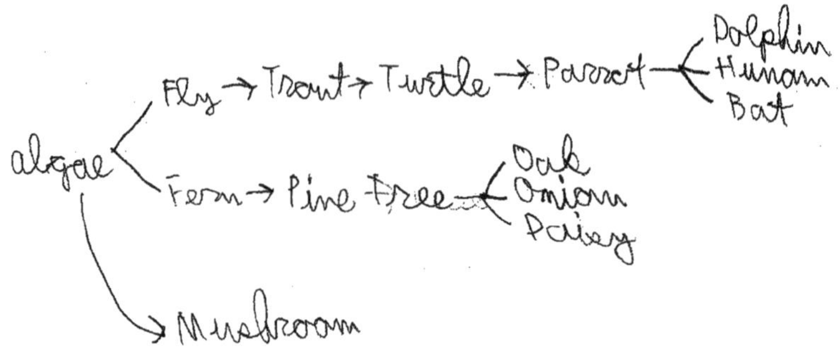

Fig. 11.6 Example of a flow chart diagram using arrow symbols to represent relationships

students at this level generate ladderized representations that more resemble phylogenetic tree representation but still symbolize progressive evolutionary histories (see Fig. 11.7).

Level 5: Conceptual Use of Representation

Tree Reading. Phylogenetic trees are viewed as 2-D illustrations of 3-D representations and their branches as being able to rotate around nodes without altering the relationships represented (see Fig. 11.8). However, students at this level do not make connections between their interpretations of phylogenetic trees and the evolutionary history represented. Furthermore, comparisons among phylogenetic representations are based upon the physical branching patterns. Similarities and differences are restricted to perceptions of how trees can be rotated, and different styles of phylogenetic representations are often excluded from consideration.

Tree Building. Generated representations begin to have hierarchical branching structures. However, these representations are flawed in that they illustrate incorrect relationships. The student has separated the organisms into different representations (see Fig. 11.9), suggesting some groups of taxa are not related to others, rather than including all taxa onto a single *tree of life*.

Level 6: Scientific Use of Representation

Tree Reading. Students are able to scientifically interpret the relationships illustrated within the topology of a phylogenetic tree based on represented common ancestry, monophyletic patterns, and implied apomorphies separating taxa. These students consistently compare phylogenetic representations based on patterns of clades regardless of the style of the representation.

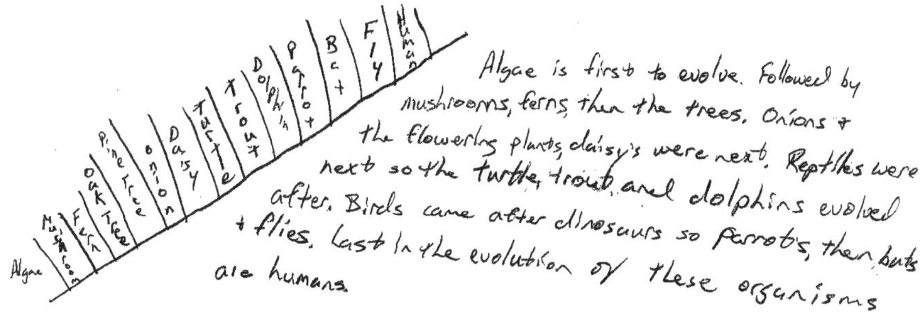

Fig. 11.7 Example of a progressive diagram illustrating taxa evolving into other taxa

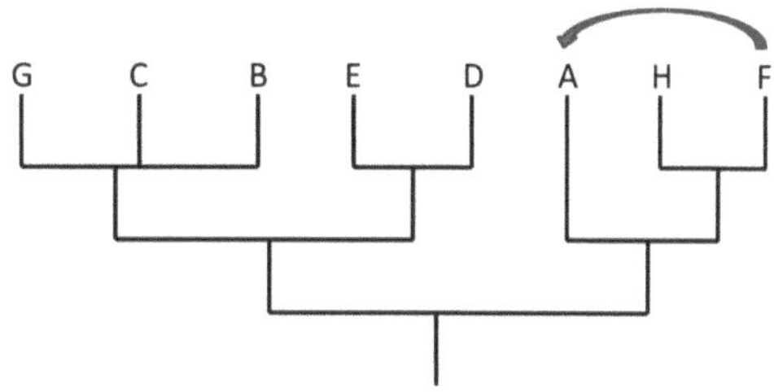

Fig. 11.8 This tree illustrates a rotated node yet depicts an identical set of relationships as the trees in Fig. 11.1

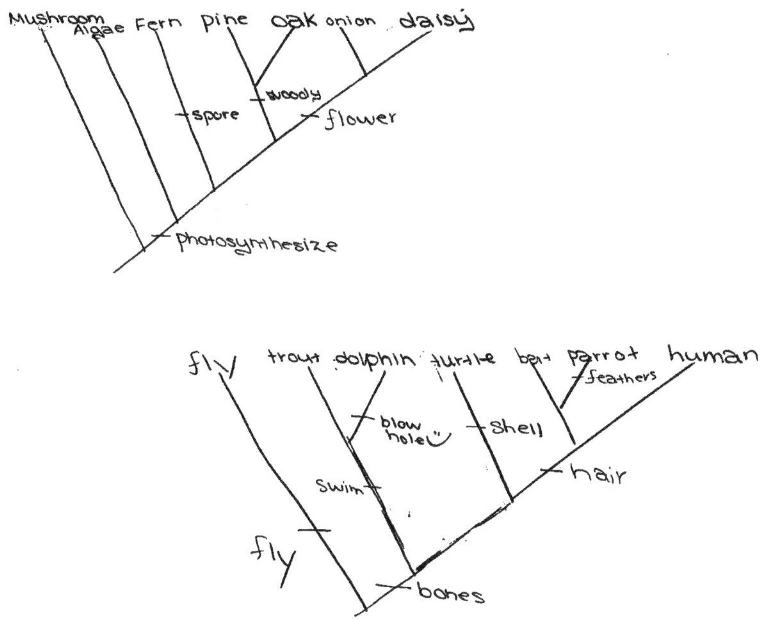

Fig. 11.9 This set of representations separates taxa into two groups: plantlike organisms and animals

Tree Building. Students at this level of competence with tree building generate scientifically correct phylogenetic representations with hierarchical branching structures and can justify/explain what their representations illustrate in terms of evolutionary content.

Level 7: Expert Use of Representation

Tree Thinking. This level is reserved for describing experts in the field of systematics and is not appropriate for beginning students. These scientists can quickly interpret representations on the underlying phylogenetic meanings that trees represent. At this level, multiple representations are used and generated consistently to solve phylogenetic problems, explain evolutionary phenomena, and make predictions. Additionally, these scientists can identify and explain why one representation is more appropriate than another when comparing or generating phylogenetic representations. Acceptance of hypotheses is positively influenced when multiple representations support similar interpretations (e.g., high bootstrapping values).

Different Levels of Representational Competence

During our study, we gauged students' levels of representational competence by the accuracy of selected answers and rationales or style of representation generated. When a student held multiple levels of representational competence, we gave credit to each competence level represented. Due to the nature of the courses and the tasks administered, we were not able to gather evidence needed to test for level 7 competency. We gathered evidence indicating that students hold different levels of representational competence dependent upon the problem and representation encountered. For example, when we assessed Aaron's representational competence prior to a plant systematics course, we found that he possessed level 1 representational competence when interpreting relationships shown on a single phylogenetic tree, level 3 competence when comparing patterns of relationships across trees, and level 2 competence when he generated a literal image of organisms to build a tree. When we examined his representational competence at the end of the course, Aaron possessed level 5 representational competence in all aspects of tree thinking.

Furthermore, we identified a developmental gap between tree reading and tree building. Trends in the data showed that most students improved from levels 2–3 to levels 5–6 tree reading by the end of each course. On average, students improved about three levels over the course of a semester. Furthermore, no students remained at a level 1 competency in tree reading after instruction. Trends also showed that most students improved from level 4 to level 5 tree building at the end of a semester—only one level worth of improvement. However, even after explicit tree-building instruction, some students remained at level 1 competence with tree building.

Discussion and Implications

We investigated the versatility of a representational competence framework in chemistry education for use with phylogenetic trees in our study. Whereas the seven core skills outlined by Kozma and Russell (2005) could be applied to tree thinking, these skills are not all inclusive. Halverson (2011) identified two unique core skill sets for tree thinking that address tree-reading and tree-building development. Additionally, unique to evolution education, this representational framework included the identification of a secondary skill set necessary to generate phylogenetic trees, a more cognitively difficult tree-thinking task. All of these skills influenced the rationales and criteria students used to make sense of phylogenetic representations as well as the styles of representations they generated. Sometimes the manners by which students make sense of a representation may lead to correct responses, but this does not mean that the students have used appropriate approaches (Tabachneck, Leonardo, & Simon, 1994). For example, Cavallo (1996) investigated the relationships among meaningful learning orientations, reasoning ability and understandings about genetics, and problem-solving abilities. She found that students were able to successfully solve genetics problems when using Punnett square representations as a tool. She also found that students with *meaningful learning orientations* were best able to understand genetics interrelationships. However, this orientation could not be used to predict problem solving with representations nor could the use of representations predict understanding of concepts. For students to become experts with representations, they must use representations correctly and as a reasoning tool when investigating problems. This idea of an essential connection between representation use and rationales was supported by a previous study (Halverson et al., 2011). Student ideas about evolution can impact upon the way in which students visualize evolutionary relationships among organisms. For example, if students viewed evolution as progressive, they tended to interpret trees in a directional manner and generate ladderized or flow chart representations.

Another aspect of learning to read and to construct representations involves determining which features are pertinent and which are not (van Fraassen, 2008). According to the literature, one reason that students struggle with making correct associations between science content and abstract representations is because students tend to rely upon superficial structures rather than use representations as analytical tools (Anderson & Leinhardt, 2002; Chi et al., 1981; Larkin, Mcdermott, Simon, & Simon, 1980). In biology education, Patrick et al. (2005) investigated how middle school-aged female students fixated upon two-dimensional and three-dimensional visualizations of DNA and the replication process. They found that purely visual characteristics such as color, shape, and complexity were important components that the students used to make sense of the images. Thus, the visualizations expressed novel information that other modes of representations could not convey. Not all of these visual characters provided informative meaning about the phenomena, and many students had difficulty distinguishing between relevant and irrelevant information. Similarly, several of the students in our study relied upon uninformative superficial structures of phylogenetic trees, such as bends in branches and proximity of tips, when making sense of the representation.

In our study, we described seven levels of representational competence (levels 1–7) that emerged from the data. We also identified more levels than previously assumed by Kozma and Russell (2005). The original model, developed in chemistry education, did not account for students choosing to ignore the representation when reasoning through phylogenetic problems. We found that students' prior knowledge about ecology interfered with students' development of higher levels of representational competence. When students were familiar with the organisms on the phylogenetic tree, they used their knowledge of morphological and ecological similarities rather than the information represented in the phylogenetic tree. Thus, we developed level 1—an additional level of representational competence to account for this lack of competence. Furthermore, the original framework stated that an initial level of competence is achieved, "when asked to represent a physical phenomenon, the person generates representations of the phenomenon based only on its physical features. That is, the representation is an isomorphic, iconic depiction of the phenomenon at a point in time" (Kozma & Russell, p. 132). This level of competency—that focuses on a person's ability to generate representations but not just on the person's ability to make sense of a representation—does not account for the competency of students who fail to generate a representation. Furthermore, whereas many chemical phenomena are readily observable (e.g., color changes), dynamic processes in biological evolution occur through generations over periods of time and cannot be observed at a single moment. Thus, synthesizing a process into a single iconic representation in biology would be a difficult and advanced skill.

The literature (Barnea & Yehudit, 2000; Kozma & Russell, 2005) has suggested that students' representational competence can change with the difficulty of the task. We presented empirical evidence of students holding differing levels of competence when facing different tasks even at the same point in the semester. We found that a majority of students were not able to generate phylogenetic trees above level 4 competency until after they had achieved at least level 5 competence in tree reading. Thus, this level of representational competence is not appropriate for understanding students' abilities to use and generate phylogenetic trees.

It is generally accepted that representations play a key role in mathematics, geography, and science (Cuoco, 2001; Gilbert, 2005b). Therefore, in order to help students achieve scientific literacy, science educators must understand how students achieve representational competency. Representations enhance learning from texts, improve problem solving, and facilitate connections between new knowledge and prior knowledge (Cook, 2006). For scientific representations to be used for their intended purpose, they must be pertinently similar to the object or phenomena represented. Gilbert stated that representations bridge scientific theory and the natural world in two ways: acting as simplified depictions of phenomena to which abstract theory can be applied and as idealizations of abstract theory comparable to observations of phenomena in the natural world. Communicating with phylogenetic tree effectively is essential to understanding evolution. By better understanding how students make sense of biological representations, particularly phylogenetic trees, we can help facilitate students developing representational competence in biology and becoming more scientifically literate.

Acknowledgments This work was supported by a Richard Wallace Research Incentive Grant from the University of Missouri. We wish to thank Sandra Abell, J. Chris Pires, and Mac Alford for their assistance with data collection and support throughout the project.

References

Ainsworth, S. (2008). The educational value of multiple representations when learning complex scientific concepts. In J. K. Gilbert, M. Reiner, & M. Nakhleh (Eds.), *Visualization: Theory and practice in science education* (pp. 191–208). Dordrecht, the Netherlands: Springer.

Anderson, K. C., & Leinhardt, G. (2002). Maps as representations: Expert novice comparison of projection understanding. *Cognition and Instruction, 20*, 283–321.

Barnea, N., & Yehudit, J. D. (2000). Computerized molecular modeling – The new technology for enhancing model perception among chemistry educators and learners. *Chemistry Education: Research and Practice in Europe, 1*, 109–120.

Baum, D. A., Smith, S. D., & Donovan, S. S. S. (2005). The tree-thinking challenge. *Science, 310*, 979–980.

Bodner, G. M., & Guay, R. B. (1997). The Purdue visualizations of rotations test. *The Chemical Educator, 2*, 1–17.

Brumby, M. (1979). Problems in learning the concept of natural selection. *Journal of Biological Education, 13*, 119–122.

Cavallo, A. (1996). Meaningful learning, reasoning ability, and students' understanding and problem solving of topics in genetics. *Journal of Research in Science Teaching, 33*(6), 625–656.

Chi, M. T. H., Feltovich, P. J., & Glaser, R. (1981). Categorization and representation of physics problems by experts and novices. *Cognitive Science, 5*, 121–152.

Cook, M. P. (2006). Visual representations in science education: The influence of prior knowledge and cognitive load theory on instructional design principles. *Science Education, 90*, 1073–1091.

Cuoco, A. A. (Ed.). (2001). *The roles of representation in school mathematics (2001 yearbook of the national council of teachers in mathematics)*. Reston, VA: National Council of Teachers in Mathematics.

Driver, R., Squires, A., Rushworth, P., & Woods-Robinson, V. (1994). *Making sense of secondary science: Research into children's ideas*. London: Routledge.

Ferk, V., Vrtacnik, M., Blejec, A., & Gril, A. (2003). Students' understanding of molecular structure representations. *International Journal of Science Education, 25*, 1227–1245.

Gendron, R. P. (2000). The classification and evolution of caminalcules. *The American Biology Teacher, 62*, 570–576.

Gilbert, J. K. (2005a). Visualization: A metacognitive skill in science and science education. In J. K. Gilbert (Ed.), *Visualization in science education* (pp. 9–27). Dordrecht, the Netherlands: Springer.

Gilbert, J. K. (Ed.). (2005b). *Visualizations in science education*. Dordrecht, the Netherlands: Springer.

Gregory, T. R. (2008). Understanding evolutionary trees. *Evolution: Education and Outreach, 1*, 121–137.

Hackling, M. W., & Lawrence, J. A. (1988). Expert and novice solutions of genetic pedigree problems. *Journal of Research in Science Teaching, 25*, 531–546.

Halverson, K. L. (2010, March). *Exploring the link between mental rotation and college student learning with phylogenetic trees.* Paper presented at the annual meeting of the National Association for Research in Science Teaching, Philadelphia, PA.

Halverson, K. L. (2011). Improving tree-thinking one learnable skill at a time. *Evolution: Education and Outreach, 4*, 95–106.

Halverson, K. L., Pires, J. C., & Abell, S. K. (2011). Exploring the complexity of tree thinking expertise in an undergraduate plant systematics course. *Science Education, 95*, 794–823.

Kozma, R. B., & Russell, J. (1997). Multimedia and understanding: expert and novice responses to different representations of chemical phenomenon. *Journal of Research in Science Teaching, 34*, 949–968.

Kozma, R. B., & Russell, J. (2005). Modelling students becoming chemists: Developing representational competence. In J. K. Gilbert (Ed.), *Visualization in science education* (pp. 121–145). Dordrecht, the Netherlands: Springer.

Larkin, J., Mcdermott, J., Simon, D. P., & Simon, H. A. (1980). Expert and novice performance in solving physics problems. *Science, 208*, 1335–1342.

Lord, T. R., & Marino, S. (1993). How university students view the theory of evolution. *Journal of College Science Teaching, 22*, 353–357.

Maroo, J., & Halverson, K. L. (2011, April). *A mental mobile: Using branch rotation to solve the puzzle, "Are these trees the same?"* Paper presented at the annual meeting of the National Association for Research in Science Teaching, Orlando, FL.

Matuk, C. (2007). Images of evolution. *The Journal of Biocommunication, 33*(3), E54–E61.

Moore, R., Mitchell, G., Bally, R., Inglis, M., Day, J., & Jacobs, D. (2002). Undergraduates' understanding of evolution: Ascriptions of agency as a problem for student learning. *Journal of Biological Education, 36*, 65–71.

National Research Council. (1996). *National science education standards.* Washington, DC: National Academy Press.

Patrick, M. D., Carter, G., & Wiebe, E. N. (2005). Visual representations of DNA replication: Middle grades students' perceptions and interpretations. *Journal of Science Education and Technology, 14*, 353–365.

Patton, M. Q. (2002). *Qualitative research and evaluation methods.* Thousand Oaks, CA: Sage.

Peterson, M. P. (1994). Cognitive issues in cartographic visualization. In A. M. Maceachren & D. R. F. Taylor (Eds.), *Visualization in modern cartography* (pp. 27–43). Oxford: Pergamon.

Reiss, M. J., & Tunnicliffe, S. D. (2001). Students' understandings of their internal structure as revealed by drawings. In H. Behrendt, H. Dahncke, R. Duit, W. Graber, M. Komorek, A. Kross, & P. Reiska (Eds.), *Research in science education – Past, present, and future* (pp. 101–106). Dordrecht, the Netherlands: Kluwer Academic.

Roth, W.-M., Bowen, G. M., & McGinn, M. K. (1999). Differences in graph-related practices between high school biology textbooks and scientific ecology journals. *Journal of Research in Science Teaching, 36*, 977–1019.

Rutherford, J. F., & Ahlgren, A. (1990). *Science for all Americans.* New York: Oxford University Press.

Tabachneck, H. J. M., Leonardo, A., & Simon, H. A. (1994, August). *How does an expert use a graph? A model of visual and verbal inferencing in economics.* Paper Presented at the 16th annual conference of the Cognitive Science Society, Atlanta, GA.

Takayama, K. (2005). Teaching visualizing the science of genomics. In J. K. Gilbert (Ed.), *Visualization in science education* (pp. 217–251). Dordrecht, the Netherlands: Springer.

Tufte, E. R. (2001). *The visual display of quantitative information* (2nd ed.). Cheshire, CT: Graphics Press.

van Fraassen, B. C. (2008). *Scientific representation: Paradoxes of perspective*. Oxford: Oxford University Press.

Williams, J. M., & Tolmie, A. (2000). Conceptual change in biology: Group interaction and the understanding of inheritance. *British Journal of Developmental Psychology, 18*, 625–649.

Zbiek, R. M., Heid, M. K., Blume, G. W., & Dick, T. P. (2007). Research on technology in mathematics education: A perspective of constructs. In F. K. Lester (Ed.), *Second handbook of research on mathematics teaching and learning* (Vol. 2, pp. 1169–1207). Charlotte, NC: Information Age.

Chapter 12
Understanding Photosynthesis and Cellular Respiration: Encouraging a View of Biological *Nested* Systems

Reneé Schwartz and Mary H. Brown

Introduction

> *....you have a sun, you have an earth, and in the trees you have cells, that have the energy, and the cells would break down into organelles, and the atoms...*Ann
>
> *...we have the sun, and it helps provide the plants with energy, so that they can photosynthesize, and then kind of goes to like the first consumer...*Jay
>
> *[Photosynthesis is] when the plants take the sunlight and they create oxygen for the plant to grow. It's their energy...*Kay

Ann, Jay, and Kay hold different conceptions regarding photosynthesis and cellular respiration, the connections between these two processes, and as being within multiple ecological levels as components in nested systems. Ann, Jay, and Kay were participants in a study which explored undergraduate education majors' conceptions of photosynthesis and cellular respiration (Brown & Schwartz, 2009). The purpose of this chapter is to explore multiple representations of photosynthesis and plant cellular respiration by presenting the cases of Ann, Jay, and a summary of students like Kay. These cases exemplify ways in which learners conceptualize the processes, their connections, and role within biological systems. Use of multiple representations that demonstrate connections and interdependencies across levels may be an effective way of helping learners develop a systems view of biology.

Photosynthesis and plant cellular respiration are challenging to learners for a number of reasons. Both processes have multiple steps and occur simultaneously.

R. Schwartz (✉)
Department of Biological Sciences, Mallinson Institute of Science Education, Western Michigan University, 3139 Wood Hall, 1903 W. Michigan Ave, Kalamazoo, MI 49008, USA
e-mail: r.schwartz@wmich.edu

M.H. Brown
Science Department, Lansing Community College, Washington Court Place, Room 030, 309 North Washington Square, Lansing, MI 48933, USA
e-mail: brownm@lcc.edu

D.F. Treagust and C.-Y. Tsui (eds.), *Multiple Representations in Biological Education*, Models and Modeling in Science Education 7, DOI 10.1007/978-94-007-4192-8_12,
© Springer Science+Business Media B.V. 2013

Learners who compartmentalize function and specialization of organelles at the cellular level may miss the significance of the plant as an independent biological system functioning at the level of the local ecosystem as well as globally (Brown & Schwartz, 2009). A compartmentalized view of the two processes isolates their function and does not consider interrelationships within and between biological systems. A systems perspective is missing.

Systems: A Theoretical Perspective

System is a unifying theme across science disciplines (Rutherford & Ahlgren, 1990). A scientific view of the living world considers a level of organization that goes beyond individual physical and chemical components to perceive life as part of a system. A system requires an integrated whole whose properties come from relationships among components. Understanding living organisms must include chemical and physical elements within their organizing relationships, including the tendency to form and be connected to multiple systems. Each system level forms a whole with respect to *its* parts but at the same time is part of a larger whole. There is a need to explore learners' conceptions of connections of biological processes within and among organizational systems (Brown & Schwartz, 2009).

Systems have been characterized as *complex dynamic processes* because of their abstractness and the multiple levels (Chi, 2001). Chi's criterion of complexity is that the *emergent mechanism* of the levels unites two systems. This is certainly the case with photosynthesis and respiration. The overall energy reaction is the emergent mechanism. Explanations of the phenomenon at the organism level or biochemical level do not account for emergent mechanisms. Capra (1996) proposed a new idea regarding life's organization which significantly changed the emphasis of biological research from individual components toward a systems approach. This view provides a level of organization beyond mere physical and chemical components. A living organism should be viewed as a system which organizes within and across multiple levels. Each system (such as the cellular level) is a whole with respect to its components and, at the same time, is part of a larger system, the organism. In this way, biological systems are organized and encompass hierarchical ecological levels and are *nested*, one within another.

> One of the fundamentals of biology is that sense can be made of the complexity of the biosphere by viewing it as a set of interrelated systems that can range in size from the subcellular to the ecosystems level. We can trace matter and energy within these systems to understand them individually and between these systems to understand their interdependence. (Wilson et al., 2006, p. 324)

Interconnections

Photosynthesis and cellular respiration in plants are interconnected processes (see Fig. 12.1). They combine to provide energy for the plant. Photosynthesis transforms the radiant energy into chemical bond energy within the carbohydrate molecule

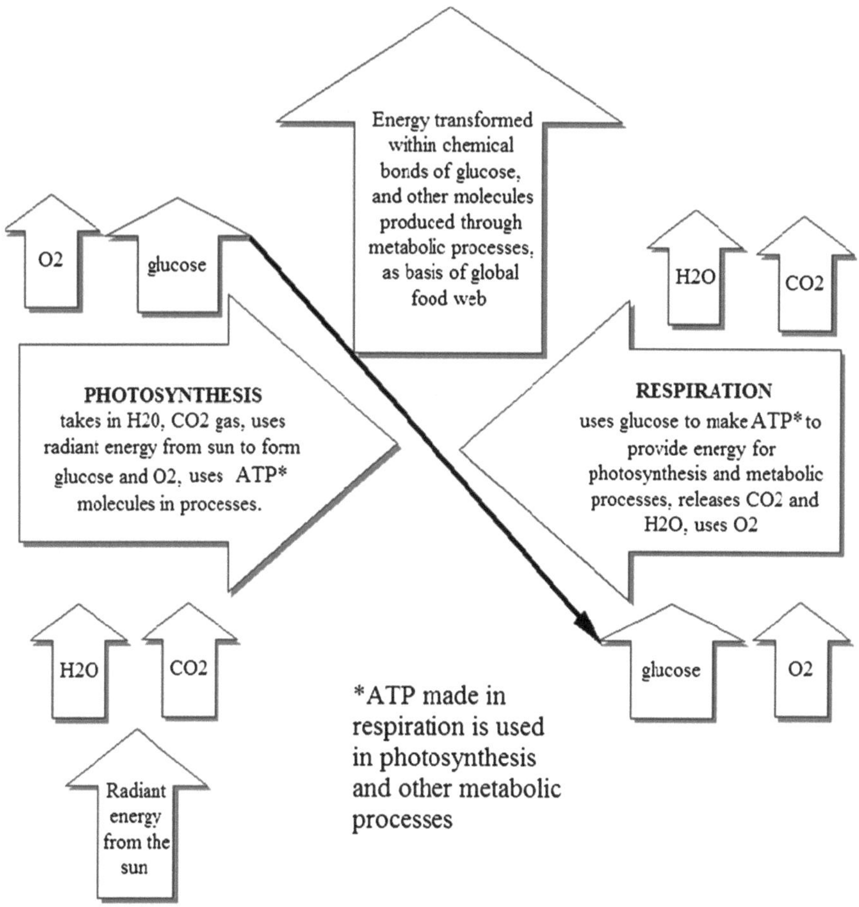

Fig. 12.1 Interconnected plant processes of photosynthesis and cellular respiration (Brown & Schwartz, 2009)

produced. This chemical bond energy is then transformed into a smaller unit of energy within the ATP molecule within cellular respiration. The energy produced during cellular respiration allows for the continuation of photosynthesis.

Research consistently shows that learners of all ages struggle to understand connections between the two processes (e.g., Canal, 1999; Haslam & Treagust, 1987; Seymour & Longden, 1991; Songer & Mintzes, 1994). Students' failure to see clear connections between the two processes indicates they see just biochemical reactions rather than processes that are components of interrelated systems (Brown & Schwartz, 2009). A more scientific view would be to consider the cell's processes on multiple ecological levels such that matter and energy are traceable within and between levels (Brown & Schwartz, 2009; Waheed & Lucas, 1992; Wilson et al., 2006) (see Fig. 12.2).

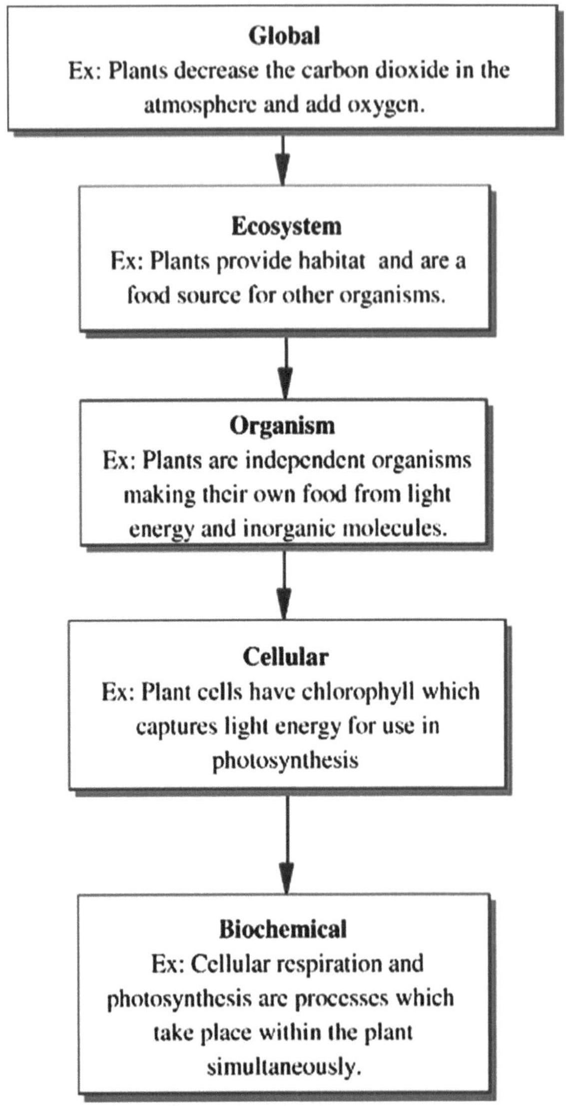

Fig. 12.2 Multiple ecological levels (Brown & Schwartz, 2009)

Plants should be considered as biological systems with processes that impact across levels (biochemical, cellular, and organism) which also interact within the larger global system. Gross anatomical structures, such as leaves and roots, bring the raw materials for photosynthesis together into the cell. Once the raw materials are assembled, the view changes, and the cellular level with the functions of the chloroplasts predominates. Plant processes in a healthy plant continue to result in both carbon dioxide and oxygen being released into and gathered from the atmosphere.

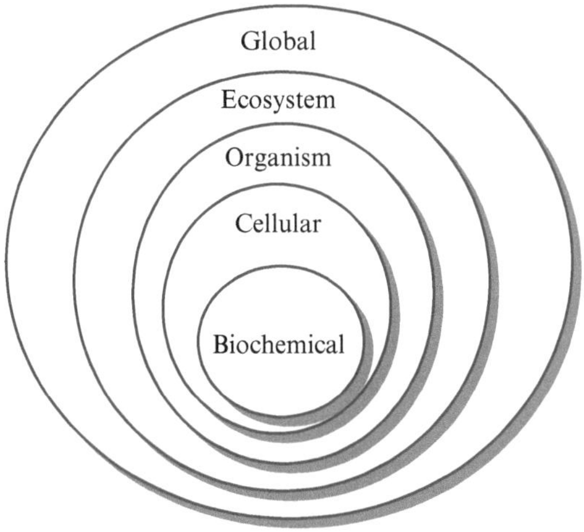

Fig. 12.3 Biological *nested systems* (Brown & Schwartz, 2009)

We consider interactions across multiple ecological levels to represent *nested systems* (see Fig. 12.3) (Brown & Schwartz, 2009). This view takes into account interconnected processes that are hierarchically organized. We want to promote conceptions of nested systems as a more complete systems view. Ideally, students recognize the importance of plants within the local and global ecosystem and view plants as a source of energy—as biological systems themselves—dependent upon cellular structures and functions.

Conceptions of Photosynthesis and Cellular Respiration: Are They Intuitive?

How do students typically understand connections between photosynthesis and cellular respiration, and how can teachers help students see plants as part of multiple and nested systems? Photosynthesis and cellular respiration remain challenging processes for students to understand. Within the last few decades, research has shown that students do not have a clear understanding of either plant process, let alone when they occur, where they occur, and how they interact or impact across systems (e.g., Anderson, Sheldon, & DuBuy, 1990; Barker & Carr, 1989a, 1989b, 1989c; Bell, 1985; Brown & Schwartz, 2009; Canal, 1999; Eisen & Stavey, 1988; Haslam & Treagust, 1987; Seymour & Longden, 1991; Stavy, Eisen, & Yaakobi, 1987). While this list of studies is by no means comprehensive, it goes to show that researchers have been aware of the longstanding issues learners have with understanding biological reactions such as photosynthesis and cellular respiration. Learners struggle with understanding biological reactions themselves and do not typically consider reactions as connected or components of systems. For example, photosynthesis and cellular respiration have been described only in terms of gas exchange with no relationship to energy (Eisen

& Stavey, 1988). Students often consider the two processes as opposite reactions, one being the reverse of the other; thus, they lose the perspective that each reaction has a specific purpose for the functioning of the cell and organism.

Perhaps these alternative conceptions are more intuitively appealing to learners, as suggested by Southerland, Abrams, Cummins, and Anzelmo (2001), when the learners provided spontaneous constructions of their knowledge that intuitively make sense. Intuitive conceptions suggest a knowledge framework that is still under construction. Such conceptions do not exist in the form of a coherent theory but rather are provided *on the spot* as the need for an explanation arises. When processes are difficult to understand, as has often been noted within the physical sciences, learners may rely on intuitive structures to provide an explanation. *Phenomenological primitives, or p-prims* (diSessa, 1993), suggest the learner is just beginning to construct a knowledge system to explain a phenomenon. Whereas diSessa suggested that intuitive conceptions may be useful in constructing scientific knowledge; others suggested the use of graphic organizers and concept maps to assess and improve biological conceptions (e.g., Kinchin, 2000; Mintzes, Wandersee, & Novak, 2001). Yet, after decades of research, it seems that a scientifically accurate conception of photosynthesis and cellular respiration still evades most learners. What are some of the typical representations learners have? We present three cases here.

How Preservice Teachers Think About Photosynthesis and Plant Cellular Respiration

The cases of Ann and Jay are developed from our larger study of 18 preservice teachers (Brown & Schwartz, 2009). We sought to answer two research questions:

1. What are preservice teachers' conceptions of photosynthesis and plant cellular respiration, and how do they conceptualize the relationships with respect to (a) interconnectedness between the processes, (b) working on multiple ecological levels, and (c) being components within *nested systems*?
2. What arguments and explanations do preservice teachers provide in support of their conceptualizations of how the two plant processes are related?

The cases presented here were part of the larger sample and were selected to represent the full range of perspectives identified. Ann was the participant with one the highest degrees of competence in her conceptions. Jay provides a different and slightly less sophisticated representation than Ann. For our third case, we present a summary of those who found it most difficult in expressing their views. Due to limited quotes beyond "I don't know" provided by participants at the lower end of the spectrum, we decided to combine information from participants who found it most difficult in expressing their ideas. The two cases and the summary case represent the range the conceptions present within this undergraduate nonmajors classroom.

Data collection comprised field notes during instruction and three separate interviews. The first interview involved *explaining sets* with tasks that challenged students to consider plant growth and functions. The two follow-up interviews served to clarify participants' responses and researcher interpretations. Full details of the data collection and analysis are found in Brown and Schwartz (2009).

Explaining Sets

The explaining set sessions took place after course instruction included photosynthesis and cellular respiration. Participants were placed into pairs and presented with four tasks. For each task, pairs were asked to take turns explaining to each other what was happening and their thought processes. After a warm-up task to become comfortable with the procedure, they were provided with a series of situations and asked to explain to each other what was happening. The specific tasks focused on eliciting conceptions at the *organism* level, the *ecosystem* level, and the *global* level. Table 12.1 includes the scenarios and questions asked for the three tasks used in our study. Later in this chapter, we will discuss the collection of strategies within Table 12.1 and how they can facilitate conceptual development. For the case studies, we only used the tasks for organism, ecosystem, and global levels; these were followed by interviews to elicit ideas related to other levels and connections.

Cognitive Interview

The cognitive interviews were conducted individually and after pairs of students completed the three explaining tasks. We asked about photosynthesis and plant cellular respiration using the plants from the organism task (plant comparison). Questions focused on the plant as a biological system and on plant growth. Participants were asked to define both processes and consider what would happen to the plants if both processes were disrupted. The participants were also asked how they had come to understand photosynthesis and cellular respiration.

Clarifying Interview

Individual clarifying interviews were conducted after preliminary data analysis. These provided an opportunity for participant validity checks and to clarify any lingering questions we had about their conceptions. We also asked two additional questions: "What do you think a life cycle or a cycle is?" and "What do you think a biological system is?" We had noted in earlier responses that their ideas of these two concepts might be quite variable.

Table 12.1 Multiple representations: strategies to elicit conceptions and scaffold student thinking about photosynthesis, cellular respiration, and biological systems

Level and task	Guiding questions
Biochemical and cellular Provide students with markers, a white board, images, models, and vocabularies from the instructional materials used in class. Images and terms can be put on individual index cards so that students can easily move them around to show relationships (e.g., concept mapping). Ask students to explain their conception of the processes using those materials	1. What do you think photosynthesis is? 2. What do you think cellular respiration is? 3. Where do these processes take place? 4. Please tell me about the two biochemical equations of photosynthesis and cellular respiration. What do you think is their relationship? 5. How do you think cells get the materials they need to do photosynthesis and cellular respiration?
Organism[a] "What are the plants doing?" Show students two plants of the same species but of different sizes. Students may be asked to explain the differences to each other, to the instructor. This could also be done as a whole class discussion or assessment 	1. Explain any differences you see in these two plants 2. Explain *how* the two plants became different from each other 3. Explain any special conditions or requirements that the plants needed in order to be the way they are now. Please tell all the details you know 4. Explain any differences you know between these plants and other organisms. How do plants differ from other organisms? 5. Explain any similarities you know between these plants and other organisms. How are plants similar to other organisms? 6. Explain how you came to know what you know about plants. How did you make sense of what you know about plants? With additional questions: 1. How do you think the leaves of one plant got larger than the leaves of the other plant? 2. What raw materials do you think the plants needed? Where did they get the materials they needed to grow larger? 3. Can you explain how they did that? What do you think the plants are currently doing? How do the various parts of the plants, such as the roots, leaves, and stem, help the plant to function? 4. Can you tell me how you think plants fit into your world? What is their role in relationship to you?
Ecosystem[a]	Tell the students that they will use their imagination a little bit with this one. This jar represents a life-sized sealed enclosure with people and plants living inside. Imagine that the top of the enclosure is completely clear,

(continued)

Table 12.1 (continued)

Level and task	Guiding questions
"Ecosystem in a Jar" 	so that sunlight can pass through the top as well as all the sides. The people have all the food and water they need. All organisms inside the enclosure are living and healthy. Imagine that the enclosure is really big, so there are no space concerns 1. Tell if you think it would be possible for a person to survive in such an environment. Why or why not? 2. Tell if you think it would be possible for the green plants to survive in such an environment. Why or why not? 3. What do you think will happen to the person as time passes? Please explain your answer 4. What do you think will happen to the plants as time passes? Please explain your answer
Prepare a clear jar with plastic greens and small figures of people inside. You can add animals too. Students can be organized into individual, pairs, or whole class	5. Explain the relationships that are going on in the enclosure. Explain your response thoroughly
Global[a]	Drawing:
"Meta-representation"	Ask students to draw a picture to explain how they think plants are part of the natural world. Their picture should demonstrate how they understand the role of plants in the world
Provide students with paper or white board and markers	Explanation:
	Ask students to explain their picture to each other, a group, or the instructor. Questions the peers or instructor might ask about the pictures include the following: (1) What is going on between the plant and the air? (or ground? or people?) (2) Where does the plant get water? (3) Is there any connection between the plants and animals? (4) What would happen to the world if all the plants were gone?
Multiple levels	1. What do you think photosynthesis and cellular respiration do for the ecosystem?
In a discussion format (pairs, groups, or whole class), ask for students' response of these questions which reflect understanding of the processes at multiple levels and to problem solve disruptions to the processes. Use any or all of the props from the other tasks and ask students to explain their ideas using the props	2. If photosynthesis were disrupted by some mechanism, what effects do you think there might be on the plant organism? And on the rest of the world? 3. If cellular respiration were disrupted by some mechanism, what effects do you think there might be on the plant organism? And on the rest of the world? 4. How do you think the chemicals, the cells, and the structure of the plant such as the roots, leaves, and stems work together within the ecosystem? How do they work on a global level?

[a]These three tasks were used in the explaining set phase of data collection for the Brown and Schwartz's (2009) study

We had also noted that participants referred to images from the course lecture or the course text. Participants seemed frustrated at being able to visualize a mental image associated with their classroom instruction but being unable to explain their conceptions without that specific image. A similar pattern emerged with their vocabulary. To offset this phenomenon in the clarifying interview, we provided copies of many of the images found in the text (which were also in the lecture presentations) and vocabulary words that participants had mentioned during the cognitive interviews or explaining sets. Participants were encouraged to use whatever they wanted to help with their explanations.

Through exploratory qualitative analysis, we reviewed all the transcripts for episodes reflecting conceptions of the processes and levels within a systems view. For each case presented here, we describe conceptions of the two plant processes in three respects: views of the connections between processes, ability to use multiple ecological levels in describing the processes, and conceptions of the processes as being nested systems.

Results: Multiple Representations Expressed in Three Cases

The cases are introduced based on the scientific acceptability of their conceptions. Ann used all ecological levels in her descriptions; although she reported misconceptions, she provided the greatest range in her justifications. Jay held conceptions of the connections of the two plant processes that were shared by at least five of his classmates. The summary case represents those participants with the most limited conceptions.

Ann

Ann was the youngest student enrolled in the course, at age 18. Ann had 4 years of high school science. She was also enrolled in college chemistry at the same time as the biology course.

Interconnections

Ann first expresses her knowledge of the interconnections between the two plant processes in the explaining set. During the organism task, Ann suggests to her partner that she sees the plant cellular processes as components of a larger system, in this case the organism:

> ...[the plant] needs all elements of photosynthesis and cellular respiration to work. If it doesn't have all the elements then it won't grow.

In the cognitive interview, Ann reasons through her problem connecting the two processes. It arises from the question, "What do you think cellular respiration is?"

> ... cellular respiration is...when ah, they make oxygen, like the cellular respiration. We did this thing in lab where we had like corn seeds and we had dead ones and live ones and we just had regular water, and we put a pH thing in there and we found ah, the corn left off CO_2...Well in the plants, they usually give off oxygen...So this Dr. Smith said that was called cellular respiration, but I think ah, she said plants do both, and I can see how they give off oxygen, but I don't really know why they give off CO_2, and I don't know if they do it just like seedlings, and that's because like growth, because they give off CO_2 because they are using... like an apple has the seeds in the middle and it uses the apple, the rest of the apple, as food to grow into a tree. So I don't know if the seeds respire too. I don't know if they give off CO_2 like we do, because they are kind of like eating, maybe, so they are breaking that stuff down.

We see here that Ann struggles to make sense of when and why plants respire. She was told they do, but she is not sure why. Ann tries to make sense by comparing respiration in plants and humans. When asked about the biochemical equations, Ann is more certain of the processes:

> ... Photosynthesis breaks down...carbons and makes energy and stuff.... to get the product back into the glucose, the six carbon sugar, then they use cellular respiration..... You need the glucose in the thing, like the energy in the food, for photosynthesis, and then you have like the products, which is energy, and then you need that to make food. So cellular respiration is taking that energy back into making food, and that's cellular respiration.

When asked what would happen if cellular respiration was disrupted, Ann is very confident of the connection between the two processes:

> ... Photosynthesis needs cellular respiration. Cellular respiration needs photosynthesis. So, I think the same effects would happen, because I think the plant needs both to live.

Multiple Ecological Levels

While there are errors or incomplete ideas in her descriptions at some of the levels, Ann at least recognizes each ecological level. Although her ideas are not always clear, she is able to discuss ATP, glucose, energy, and food. Ann explains how plants use glucose:

> I think that's what it does for food, is that the carbon in it, like it breaks it down. That's how it has oxygen as a by-product, as it takes the oxygen out of it, the glucose.

When asked, "What does it use the carbon for," Ann responds, "Well, I'm guessing that it makes food." Ann can talk about ATP and knows it as an energy molecule. She knows why the plant, as an organism, needs ATP:

> I think it uses it for everything. That's why um, it needs a lot of ATP. It uses it to, um, grow, and it goes through the food, and it uses it to make the food...

Ann struggles to understand the source of carbon. She does not make the association between the carbon she talked about as being in glucose and the carbon in carbon dioxide. She also does not connect carbon dioxide as an atmospheric gas

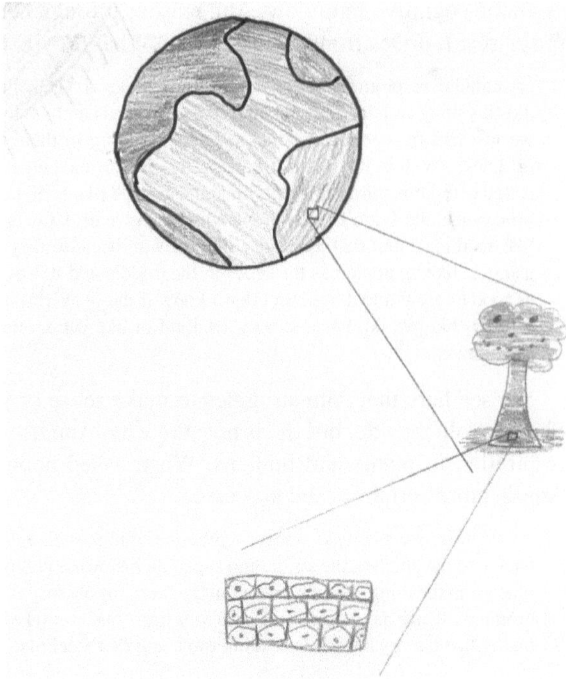

Fig. 12.4 Ann's meta-representation of a plant's role in the world (Brown & Schwartz, 2009)

and suggests that plants take up carbon through the roots. We find that she is unable to accurately trace carbon across systems.

> I think they [plants] just like soak it [the CO_2] up. I don't know how they get the carbon dioxide. I think they just must soak it up from the leaves, or the roots or somewhere.

Ann also has difficulty with the concept of energy. Her global understanding of plants includes interactions; and she recognizes ATP as a necessary transformation from the global form of radiant energy to energy for the cell, but she still struggles with the subtle distinctions between food and energy. She states that photosynthesis is a way to make food and energy. Then, when asked to clarify how she understands "food" and "energy," she says:

> Well, I guess, 'cause energy, food is something [plants] use, I guess to grow. I don't know how to explain food and energy. It's just we eat for energy, and then there is like energy of ATP, and it makes like food, so. And, then food is used to make energy...

Nested Systems

Even though Ann struggles with her understanding of connections across levels, she recognizes that connections exist and the levels rely on one another. Ann's meta-representation (see Fig. 12.4) depicts her idea of how plants play a role at multiple levels and connected systems. She describes her picture while pointing to the cells and to the globe she has drawn:

> . . .and the littlest part has to work for the biggest part to work. The biggest part has to work for the littlest part to work. I could even break this down further, with one little organelle and show all the parts. . .

Misconceptions

In addition to the issues mentioned above, Ann reveals other misconceptions. For example, she has a broad interpretation of the word *cycle*. She explains the relationship between plants and people during the “ecosystem task”:

> So the green plants will live because they have sunlight. They have oxygen that the humans are giving off, kind of like a cycle. . . I think this is the cycle of life. I mean organisms are there, like humans. They eat the plants. They die. They hold the soil. They create carbon and nutrients in the soil for the plants to eat. The plants grow. The plants die, get decomposed, or we eat them, whatever the case may be. They grow. We have offspring. I’m guessing there is a boy and girl and they can have little babies. And then, they grow. It’s just like the cycle of life in there.

She seems to be talking about an oxygen cycle, where humans are a necessary source of oxygen for plants to live, and plants are a source of oxygen for humans to live. She reveals a misconception about what humans “give off” and the source of carbon for plants. She is consistent in her view that humans and plants are reliant upon each other for what she considers a “cycle of life.”

Jay

Jay had been out of high school for 4 years. He had taken two science classes during high school and one college level nature study class prior to taking the current biology course. Jay admitted he knew little of the two processes.

Interconnections

Jay’s idea of photosynthesis is an energy reaction, whereas cellular respiration is a gas exchange.

> . . . I think [photosynthesis is] taking energy from the sun’s light and transferring it to energy to be used by the plant which can be done through glucose and all that.

> I think [cellular respiration is] taking carbon dioxide, using it for the photosynthesis process and in the end comes oxygen, which is given off, and used by others, such as humans.

He then combines the reactions to form one equation:

> In the presence of light energy, CO_2 (of cellular respiration) and H_2O (of photosynthesis) yields O_2 (of cellular respiration) and carbohydrate (of photosynthesis).

In his view, one equation involves both processes, with the gas portions being the respiration and the water and carbon portion (carbohydrate manufacturing) being the photosynthesis segment.

Multiple Ecological Levels

We found that Jay's focus is mainly on the connection of the processes to the organism and ecosystem and, to a lesser degree, the global level. He does not mention clear biochemical equations or cellular details. He reveals misconceptions on each level. In describing growth, Jay starts with a seed and focuses on the organism level:

> ...[the plant] needs the resources like water, sunlight, and they started to grow, obviously. With the resources there, they just keep getting bigger and bigger.

It becomes clear that Jay confuses the terms nutrients, food, glucose, and carbohydrates. When asked what the plant is getting from the soil, Jay originally says "nutrients" which he says includes water. He then explains:

> Jay: 'Cause I thought, if I remember correctly, is it something that [plants] use carbohydrates, and they transfer it into glucose?...I'm wondering if possibly there is oxygen in the soil too, from other plants living on them, decomposing.
> Researcher: Does it take that oxygen from the soil do you think?
> Jay: Possibly, and then transfers it to carbon dioxide, or carbon dioxide, no it takes the carbon dioxide and puts it into oxygen.

Jay suggests a connection between the ecological and organism levels, but he still struggles to explain it. Perhaps Jay is tapping into his intuitive conceptions while trying to make sense of the source of oxygen and carbon dioxide. A subsequent response suggests Jay is certain the plant uses carbon dioxide but is unsure of its source. He knows carbon dioxide is in the air, but he does not want to give up on the soil idea:

> Researcher: Where would it get the carbon dioxide from then?
> Jay: In the soil.
> Researcher: Well, do you think it's getting in from the soil?
> Jay: I think most of it that it gets is from the air.
> Researcher: So, it's taking most of the carbon dioxide from the air?
> Jay: I'm just not sure if like most of it seeps into the ground too, a little bit, like maybe there is a trace of it inside the soil?

Jay's conception of the plant's role in the ecosystem is one of *use*. He talks about the medicines that are harvested from the rainforest, oxygen and food, and the prevention of soil erosion. Jay also holds a view that plants are dependent on humans and humans are dependent on plants.

> . . . I think the leaves take in the carbon dioxide, which is given off by humans. For example, that's poisonous for us. We can't survive if we have too much CO_2, um, so they take in that. The plant uses that for cellular respiration, which is also used for photosynthesis to make energy, for the plants, and then it gives off oxygen. The plant obviously is not going to use that [oxygen], but we can use that, and that's what we need to survive.

Nested Systems

We get a glimpse into Jay's notions of nested systems from the above quote. He sees a connection and reliance between the ecosystem and organism and among organisms. He hesitates, though, when describing biological systems.

> . . .I guess it's everything within like biology and stuff, going hand-in-hand. . . ..like basically, how everything works and interacts with one another.

Jay is one of the few participants to suggest that systems involve interactions. In Jay's conception, photosynthesis directly takes radiant energy from the sun to make carbohydrates. Logically, but erroneously, Jay's conception of photosynthesis needs no cellular respiration. The connection is made with one equation. The extent of recognizing the nestedness of systems is difficult to determine, but his one-equation approach suggests a limited perception.

Misconceptions

Jay places considerable importance on the soil. He considers materials moved by the roots of plants as the source of "plant's food." His confusion of food, nutrients, carbohydrates, sugar, and glucose allows him to reason that carbon dioxide is present in sufficient quantities in the soil. He also fails to recognize that plants use oxygen for cellular respiration.

Summary Case of Most Limited Conceptions

We combined results from several participants to describe the typical conceptions of those who struggled the most to verbalize and rationalize their ideas. These participants considered the processes to be too abstract to fully understand. For example, when asked what plants do and what they need to grow, responses included:

> I don't know. You probably can't see what [the plant] does. . . .When they absorb the light, you can't really see when they do that, but they do.

> [The plants] are possibly growing, a little bit. They could grow a little bit each day, and that would be hard for us to see. . .Growing by, you know, just looking at it by our eyes, but it's growing. It's giving off energy as well.

Interconnections

The following describes one view of process connections. This view was typical of those who saw a linear progression from photosynthesis to cellular respiration.

> . . .I think you need the photosynthesis to be able to go on to cellular respiration. Because it uses the light as energy and without energy, you couldn't do cellular respiration.

Students at this level did not freely discuss plant cellular respiration. One offered a guess as to the meaning of cellular respiration and assigned it a gas exchange function.

> Isn't that when the by-product is oxygen? It takes the CO_2, and it gives to ATP and then comes out of the oxygen and gives it to the Calvin cycle.

In this view, photosynthesis provides energy for cellular respiration to continue. Although we do not get a good sense of how these participants understand cellular respiration, they do not seem to consider it to be an energy reaction. These participants tended to confuse oxygen and energy:

> Kay: [Photosynthesis is] when the plants take the sunlight and they create oxygen for the plant to grow. It's their energy.
> Researcher: Which is their energy? The oxygen is their energy?
> Kay: . . .Yeah, I think.

Multiple Ecological Levels

Those with most limited conceptions described the two processes at the organism level. One participant shows her lack of understanding at the organism level when she describes the function of leaves:

> . . .the leaves would be a deterrent from bugs, because they would eat the leaves, instead of digging down to the roots.

These participants would make connections to other levels only when asked and always inaccurately. The strongest connection was made across organisms rather than across levels. They held a consumer role for plants. That is, plants exist to provide for humans.

> . . .plants give us oxygen; they give us food; they give us shade. If you look at plants, as if like a whole tree, they give us wood, fuel for heat, homes, houses. . . They give homes to animals and food.

Nested Systems

It is not surprising that, with such little awareness of different ecological levels, there is no evidence that these participants consider nested systems. They indicate plants are in service to humans, a very egocentric, yet common, view. When asked how plants function within the ecosystem, they tie plants to human survival and evolution:

> Well, without them, we wouldn't survive. I don't think we would have even made it to be a little cell, because we use them to eat. . .I think it's just a chain reaction.

Misconceptions

These participants hold the common misconception that plants take most of the food from the soil. They suggest that the function of the leaf is one of protecting the root and, if they consider cellular respiration at all, it is for gas exchange. They tend to equate oxygen and energy, which arise from the process of photosynthesis. Plants exist to provide resources for humans.

Summary of the Three Cases

These cases represent learners who could benefit from opportunities to trace matter and energy across multiple levels. Instruction needs to explicitly highlight connections across and within systems. The role of the processes in the broader spectrum of living systems is missing. Misconceptions remain prevalent. Even Ann thought that the plant obtains food from the soil. She, like the others, viewed the plant as dependent on humans for survival. Ann had partial recognition of cellular respiration as an energy reaction but was unable to connect accurately the two biochemical equations, nor could she connect an accurate purpose to cellular respiration. Jay combined the two processes on the biochemical level into one equation and gave each process a separate purpose. He viewed the two reactions as being a simplistic need to obtain a source of food and discounted all classroom discussions of ATP and NADH as being insignificant to the purpose of food attainment. The summary case demonstrated only a superficial knowledge of cellular respiration and could only discuss the processes at the level of the organisms. Explanations were based on personal experiences, discounting classroom instruction. The misconceptions reported here are consistent with those documented by Bell (1985), Wood-Robinson (1991), Haslam and Treagust (1987), and others. In addition, these cases portray misconceptions about cycles and systems, two concepts that are central to biological literacy.

Ann was the only participant who considered nested systems. She was able to see connections from the organism, ecosystem, and global levels. The other participants compartmentalized the two plant processes, maintaining an organism view that rarely extended connections. Underlying sociological and egocentric perspectives correlated with their conceptions of plants. Levels other than the organisms may be too abstract to be clearly connected. However, their sociological and egocentric perspectives connected organism (plants) to organism (humans). Biochemical and cellular levels did not fit within their perspective of egocentrism and social needs. Their use of the intuitive conception *self as first referent* only permitted a limited view of cellular or chemical systems, as it focused their conceptions on self as a member of human society in which plants serve humans. All three cases used human analogies to describe plant processes. They might have been targeting their own human attributes in a *self as first referent* intuitive conception similar to the *need for change* biological p-prims suggested by Southerland et al. (2001). Our results suggested that the three cases, although variable in their conceptions, all used *self as first referent* to explain their conceptions. With the knowledge that students may rely on this intuitive conception, how can teachers reduce confusion of the processes and address misconceptions such that students better understand the context of ecological levels and nested systems?

Scaffolding Connections Across Levels

We recommend several pedagogical approaches to improve student understanding of connections between the processes of photosynthesis and plant cellular respiration and connections across biological systems. Table 12.1 details the levels, tasks, and instructional guides that can scaffold student thinking about photosynthesis and cellular respiration within and across biological levels. Our research used the organism, ecosystem, and global level exercises to elicit learners' conceptions. These, and the tasks targeting the other levels, can be used as instructional tools as well as research tools. Teachers can use them to identify preconceptions, develop more coherent systems views, and as formative assessments. The tasks are easily prepared from common materials and fairly quick to administer to a group, pairs, or even individual students. Instructional episodes can target specific ecological levels while also using intuitive ideas and guideposts to help students transfer from one level to another. We recommend these strategies in response to the conceptual problems identified in our study and other studies.

Use Self as First Referent

The *self as first referent* intuitive conception may be tapped during instruction to emphasize connections across systems. Intuitive conceptions are used within physical science instruction to scaffold student learning toward more accurate scientific conceptions. In the case of photosynthesis and cellular respiration, *self as first referent* could help build analogies between the human need for energy and its sources, and the plant's need for energy and its sources. Learners already see similarities between plant and human functions regarding energy. Both humans and plants have a need for energy. Both use the cellular respiration process to package the energy into units useful at the cellular level. Both have mitochondria which perform this process. Both species exist in multiple ecological systems. Using a series of tasks from Table 12.1 allows for discussion of how plants and humans are similar, different, and connected (or not) within the various levels. Instruction can focus at the level of the organism during the plant comparison and ecosystem jar tasks. The organism level is easily accessible for the learner. Is cellular respiration really like *breathing*? How is cellular respiration similar in plants and humans? How is breathing (as in the cardiovascular system) different from cellular respiration? Such an approach may help reduce the misconception of cellular respiration as *breathing*. Cellular respiration is clearly not *breathing* at the organism level of the humans, and an analogical comparison may help students understand that it is not breathing for the plant either.

Use Guides and Signposts

Language

Our participants focused on the organism level primarily although instruction focused on the biochemical and cellular levels. At the organism level, *air* is necessary. At the cellular level, molecules of *oxygen* and *carbon dioxide* must be distinguished. Learners confused oxygen with carbon dioxide (they are both *air*); and this confusion may be a barrier to understanding the more abstract biochemical and cellular processes and connections to the broader levels. The biochemical level has a great deal of symbolism needed to associate CO_2 with carbon dioxide. Such chemical knowledge is another barrier. Teachers need to recognize that images, chemical symbols, and text may all be idiosyncratically interpreted by students based upon their previous experiences. Because instruction has to be explicit in terms of language and connections between symbols, teachers should use all these representations and guide students from one type to another. Again, using a series of the tasks in Table 12.1 provides these opportunities.

Teachers should consider the prerequisite concepts and assumed common language when teaching about any new concept. For example, do the students know what chlorophyll is? Do they know electrons and energy relationships? How do they understand *energy* as a biological concept, chemical concept, or neither? Do they know the difference between food, energy, and nutrients? Do they know the function of leaves? Teachers can scaffold their own language use and create opportunities for students to demonstrate their understandings of prerequisite and assumed knowledge, including common vocabulary that might seem basic but can be problematic (like food and energy). Further study is needed to better understand the prerequisite conceptual knowledge for these learners.

Tracing, Organizing, and Mapping

The undergraduate course where Ann, Jay, and the others were introduced to biological processes did not include explicit cross-system cues. The instructor mentioned multiple ecological levels; however, there was rarely explicit reference to interactions or dependencies across levels. There was no tracing of elements, molecules, or products from one level to another. The concept of *nested systems* was implicit. Instruction could be enhanced if the instructor provides signposts for the learners when transitioning from one level to another. Clear tracing of matter and energy across levels needs to be modeled during instruction and then reinforced through formative and summative assessments. Connections can be made explicit as instruction utilizes various learning tasks, such as those in Table 12.1. As seen in our study, leaving the learner to make his/her own connections may result in a perspective focused on the most accessible and comfortable level, such as the organism.

Graphic organizers and concept maps can help student make connections across system levels. Organizers should emphasize cause and effect relationships and illustrate contrasts and comparisons (Trowbridge & Wandersee, 1998). Flowchart organizers may help illustrate the biochemical connections within cellular processes. Telescoping circles from a central point could show the subsystems within and the interconnectedness of the systems (as in Fig. 12.3). A side-by-side comparison of plant system and self-system could help the learner use their intuitive conception of plant processes as a foundation for a more scientifically accurate conception. Finally, instruction can utilize concept maps as signpost references while moving within and across multiple ecological levels. Concept map exercises, introduced at critical junctions, can serve as formative assessments to check students' abilities to trace matter and energy. Using the exercises in Table 12.1 in combination with concept maps for each level may be helpful to convey a *nested system* view, visually illustrating that interactions within the subsystems of multiple ecological levels lead to global consequences. Further research into the effectiveness of these approaches is warranted.

We recommend using multiple representations to teach students about photosynthesis, plant cellular respiration, and nested biological systems. Providing multiple opportunities, perspectives, signposts, and assessments gives priority to a *systems* view and can make seemingly abstract and disconnected biological processes relevant and more accessible to all learners.

References

Anderson, C., Sheldon, T., & DuBuy, J. (1990). The effects of instruction on college nonmajors' conceptions of respiration and photosynthesis. *Journal of Research in Science Teaching, 27*(8), 761–776.

Barker, M., & Carr, M. (1989a). Teaching and learning about photosynthesis. Part I: An assessment in terms of students' prior knowledge. *International Journal of Science Education, 11*(1), 49–56.
Barker, M., & Carr, M. (1989b). Teaching and learning about photosynthesis. Part II: A generative learning strategy. *International Journal of Science Education, 11*(2), 141–152.
Barker, M., & Carr, M. (1989c). Photosynthesis – Can our pupils see the wood for the trees? *Journal of Biological Education, 23*(1), 41–44.
Bell, B. (1985). Students ideas about plant nutrition: What are they? *Journal of Biological Education, 19*(3), 213–218.
Brown, M., & Schwartz, R. (2009). Connecting photosynthesis and cellular respiration: Preservice teachers' conceptions. *Journal of Research in Science Teaching, 46*(7), 791–812.
Canal, P. (1999). Photosynthesis and 'inverse respiration' in plants: An inevitable misconception? *International Journal of Science Education, 21*(4), 363–372.
Capra, F. (1996). *The web of life: A new scientific understanding of living systems*. New York: Anchor Books.
Chi, M. (2001, April 10–14). *Why do students fail to understand complex dynamic concepts*? Paper delivered to American Education Research Association, Seattle.
diSessa, A. (1993). Towards an epistemology of physics. *Cognition and Instruction, 10*, 105–225.
Eisen, Y., & Stavey, R. (1988). Students' understanding of photosynthesis. *The American Biology Teacher, 50*(4), 208–212.
Haslam, F., & Treagust, D. (1987). Diagnosing secondary students' misconceptions of photosynthesis and respiration in plants using a two-tier multiple choice instrument. *Journal of Biological Education, 21*(3), 203–211.
Kinchin, I. (2000). From 'ecologist' to 'conceptual ecologist' the utility of the conceptual ecology for teachers of biology. *Journal of Biological Education, 24*(4), 178–183.
Mintzes, J., Wandersee, J., & Novak, J. (2001). Assessing understanding in biology. *Journal of Biological Education, 35*(3), 118–124.
Rutherford, E. F., & Ahlgren, A. (1990). *Science for all Americans*. New York: Oxford University Press.
Seymour, J., & Longden, B. (1991). Respiration-that's breathing isn't it? *Journal of Biological Education, 25*(3), 177–183.
Songer, C., & Mintzes, J. (1994). Understanding cellular respiration: An analysis of conceptual change in college biology. *Journal of Research in Science Teaching, 31*(6), 621–637.
Southerland, S., Abrams, E., Cummins, C., & Anzelmo, J. (2001). Understanding students' explanations of biological phenomena: Conceptual frameworks or p-prims? *Science Education, 85*(4), 328–348.
Stavy, R., Eisen, Y., & Yaakobi, D. (1987). How students aged 13–15 understand photosynthesis. *International Journal of Science Education, 9*(1), 105–115.
Trowbridge, J., & Wandersee, J. (1998). Theory driven graphic organizers. In J. Mintzes, J. Wandersee, & J. Novak (Eds.), *Teaching science for understanding* (pp. 95–131). New York: Academic Press.
Waheed, T., & Lucas, A. (1992). Understanding interrelated topics: Photosynthesis. *Journal of Biological Education, 20*(3), 193–198.
Wilson, C., Anderson, C., Heidemann, M., Merrill, J., Merritt, B., Richmond, G., et al. (2006). Assessing students' ability to trace matter in dynamic systems in cell biology. *CBE Life Sciences Education, 5*(4), 323–331.
Wood-Robinson, C. (1991). Young people's ideas about plants. *Studies in Science Education, 19*, 119–135.

Part III
Assessment of Learning and Teaching with Multiple Representations

Chapters in Part III address research approaches using a range of methodologies and methods for assessing students' conceptual understanding of biology in terms of reasoning, problem solving and other higher order learning. The systems thinking highlighted in Chap. 18 points to a new direction for assessing biological education in the twenty-first century.

Chapter 13
Scientific Models in the Severe Acute Respiratory Syndrome (SARS) Research and in the Biology Curriculum

Alice Siu Ling Wong, Maurice M.W. Cheng, and Valerie W.Y. Yip

New Curriculum Goals of Hong Kong Science Education

In response to the rapid advancement of science and technology, science education in Hong Kong has seen a shift from predominantly content-focused goals to a wider goal of promoting scientific literacy. Appreciation of nature of science (NOS) is often regarded as an important component for scientific literacy.

The importance of promoting students' understanding of nature of science (NOS) has been explicitly spelt out in the Curriculum and Assessment Guides of the science subjects in Hong Kong (CDC-HKEAA, 2007). Such goals are in line with the science curricula in many other countries (e.g., American Association for the Advancement of Science, 1993; Council of Ministers of Education, 1997; Millar & Osborne, 1998). Earlier studies reported the disappointing findings that both students and science teachers have inadequate understanding of NOS (Lederman, 1992); however, there is encouraging empirical evidence that can inform initiatives to improve NOS understandings. Explicit and reflective approaches in teaching NOS can support learner development of sophisticated NOS ideas (Abd-El-Khalick & Lederman, 2000; Khishfe & Abd-El-Khalick, 2002). The critical review of Abd-El-Khalick and Lederman suggested that "approaches that utilize elements from history and philosophy of science and/or direct instruction on NOS are more effective in achieving that end than approaches that utilize science process-skills instruction or non-reflective inquiry-based activities" (p. 694).

Being cognizant of the challenges about teachers' general inadequate understanding of NOS and pedagogical skills in teaching NOS in Hong Kong, our preservice and in-service science teacher education programs were restructured to align with the direction of the curriculum reform. Since the early 2000s, we have

A.S.L. Wong (✉) • M.M.W. Cheng • V.W.Y. Yip
Faculty of Education, The University of Hong Kong, Pokfulam Road, Hong Kong SAR, China
e-mail: aslwong@hku.hk; mwcheng@hku.hk; valyip@hku.hk

D.F. Treagust and C.-Y. Tsui (eds.), *Multiple Representations in Biological Education*, Models and Modeling in Science Education 7, DOI 10.1007/978-94-007-4192-8_13,
© Springer Science+Business Media B.V. 2013

made use of science stories, such as the discovery of penicillin, the development of cowpox, Newton's proposition of law of universal gravitation, and the treatment of stomach ulcers (Tao, 2002) as a medium through which NOS could be introduced. However, due to the lack of both understanding of NOS and experience in learning and teaching NOS, many teachers who have not formerly participated in our teacher education program only saw these stories as a good means of arousing students' interest without having noticed the intended learning outcomes of NOS understandings. Such a situation was reflected in the comment made by a junior science teacher who had been telling the interesting science stories to his students. He came to realize his oversight of not having made good use of the stories for teaching NOS after he had attended our NOS in-service education workshop:

> I found the story on stomach ulcers very interesting.... Marshall tested his hypothesis by trialing out himself.... Students all enjoyed the story... I only realized now that there are deeper meanings behind the story and other important learning outcomes to be achieved through it and other stories.

There were further inadequacies of these relatively *old* stories. Teachers and students expressed the view that though these stories aroused their interests, they happened quite a while ago. Those who did not have the historical and cultural backgrounds of the scientific discoveries and inventions would fail to develop an in-depth understanding of, and hence appreciate, the thought processes of the scientists related to what they encountered at their time.

Bell, Blari, Crawford, and Lederman (2003) and Schwartz, Lederman, and Crawford (2004) investigated the effectiveness of promoting NOS understanding among high school students and preservice teachers by providing the authentic research experience of working with practicing scientists. The results showed that better understanding of NOS does not necessarily result from doing science per se (Bell et al., 2003). There also needs to be frequent opportunities for reflection on NOS in the context of that authentic scientific research experience through journal writing and seminars (Schwartz et al., 2004). In their comparison of the epistemic beliefs of chemistry students and research chemists, Samarapungavan, Westby, and Bodner (2006) came to the broadly similar conclusion that apprenticeship experiences are no automatic guarantee of epistemic development in students. They suggested that engaging students and expert researchers in conversation and reflection on epistemic issues related to research work will have a greater chance of success. While apprenticeship or internship experiences offer enormous potential for enhancing NOS understanding, they created major logistic problems—especially in East Asian classrooms, where class size is routinely about 40. Our response has been to present students with insights into authentic scientific practice through the development of a case study of contemporary scientific practice that shows the importance of developing models to understand the phenomena—a procedure essential in the understanding of NOS.

In the summer of 2003 when the crisis due to the severe acute respiratory syndrome (SARS) in Hong Kong was coming to an end, we saw a golden opportunity to turn the crisis into a set of instructional resources which aimed to address the issues raised above.

The SARS incident was a unique experience through which everyone in Hong Kong had lived and the memories of which would stay for years to come. At the beginning of the outbreak, the causative agent was not known, the pattern of spread was not identified, mortality was soaring, yet an effective treatment regimen was uncertain. It attracted the attention of the whole world as scientists worked indefatigably to understand the biology of the disease, develop new diagnostic tests, and design new treatments. Extensive media coverage kept people up to date on the latest development of scientific knowledge generated from the scientific inquiry about the disease.

The details on how we made use of the news reports and documentaries on SARS, together with episodes from the scientists' interviews—to develop a set of instructional materials and to explicitly teach a wide range of prominent features of NOS identified in the authentic scientific research—can be found in Wong, Hodson, Kwan, and Yung (2008). Since January 2005, we have been using the SARS story in developing NOS understanding among hundreds of preservice and in-service science teachers. The contextual approach which situated the learning of NOS in the authentic scientific research during the SARS epidemic was found to be particularly successful in promoting teachers' understanding of NOS in terms of (1) the realization of inseparable links between science and the social, cultural, and political environment; (2) deeper understanding of how science and technology impact on each other; and (3) a richer appreciation of the processes of authentic scientific inquiry and the humanistic character of scientists (Wong, Kwan, Hodson, & Yung, 2009). We have also recently reported some exemplary classroom practice translating teachers' own effective learning involving these NOS aspects in their science lessons (Wong, Wan, & Cheng, 2011). However, we are mindful not to be complacent with the learning of the three areas of NOS aspects as we noted less sophisticated discussion on the role and nature of scientific models and modeling in classroom practice (Cheng, Wong, & Yung, 2007), let alone the discussion of multiple levels of representations in science.

In this chapter, our discussion is centered on the role of models and the prominent activity of model building as shown in authentic scientific inquiries during the SARS epidemic. By following more subsequent research findings and reports related to SARS, we have recently enriched our teacher education materials to strengthen the discussion of these important aspects of NOS. We then compare and contrast the nature of models and modeling as reflected in authentic scientific practice and that appear in the Biology Curriculum and Assessment Guide (CDC-HKEAA, 2007). In Hong Kong, biology is arguably the science subject that has placed greatest attention to NOS. Many local biology teachers also believe that understanding of NOS will be assessed in the public examination for the reformed curriculum (Kwan, 2011).

Models and Modeling in Research on SARS

This section elaborates the series of events that occurred in four key scientific inquiries during the SARS epidemic, namely, (1) the identification of the transmission mode of SARS and (2) the hunt for the causative agent of SARS, (3) the search for the natural host of the SARS-related coronavirus, and (4) the modeling of the mysterious transmission rate and infection pattern in the tragic outbreak at Amoy Gardens. For each scientific inquiry, we highlight the important roles and characteristics of models, modeling, and the multiple levels of representations of science.

New Infectious Disease: Identification of Transmission Mode of SARS

The first scientific inquiry during the beginning of the SARS crisis was prompted by the urgent societal demand for the understanding of the transmission mode of SARS.

Starting from November 2002, there had been rumors that a mysterious atypical respiratory illness (later known as SARS) had occurred in Guangzhou, southern China. Around mid-February 2003, Dr. Liu from Guangzhou, who was infected with SARS virus but not knowing its morbidity, visited Hong Kong. He stayed at the Metropole Hotel, from where SARS started to creep into the Hong Kong community. A number of residents living in the hotel were infected. Most of them left Hong Kong by air to other countries and quickly spread the disease to the rest of the world without anyone or even the World Health Organization (WHO) noticing. By 4 March 2003, a Hong Kong young man visited a friend staying on the same floor at the Metropole Hotel as Dr. Liu got infected. He was admitted to the Prince of Wales Hospital. He then became the index patient of the outbreak in this hospital where over 100 medical doctors and other healthcare workers got infected within days and the following weeks.

By 15 March 2003, WHO was then aware of the severity of the disease and formally named the disease as SARS. A list of symptoms and a set of preventive measures and guidelines were disseminated to hospitals all over the world. It started with symptoms including high fever, headache, and dry cough. Most cases developed into pneumonia. Cases with serious buildup of fluid and inflammation of the lungs were admitted into hospitals.

The Hong Kong government started to disseminate guidelines and advice related to the likely transmission means of SARS disease through various media. A number of preventive measures were quickly put in place in Hong Kong, including cleaning of lift buttons every 2 h with diluted bleach and taking body temperature before going to school and work.

Models/Modeling/Multiple Levels of Representations Related to Identification of Transmission of SARS

As SARS spread in Hong Kong and in different parts of the world, epidemiologists (who study transmission and control of diseases) would have to review and screen a

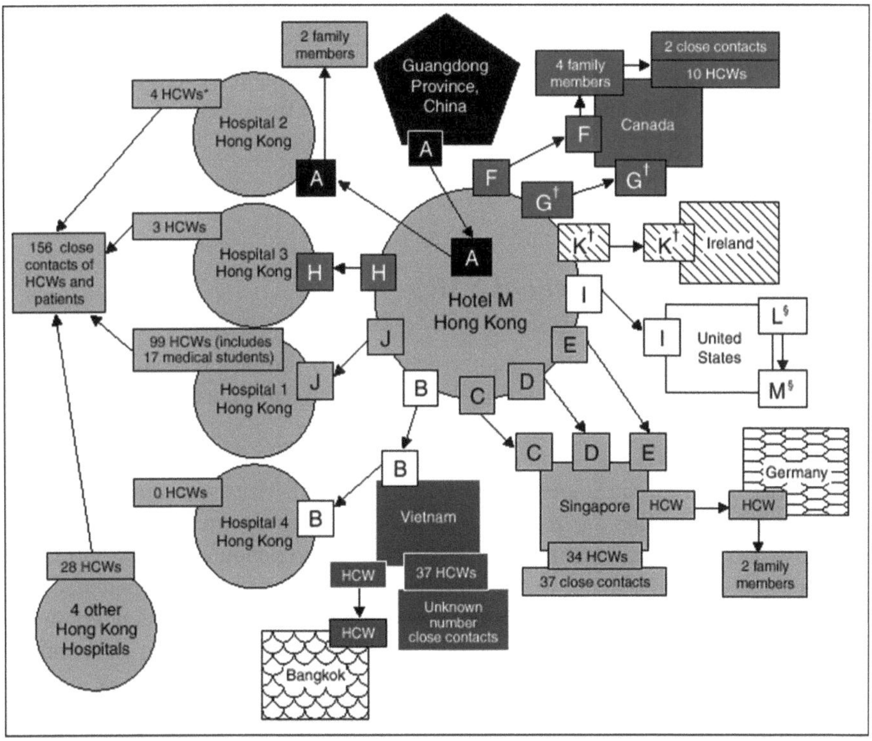

* Health-care workers.
† All guests except G and K stayed on the 9th floor of the hotel. Guest G stayed on the 14th floor, and Guest K stayed on the 11th floor.
§ Guests L and M (spouses) were not at Hotel M during the same time as index Guest A but were at the hotel during the same times as Guests G, H, and I, who were ill during this period.

Fig. 13.1 Chains of transmission showing how SARS spread from the Metropole Hotel to other parts of Hong Kong and the world as of late March in 2003 (Reproduced with permission from Centers for Disease Control and Prevention, Altanta 2003, p. 243)

massive amount of information, for example, the profile of SARS patients, clusters where people got infected, and infectivity of the disease. They could then model the chain of infection, patterns of spread, and the speed of transmission and hence pinpointing Dr. Liu and Metropole Hotel as the origin of the initial cases. As shown in Fig. 13.1, such a systematically organized diagrammatic representation of the pattern and sequence of infection at the macro level[1] (Fig. 13.1) led to their further proposal of a model at the micro level which suggested that the key means of transmission of SARS disease was through close contact with respiratory droplets containing the SARS viruses (e.g., through aerosols from coughing leaving viruses on public facilities or surroundings like lift buttons, door handles).

[1] We adopt the labels proposed by Gilbert and Treagust (2009) for different levels of representations (i.e., the macro, the submicro, and the symbolic).

With the most probable transmission mode and spreading rate of SARS virus (at the micro level), a set of preventive measures for the public and healthcare personnel working in the hospitals could then be recommended. The set of preventive measures stand as important products which result from the predictive power of a scientific model (in this case, the mode of transmission of SARS) deduced by modeling of the available data.

Hunt for Causative Agent of SARS

The second important scientific inquiry was the hunt for the causative agent. With the knowledge of the causative agent, diagnosis of SARS and a possible cure could be found.

On 18 March 2003, the virologists at the Chinese University of Hong Kong first announced that they had found evidence that the SARS virus was a member of the paramyxovirus family, a human metapneumovirus. Immediately afterward, scientists from Germany, Singapore, and Canada also announced they had found evidence of paramyxovirus in the samples collected from SARS patients. The announcement by the first research group and the immediate subsequent confirmation by the other laboratories came as exciting news for the world as it gave hope of prompt actions to cure the disease.

In less than 3 days' time, the University of Hong Kong found evidence suggesting that coronavirus is the primary cause of SARS. A team of microbiologists had isolated the virus from a SARS patient. The halo of dots surrounding the virus observed through an electron microscope was strongly suggestive of coronavirus, and it was further confirmed by the genetic analysis that showed fragments of genetic materials that was distinctive to the coronavirus family. After their announcement, scientists from Rotterdam and CDC in Atlanta also quickly announced that they had also found evidence in favor of coronavirus as the causative agent of SARS.

Subsequent stronger evidence was further provided by scientists in Netherlands showing that the SARS coronavirus fulfilled Koch's postulates by the experiments in which monkeys infected with the virus developed the same symptoms as human SARS victims. On 12 April 2003, the first genomic sequence of the SARS coronavirus was mapped just 20 days after its discovery. Never in the history of science had the genome of a new disease-causing agent been sequenced in such a short time.

Models/Modeling/Multiple Levels of Representations Related to Hunt for Causative Agent of SARS

Scientists had a tendency to accept newly proposed models or explanations if they tied in with their expectation. It was the case in the identification of the causative agent for SARS when they could also find the same type of virus in most samples collected from the SARS patients.

It is commonplace that scientists' observations are influenced by their knowledge and the theoretical framework they employ, that is, their observations may be affected by what they expect to see based on some initial scientific models in their mind. Coronavirus is well known to cause mild common cold, and hence many scientists did not make any linkage to it as the causative agent of SARS, not until it was later believed to be mutated into a more severe form of pneumonia. Development of scientific models is a prominent and important activity in the scientific community in the pursuit of understanding and appreciation of the neatness and beauty of the natural phenomena. Scientific models with stronger supportive evidence and greater explanatory and predictive power possess higher status. The experiment based on Koch's postulates made coronavirus the more likely candidate as the causative agent.

As biology has advanced into the molecular regime, a disease can now be understood comprehensively at different levels of representations. Using the case of SARS as an example, it could be understood at the macro level in terms of the symptoms expressed by the host of the disease. Indeed, the submicro level of representation of a disease only became available after the invention of microscope. The form of representation also evolved from hand-drawn figures in the past to the current high-resolution digital photos. At the submicro level, the SARS coronavirus would show a typical crown-like halo of spikes on the outer shell of the virus under an electron microscope. At the molecular level, the whole genome can now be obtained and often expressed in symbolic representation.

Search for Natural Host of SARS-Like Coronaviruses

Finding the natural reservoir of SARS-like coronaviruses is important for preventing and controlling future outbreaks of SARS. This search had begun ever since the human SARS coronavirus was identified as the causative agent of SARS in mid-April 2003. Such a search was performed by genetic analysis of the viral samples collected from SARS patients and other possible hosts.

Evidence showed that the early SARS patients in southern China were mostly chefs and restaurant workers who handle wild animals and serve exotic food like civet cats. It prompted researchers—from the University of Hong Kong and Shenzhen Center for Disease Control and Prevention—to collect samples from animals and animal traders for testing if there were SARS-like coronaviruses. In May 2003, the researchers found that civet cats carried a coronavirus that was 99.8% genetically identical to the human SARS coronavirus. They also found some animal traders who were involved in slaughtering the animals had antibodies of the virus carried by civet cats. These data indicated that the virus was passing between animals and humans. Yet without concrete data, they consciously commented that civet cats might have been infected from yet another unknown animal source.

Nevertheless, as this was a concern for global health, the researchers immediately reported the data to the officers in Guangzhou. The provincial authorities across China banned the sale of civet cats and wild animal species and tightened up regulations on animal trade from late May 2003. This directly affected the livelihood of those who sold wild animals. The findings had also impacted on the habit of eating wild animals in China. People are now less keen on eating wild animals as before. These events represent an example where science impacts on political decision, social, and cultural practices. Subsequent studies reported in 2004 and early 2005 revealed no widespread infection in wild or farmed civets. Such findings indicated that civet cats were more likely an intermediate host which got infected in the markets where they were caged in close proximity with other animals carrying the virus.

A research team then turned their attention to bats which had been found to be reservoir hosts of several types of viruses. The increasing presence of bats and bat products in food and traditional medicine markets in southern China and elsewhere in Asia was also a contextual factor which pointed to bats as their next target (Li et al., 2005). They did found SARS-like coronaviruses when they started surveying different species of bats in the search for the natural reservoir of SARS coronavirus. They then generated the genomic sequences of the viral samples from bats carrying the SARS-like viruses, infected civet cats, and infected human beings to study the evolutionary history of the SARS-like coronavirus. In October 2005, the scientists reported the phylogenetic trees (or evolutionary trees) constructed after a series of tedious comparison of the genomic patterns and logical reasoning in the deduction of the order of changes that happened to the genomic sequence of the strains during the evolution of the viral strains. Figure 13.2 shows a simplified form (for ease of illustration) of a typical phylogenetic tree, essentially scientific model, constructed by the scientists. The tree indicated bats as the likely natural host of the SARS-like coronaviruses. This finding would be welcomed by people who would benefit from the understanding and knowledge of the natural host and the interaction between the host, intermediate host, and human beings. Prevention of future outbreak thus becomes more feasible. Of course employers and employees of the restaurants serving exotic animals would not be too pleased to know that the ban of sale of civet cats (political decision) which had affected their livelihood was based on less than robust scientific evidence.

Models/Modeling/Multiple Levels of Representations Related to Search for Natural Host of the SARS-Like Coronavirus

Similar to the hunt of the causative agent of SARS coronavirus, it is essential that the modeling and models (evolutionary trees) generated in the identification of the natural reservoir should be consistent with the observations and guided by logical reasoning governed by some known criteria, for example, the variation of the genomic sequence from one stage to the next will not be drastically different from the previous one to the

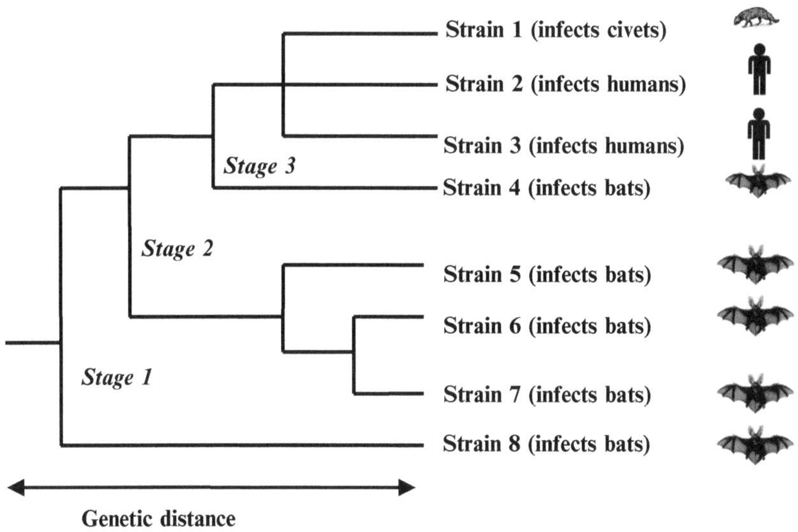

Fig. 13.2 A phylogenetic tree showing the evolutionary pattern of the bat SARS-like coronavirus mutated into human SARS coronavirus strains 2 and 3

next sequence. In the search of the natural host of the SARS-like coronaviruses, phylogenetic trees or evolutionary diagrams should illustrate the proximity of the viruses from bats, civets, and humans in terms of the genetic distance (see Fig. 13.2). As shown in the left of the phylogenetic tree, all these viruses share a common ancestral strain found in bats. Given the time for mutation, two strains were formed (stage 1[2]). One strain shares similar genetic composition as the ancestor (strain 8), whereas the other evolved (stage 2) to form several new species that only infect bats (strains 5–7), and another evolved to form viruses infects bats, civets, and humans (strains 1–4). The reason why the coronavirus found in civets (strain 1) are so similar to the human strains (strains 2 and 3) is that they share the same recent ancestor evolved in stage 3. Moreover, viral strain 4 (infecting bats) and strains 1–3 are placed at the same level in the diagram, indicating the high possibility of the bat virus spreading to humans either by direct contact or through civets (or the animals) sold in the markets.

A phylogenetic tree, essentially a scientific model, is constructed based on careful comparison of genomic sequence of each viral strain, that is, thorough understanding of the molecular level (or submicro level) of the different viral strains obtained from the genomic sequence of each viral strain. Scientists frequently have to communicate through representations of a part of or the whole genome to develop the evolutionary diagram which is at the symbolic level. In other words, modeling and models in constructing a phylogenetic tree also require knowing how to represent and communicate the models at the symbolic level.

[2] The stages written here are for illustration purpose. There should not be any labels such as *stages* in a typical phylogenetic tree. Similarly, the arrow for genetic distance is optional in this diagram.

Tragic Outbreak at Amoy Gardens

Amoy Gardens, a residential complex comprising 19 blocks, was found to have an alarming number of cases of infection. New infected cases rose from 7 to 185 within 4 days. By 31 March 2003, most of the new cases in Hong Kong were from Amoy Gardens and most of the cases from Amoy Gardens were from Block E. Among them, most were residents from Flat 7 and Flat 8 (see Fig. 13.3).

On 31 March 2003, the Department of Health imposed quarantine on Block E of Amoy Gardens—an unprecedented order from the Hong Kong government to move all residents of Block E to isolation camps. This quarantine allowed a thorough investigation (by a cross-disciplinary investigation team consisting of epidemiologists, engineers, virologists, and other experts) to find clues to the causes of the devastating outbreak and the puzzling infection pattern.

The quarantine order sped up the scientific inquiry by the investigation team for the possible causes of the infection happened in Amoy Gardens, especially the widespread infection in Block E. From early to mid-April, scientists obtained the following crucial findings through epidemiological and environmental investigations as documented in their report to the government on 17 April 2003 (Department of Health, Government of Hong Kong Special Administration Region, 2003):

- The index patient (first case of the Amoy Gardens outbreak) visited his relatives on 14 and 19 March 2003 in a flat of Block E in mid-March around the time he developed SARS. He was having diarrhea at that time and he used the toilet there.
- Scientists collected every possible type of sample including the air, the water stored in the tank for the use of residents of Block E, the sewage system, as well as cats, dogs, rats, and cockroaches in and around Block E. They quickly identified the presence of SARS coronavirus in rats, cockroaches around the residential area, and the sewage from the drainage system of the building.
- Many Amoy Gardens residents reported foul smell in their bathrooms which suggested that the U-shaped water trap (U-traps) of the floor drainage system might not be filled with water to perform the proper function of preventing foul smell and insects from entering the bathrooms. (As the toilets, the basins, and the bathtubs were frequently used, their U-traps should be charged with water and should have been functioning properly. However, most households had the habit of cleaning the bathroom floor by mopping instead of flushing it with water, and the U-traps of the floor drains were likely to be dry and not functioning properly.)
- The pattern of vertical spread of the infection (most of the infected residents lived in Flat 7 and Flat 8, with more cases on higher floors) suggested a close connection to the sewage system that is connected to the same drainage system.
- However, the drainage system alone could not explain the high infection rate of the higher floors. It was then postulated that habit of the use of exhaust fan during the use of the bathroom might explain such a pattern.
- [Later epidemiological data revealed that "those who had used their exhaust fans while taking a shower had a five times greater chance of getting SARS" (Abraham, 2004, p. 75).]

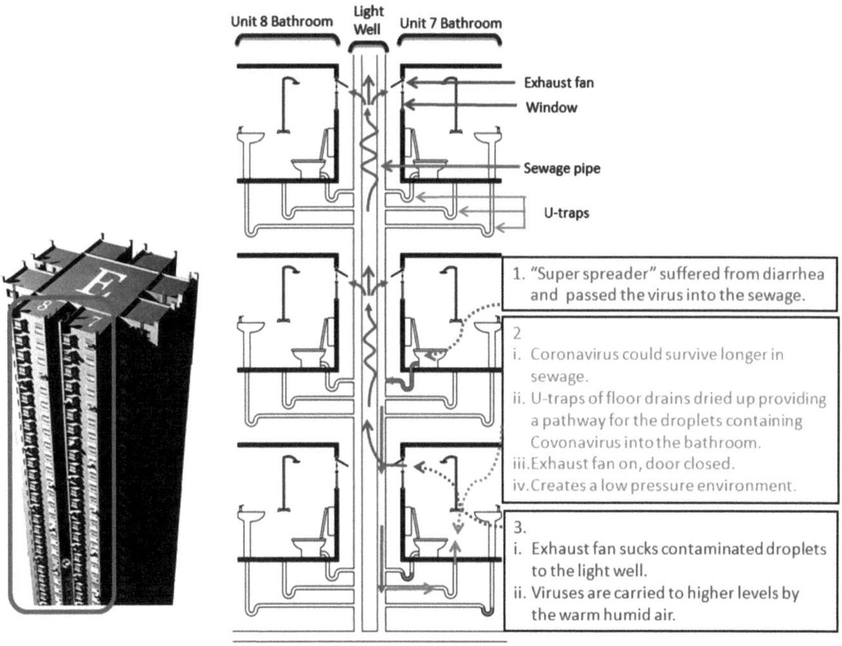

Fig. 13.3 Diagrammatic representation of the model which explains the peculiar infection pattern and its fast transmission rate

The newly constructed model was widely reported in different media in Hong Kong and quickly drew people's attention to an overlooked hygienic measure of proper use of U-traps in the bathrooms.

WHO initially had reservation about the proposed transmission model as the explanation of the infection rate and pattern. The model proposed by the Hong Kong scientists was only accepted by the scientists of the WHO after they had conducted an independent investigation in Amoy Gardens during the visit to Hong Kong in late April. Better understanding of the building structure, the drainage systems, and the overpacked conditions of the neighboring flats enabled them to appreciate the investigation and conclusions by the Hong Kong investigation team.

Models/Modeling/Multiple Levels of Representations Related to Tragic Outbreak in Amoy Gardens

Development of scientific models is sometimes prompted by an urgent demand by society in tackling societal and global problems instead of just being driven by curiosity or competition among scientists in their understanding certain aspects of the natural world. Such a demand together with unreserved funding and resources from the government could hugely speed up the whole process.

The diagrammatic representation of the model proposed by the cross-disciplinary investigation team is given in Fig. 13.3. It depicts how the unfortunate outbreak in Amoy Gardens occurred due to a combination of a series of rare events: (1) The index patient who turned out to be a superspreader had introduced considerable amount of SARS coronavirus into the sewage drainage system which was shared by residents of Flat 7 and Flat 8. (2) The U-traps of the floor drains in the bathrooms of some flats were dried up and opened a pathway for small droplets containing coronavirus into the bathrooms. (3) The exhaust fan which was too powerful for a small bathroom typical in Hong Kong then sucked the contaminated droplets to the light well.[3] (4) The viruses were carried to floors higher up by the warm humid air (the so-called chimney effect) through open windows.

Modeling is a typical process in scientific research that offers an explanation to observations. It is noteworthy that parts of the models could be grounded on available empirical data but some could only be derived based on logical deduction when the availability or accessibility of the data is limited.

Although *science in the making* is more likely to be subject to changes, the government decided the risk of slow action was not affordable. Even if the model might not be fully correct, preventive actions based on the newly constructed model should cause no harm. It is not uncommon to rely on the most recent scientific models to make decision. It is always a balance of pros and cons of the different consequences that would be incurred if a model is adopted or not and if it is valid or not.

Good understanding of the contextual and environmental conditions as well as the social practice is crucial in the identification of the bits and pieces for formulating the explanatory model. The combined effort from the local team and the WHO team could capitalize on the best of both teams in terms of their developmental and confirmatory roles in the proposed model. The Hong Kong team's familiarity of local practices (e.g., habit of cleaning floor by mopping instead of flushing it with water might result in dried-up U-traps) and the evaluation of the model by the WHO team as an independent reviewer gave weight to the proposed model.

Prominent Roles and Functions of Models and Modeling in Authentic Research

From the above four episodes, we could identify some prominent roles or functions served by models. First, some models serve to organize some complex data in a systematic manner so that patterns, trends, or relationships could be more easily

[3] A light well is an architectural design of an open area or vertical shaft in a building bringing natural light to the lower floors. In Hong Kong, the long and vertical space between neighboring flats from the bottom to the top floors is typically referred to as a light well. The long and narrow space (almost like a long chimney) between Flats 7 and 8 of Block E (see Fig. 13.3) is a typical example in a crowded residential complex.

identified, for example, the diagrammatic representation in Fig. 13.1. Second, some models are constructed for the explanation of some observations, for example, the mysterious outbreak in Amoy Gardens. Oftentimes, such models are explaining certain patterns, trends, or relationships (first types of models). Third, due to the predictive power of a model, it can often be subsequently applied, for example, the preventive measures in reducing the spreading of SARS are a product resulting from the transmission model of SARS.

It is also worth noting that many different levels of representations were evoked in the scientific research and the dissemination of findings. The phylogenetic tree and the diagram representing the transmission of the disease are symbolic. Yet the phylogenetic tree represents the evolution of SARS-like coronavirus which is at the submicro level; and the diagram representing disease transmission represents phenomena that are at the macro level. In the virus hunt, electron micrographs (at the submicro level) were useful for the identification of the likely causative virus (which needs further experiment based on Koch's postulates for confirmation). The fuller details of the coronavirus were obtained through genetic analysis (at the submicro level or more specifically the molecular level). The above examples may seem to suggest that contemporary scientific research does not make use of representations that are iconic. However, the pattern of spread in Amoy Gardens reveals that iconic diagrams at the macro level did play an essential role in modeling the way in which the disease was spread in the building. In different contexts and in fulfilling different needs, models at different levels of representations were evoked. In other words, scientific modeling is a purposeful activity and is conducted to fulfill the needs of particular contexts.

Modeling is unavoidably guided by the prior knowledge of and the preconceptual framework adopted by the researcher; for example, certain symptoms are more likely associated with certain family of viruses. In a way, while expert knowledge could help swiftly eliminate many possible unlikely causes through logic or evidence (e.g., sewage system rather than rodents is a more likely culprit for the infection pattern in Amoy Gardens), it would also inevitably lead to the possibility of missing a target or a breakthrough.

Construction of models could be prompted by curiosity or in fact is more often driven by social expectations and demands. When there is more than one model, the one with the greater descriptive/predictive/explanatory power will normally prevail.

Models and Modeling Represented in Curriculum Guide

To compare and contrast the models and modeling as represented in our newly reformed curriculum to those reflected in the authentic research as illustrated in several episodes of scientific inquiries in SARS, we conducted a simple content analysis of the Hong Kong Biology Curriculum and Assessment Guides (CDC-HKEAA, 2007). We found 25 places where the document mentioned model(s)/modeling, of which there was one instance that *model* referred to the behavior of teachers from whom their students should learn (p. 77). Such a use was not deemed relevant to our analysis. For the other 24 relevant places, we observed that there were three distinctive ways in which the idea of model/modeling was made use of in the document as described in the following sections. We then put forward our suggestion on how NOS could be better included in the new curriculum.

Models as Physical Artifacts (n = 12 places)

Models are regarded as physical artifacts through which structures are observed (CDC-HKEAA, 2007). As teaching and learning activities, students were expected to "construct models of DNA and RNA" (p. 27), "examine models of the human brain, eye, ear and arm" (p. 36), and "examine prepared slides or models to identify features of mammalian skin that are related to body defence" (p. 42). That is, models are regarded as aids through which students learn target scientific ideas. In this connection, they are *teaching models* based on Gilbert's (2005) taxonomy of models.

The use of physical models extended from teaching and learning to school-based assessment (CDC-HKEAA, 2007). According to the document, teachers were recommended to use "a variety of assignment tasks—such as exercises, essays, designing posters or leaflets, and model construction. . .to allow students to demonstrate their understanding and creative ideas" (p. 97).

Models as physical artifacts were represented not only as the most frequently used models among the other models but were also most widely made use of in the classrooms. The curriculum was meant to be taught from grade 10 to grade 12 and was allocated 270 h for a complete coverage (p. 14). Teachers were expected to spend 200 h to cover the compulsory part that was composed of four topics, namely, *cell and molecules of life*, *genetics and evolution*, *organism and environment*, and *health and diseases*. It was observed that examining or building physical models were suggested for all of these topics. For example, they included building fluid mosaic model (p. 22) and DNA model (p. 27), and examining mammalian skin model (p. 42) and kidney model (p. 46).

Models as Virtual Artifacts (n = 6 places)

Models were regarded as computer simulations through which physical phenomena could be represented, tested, and manipulated (CDC-HKEAA, 2007). In this regard, models were introduced in the context of using the Internet and technology in facilitating students' learning. The document suggested that "modeling software, which allows students to test their proposed models through virtual experiments, is useful in helping students to develop conceptual understanding. . ." (p. 108).

Virtual experiments could have been applicable in different biological topics. Throughout the curriculum guide, however, there was only one single suggestion that such strategies were made use of. The *Suggested Learning and Teaching*

Activities section of the topic *Genetic and Evolution* listed "Use computer simulations or other simulations to model natural selection" (p. 27). In another part of the document which advised teacher how to use information technology to exercise interactive teaching, the same statement was reiterated (p. 85). In short, although there were six suggestions when models could be used as virtual artifacts, their proposed use was very limited.

Models as Exemplar Phenomena (n = 1 place)

Given the complexity of the physical world and biological organisms, scientists have to idealize or simplify the phenomena to be studied. For example, the complexity of inheritance of characteristics of organisms was simplified to the study of single features (hence Mendelian genetics). In school science, typically, it is the study of the flower color of the parent plants and their offspring. Atkins (2003) and Gilbert (2005), respectively, called such simplified but representative phenomena as "core phenomena" (p. 2) and "exemplar phenomena" (p. 10) regarding the study of the core/exemplar phenomena as a key part of scientific activities and scientific method.

Materials to be included in a curriculum inevitably would have to be selective. Given the myriad number of animals, their characteristics, and how they survive in their environment, human beings were chosen as the exemplar phenomena. The preamble to the topic *Organism and Environment* in CDC-HKEAA (2007) stated that "[s]tudents will study reproduction, growth and development to understand how organisms perpetuate and proliferate in the environment. The human being is used as a model for students to understand the essential life processes of animals" (p. 28).

In this topic, among other biological concepts, students were expected to study life processes of animals, which included nutrition, gaseous exchange, growth and reproduction, nervous and hormonal coordination, movement, and homeostasis. Based on these contents to be covered, it is unlikely that the selection of human beings as a model/exemplar phenomenon reflected the use of a simplified phenomenon (for human beings are highly complex and sophisticated). Other than the statement quoted above, the document did not elaborate on how human beings could be a model of animals, which include insects, fishes, birds, and so on. It is likely that the selection was (justifiably) based on the familiarity and relevance of human beings to students.

Models as Processes/Outcomes of the Scientific Enterprise (n = 5 places)

Models were regarded as activities and as products of the scientific community (CDC-HKEAA, 2007). It could be found under the "curriculum emphases" section of the document that "[the curriculum] should enable students to...formulate and revise scientific explanations and models using logic and evidence" (p. 12). Elsewhere, a statement that "students should be able to... appreciate the uses and limitations of scientific models" (p. 19) was indicated.

Based on these two statements, it might seem that the curriculum adopted a model-based approach to science teaching and learning. Nevertheless, a scientific model was addressed in only one model: "Use the fluid mosaic model to explain the properties and functions of cell membrane" (p. 19). Also students were expected to "be aware that biological knowledge and theories are developed through observations, hypotheses, experimentations and analyses (e.g., fluid mosaic model of cell membrane structure)" (p. 18).

Without going to the detailed discussion on whether *logic and evidence* were the key to the formulation to a model, or scientific models were developed through *observations*, *hypotheses*, *experimentations*, and *analyses*, we argue that the inclusion of only a scientific model in the content specification was inconsistent with the overarching curriculum emphasis. This situation leaves a big challenge for teachers if they are to achieve the curriculum aims based on a single scientific model.

Summary

We observed that the curriculum made different uses of the idea of *models*. In general, models—be they physical or virtual—were taken to be teaching and learning aids through which the students would learn target biological concepts. Such a use of models was evident in the frequency of their use across different biological topics. Models as exemplar phenomena were also referred to in the curriculum. As argued above, due to the complexity of human beings, it is debatable whether humans can be regarded as a model of other animals.

The curriculum document (CDC-HKEAA, 2007) was written in line with the advocacy of the science education literature. *Nature and history of biology* was taken to be one of the three curriculum emphases. We have argued elsewhere that the focus on NOS might be more of paying lip service than of having substantial commitment (Wong, Yung, & Cheng, 2010). An issue which further confounds the problem was that scientific models/modeling as outcomes and processes of the scientific enterprise were underrepresented and were used interchangeably with teaching and learning aids.

We argue that the use of models as physical and virtual artifacts has been a daily practice of biology teachers; how they are to be used have been widely discussed in the existing literature. In contrast, how teachers could help students learn the notion of *modeling and models*—as a process and outcomes of scientific activities (rather than simply as artifacts or *copies of reality*) while covering curriculum contents that would be assessed in high-stake public examination—remains a challenge for biology teachers and science educational researchers. However, we believe that the in-depth analysis of the four scientific inquiries during the SARS crisis has provided convincing evidence that modeling and scientific models could be vividly illustrated by appropriate episodes in the authentic scientific inquiries in SARS as other aspects of NOS reported earlier (Wong et al., 2009). If a similar effort is put into developing instructional materials with a focus on promoting the understanding of modeling/models/multiple levels of representations as we did for the other aspects of NOS (Wong et al., 2008), we anticipate that similar favorable learning outcomes could be achieved.

Concluding Remarks

This chapter discusses how scientific models at different levels, namely, the macro, the submicro, and the symbolic, were developed in the SARS-related scientific research. The discussion serves to exemplify the roles of scientific models and modeling in knowledge construction and representations in biological sciences. In the authentic practice, scientific models are developed to describe, explain, and predict physical phenomena. In solving different problems, scientific models of different levels are made use of in ways that were fit for their purposes.

Compared with the authentic scientific practice, we argue, based on our observation of a school biology curriculum, that the scope of *models* advocated in school biology was rather limited. Also, the roles of models as teaching tools and as outcomes/process of scientific research are not differentiated. There might be an issue that such an unspecific use of models would affect the quality of teaching and learning of school science/biology. We suggest that a way forward for research is to further investigate how biology teachers handle *scientific models* (rather than merely *teaching models*) in their classrooms and to study students' learning of *scientific models*. Meanwhile, as far as curriculum material development is concerned, we support the view of van Dijk (2011) that exemplars based on authentic and contemporary scientific practice could be further developed and the SARS crisis could again be turned into opportunity as an excellent local exemplar for a rich discussion and illustration of the nature and roles of scientific models and modeling as part of NOS.

References

Abd-El-Khalick, F., & Lederman, N. G. (2000). Improving science teachers' conceptions of the nature of science: A critical review of the literature. *International Journal of Science Education, 22*, 665–701.

Abraham, T. (2004). *Twenty-first century plague: The story of SARS*. Hong Kong: Hong Kong University Press.

American Association for the Advancement of Science. (1993). *Benchmarks for scientific literacy*. New York: Oxford University Press.

Atkins, P. (2003). *Galileo's finger: The ten great ideas of science*. Oxford: Oxford University Press.

Bell, R., Blari, L., Crawford, B., & Lederman, N. (2003). Just do it? Impact of a science apprenticeship program on high school students' understandings of the nature of science and scientific inquiry. *Journal of Research in Science Teaching, 40*(5), 487–509.
CDC-HKEAA. (2007). *Physics/chemistry/biology/integrated science curriculum guide and assessment guide (secondary 4–6)*. Hong Kong: Curriculum Development Council and Hong Kong Examinations and Assessment Authority.
Centers for Disease Control and Prevention, Atlanta. (2003). Update: Outbreak of Severe Acute Respiratory Syndrome – Worldwide, 2003. *Morbidity and Mortality Weekly Report, 52*(12), 241–248.
Cheng, M. M. W., Wong, S. L., & Yung, B. H. W. (2007, April). *Students' understanding of scientific models in different contexts: The impact of teaching on the nature of models*. Paper presented at the National Association for Research in Science Teaching (NARST) annual meeting. New Orleans, LA.
Council of Ministers of Education. (1997). *Common framework of science learning outcomes*. Toronto, ON, Canada: CMEC Secretariat.
Department of Health, Government of Hong Kong Special Administration Region. (2003). *Outbreak of severe acute respiratory syndrome (SARS) at Amoy Gardens, Kowloon Bay, Hong Kong: Main findings of the investigation*. Retrieved November 18, 2011, from http://www.info.gov.hk/info/ap/pdf/amoy_e.pdf
Gilbert, J. K. (2005). Visualization: A metacognitive skill in science and science education. In J. K. Gilbert (Ed.), *Visualization in science education* (pp. 9–27). Dordrecht, the Netherlands: Springer.
Gilbert, J. K., & Treagust, D. (2009). Macro, submicro and symbolic representations and the relationship between them. In J. K. Gilbert & D. Treagust (Eds.), *Multiple representations in chemical education* (pp. 1–8). Dordrecht, the Netherlands: Springer.
Khishfe, R., & Abd-El-Khalick, F. (2002). The influence of explicit/reflective versus implicit inquiry-oriented instruction on sixth graders' views of nature of science. *Journal of Research in Science Teaching, 39*, 551–578.
Kwan, J. (2011). *Interactive relationships among teachers' intentions, beliefs, pedagogical content knowledge and classroom instruction on the nature of science*. Unpublished PhD dissertation, The University of Hong Kong, Hong Kong.
Lederman, N. G. (1992). Students' and teachers' conceptions of the nature of science: A review of the research. *Journal of Research in Science Teaching, 29*, 331–359.
Li, W., Shi, Z., Yu, M., Ren, W., Smith, C., Epstein, J. H., et al. (2005). Bats are natural reservoirs of SARS-like coronaviruses. *Science, 310*(Science), 676–679.
Millar, R., & Osborne, J. (Eds.). (1998). *Beyond 2000: Science education for the future*. London: King's College.
Samarapungavan, A., Westby, E., & Bodner, G. (2006). Contextual epistemic development in science: A comparison of chemistry students and research chemists. *Science Education, 90*(3), 468–495.
Schwartz, R. S., Lederman, N. G., & Crawford, B. (2004). Developing views of nature of science in an authentic context: An explicit approach to bridging the gap between nature of science and scientific inquiry. *Science Education, 88*(4), 610–645.
Tao, P. K. (2002). A study of students' focal awareness when studying science stories designed for fostering understanding of the nature of science. *Research in Science Education, 32*, 97–120.
van Dijk, E. M. (2011). Portraying real science in science communication. *Science Education, 95* (6), 1086–1100.
Wong, S. L., Hodson, D., Kwan, J., & Yung, B. H. W. (2008). Turning crisis into opportunity: Enhancing student teachers' understanding of the nature of science and scientific inquiry through a case study of the scientific research in Severe Acute Respiratory Syndrome. *International Journal of Science Education, 30*, 1417–1439.

Wong, S. L., Kwan, J., Hodson, D., & Yung, B. H. W. (2009). Turning crisis into opportunity: Nature of science and scientific inquiry as illustrated in the scientific research on Severe Acute Respiratory Syndrome. *Science Education, 18*, 95–118.

Wong, S. L., Yung, B. H. W., & Cheng, M. W. (2010). A blow to a decade of effort on promoting teaching of nature of science. In Y.J. Lee (Ed.), *The world of science education: Handbook of research in science education research in Asia* (pp. 259–276). Rotterdam, the Netherlands: Sense.

Wong, S. L., Wan, Z., & Cheng, M. M. W. (2011). Learning nature of science through socio-scientific issues. In T. D. Sadler (Ed.), *Socio-scientific issues in the classroom: Teaching, learning and research* (pp. 245–269). Dordrecht, the Netherlands: Springer.

Chapter 14
Supporting and Assessing Complex Biology Learning with Computer-Based Simulations and Representations

Barbara C. Buckley and Edys S. Quellmalz

Introduction

Biology learning is, by its very nature, complex. Living organisms are composed of systems nested within systems, each of which has components that interact to produce the emergent behavior of that system and interact in the next larger system. The components of living systems can be as small as ions and can participate in systems as large as the biosphere of Earth. The mechanisms for these systems are dictated by the evolutionary pressures that have enabled an array of structures and behaviors to survive. This systems view of biology is very different from the fragmented, inert, and inaccurate knowledge that too often results from *taking biology* in school.

Numerous studies have documented the ways in which US science curricula are failing today's students. Studies repeatedly report that teachers are typically required to cover a daunting number of standards, often resulting in a focus on superficial recall with insufficient attention to deep understanding (Weiss & Pasley, 2004). Standard science instruction has been characterized as requiring that students read sections in a textbook, take notes on definitions of key terms, and take examinations that test recall, thereby leaving students without experiences in what it means to know and do science. Analyses of American science textbooks indicate that they cover too many topics, use difficult vocabulary, make few connections with students' background knowledge, and do not address commonly held misconceptions (Stern & Roseman, 2004). Textbooks also often do not coherently develop and relate concepts (Shymansky, Yore, & Good, 1991). Even in curricula with hands-on laboratories, students tend not to address authentic problems but to simply replicate standard experiments. These textbooks and

B.C. Buckley (✉) • E.S. Quellmalz
Science, Technology, Engineering and Mathematics, WestEd, 400 Seaport Court, Suite 222, Redwood City, CA 94063, USA
e-mail: bbuckle@wested.org; equellm@wested.org

D.F. Treagust and C.-Y. Tsui (eds.), *Multiple Representations in Biological Education*, Models and Modeling in Science Education 7, DOI 10.1007/978-94-007-4192-8_14,

© Springer Science+Business Media B.V. 2013

associated laboratory activities are critiqued as being limited to transmitting science rather than learning its practices (Duschl, Schweingruber, & Shouse, 2007).

For example, in high school biology textbooks, the major human body systems are typically presented in separate chapters and depicted by static images. While many publishers now include online resources such as animations to present concepts, these resources tend to be short, decontextualized fragments. As a result, many students do not understand the dynamic complexity of human body systems or how these systems work together to ensure sufficient energy and building materials to sustain life. In recent field tests of assessment items on human body systems designed for middle and high school levels, Project 2061 researchers at the American Association for the Advancement of Science (AAAS) found that many students are unfamiliar with how the multiple systems of the body are connected (AAAS, 2011). Other studies have documented that students have difficulty understanding the dynamic nature of biological systems as well as the structures that enable them to interact (cf. Buckley & Boulter, 2000; Feltovich, Spiro, & Coulson, 1988; Hmelo-Silver & Pfeffer, 2004; Patel, Kaufman, & Magder, 1991). To understand biology one cannot rely on a small set of laws for reasoning. Rather, the understanding must be grounded in the interacting structures that provide the mechanisms for biological functioning. Understanding how the different systems work together is critical for a solid understanding of how living organisms work. When this understanding is missing, students struggle in their learning. Without an understanding of the structures that enable the interactions of systems, even medical students are unable to reason about phenomena (Patel et al., 1991).

This chapter describes three projects designed to foster richly connected, extensible systems views among biology students. For each project I describe the representations employed, the nature of student interactions with those representations, and the lessons learned. I summarize across the projects to nominate several components for inclusion in a systems view of biology learning that can be used to guide both instruction and assessment.

Science for Living: The Circulatory System (1986–1992)

The *Science for Living* (SFL) project at Stanford University developed an interactive multimedia resource intended to help secondary students learn about the circulatory system using the then state-of-the-art technology. SFL employed two screens—one for the computer, another for the laser disc player. Although the specific technologies are obsolete, the conceptual and representational aspects of the project are still salient.

Representations in *Science for Living: the Circulatory System* included video of live phenomena, animations, and simulations as well as videos of experts who presented information on a variety of topics. SFL did not provide instruction or direction for the students. These representations were organized around an anatomical hierarchy. At the main level, information was available about the

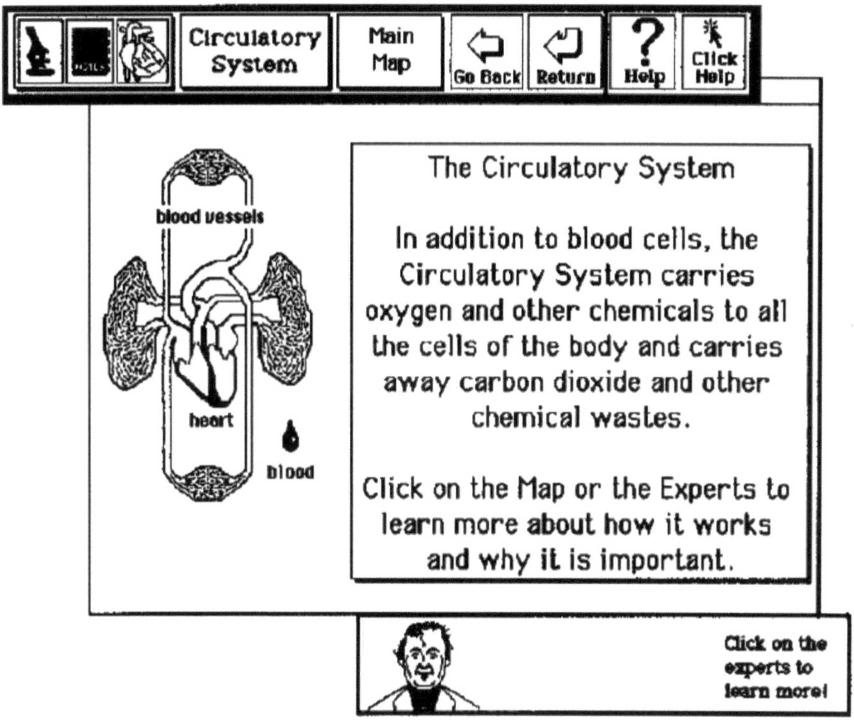

Fig. 14.1 Main interface screen for *Science for Living: the Circulatory System*

circulatory system as a whole. At a lower level, information was available about the heart, blood vessels, or blood. As shown in Fig. 14.1, the diagram of the circulatory system on the left side of the screen served as a map of and route to the different levels.

When a level such as the heart was selected (see Fig. 14.2), the heart was highlighted on the map and an expanded diagram of the heart was displayed while the first video for that level played on the video screen.

At each level the user could access multiple pieces of information that included video of live phenomena, photographs, slides, or drawings. Each video segment was accompanied by a text caption and playback was controlled by the user. Users could also access additional information about the image by clicking on a predefined portion of the image, which caused that portion of the image to be highlighted and a text box to be displayed indicating the name of the part and/or what was happening. The ability to highlight parts of the images and thus to see boundaries as well as names helped overcome some of the challenges presented by representations of natural phenomena (Goldsmith, 1984). SFL's experts were available on demand to provide lectures or demonstrations on topics related to the selected part of the circulatory system. Care was taken to ensure that the

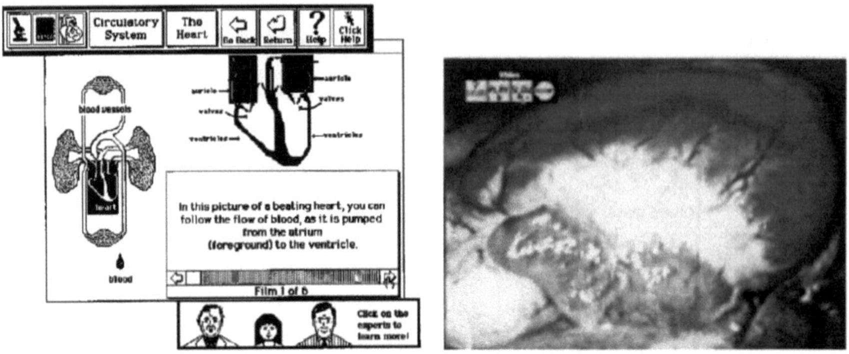

Fig. 14.2 Example of two screen views at middle level (the heart). Screen on the *left* is the interface for navigation and information; the screen on the *right* is the laser disc display

information and representations were tightly linked to the anatomy of the circulatory system and accurate terminology.

Users could, at any time, access the laboratory simulations or the NoteBook provided in SFL. The simulations included the Pump Lab for exploring and experimenting with valves in a single-chamber pump and the Life Lab for exploring and experimenting with risk factors associated with heart disease. The NoteBook allowed users to select video segments and create note cards linked to the segment. Note cards could then be annotated with text and/or graphics and sequenced for linear presentation. SFL generated a log that traced the user's path through the information indicating lapsed time and buttons clicked (for a more detailed analysis of the representations in SFL, see Buckley & Boulter, 2000).

Methods

The naturalistic, cognitive case study design employed a variety of data collection and analysis techniques used in case studies (Merriam, 1991) and cognitive psychology (Cronbach, 1985; Ericsson & Simon, 1993). It was conducted in a technology-rich classroom of 15–16-year-old students participating in the Apple Classroom of Tomorrow (ACOT) long-term research project. Both training time and novelty effects were minimized because the classroom had sufficient hardware to support the research and its teacher and students were experienced users of the technology and accustomed to the presence of researchers. The composition of the class mirrored the composition of the large urban Midwestern US secondary school in ethnicity and socioeconomic status.

Diverse data sources (see Fig. 14.3) were gathered to provide evidence of learning and learning activities, including 160 h of videotape that captured classroom activities, group work at the multimedia workstations, presentations, and

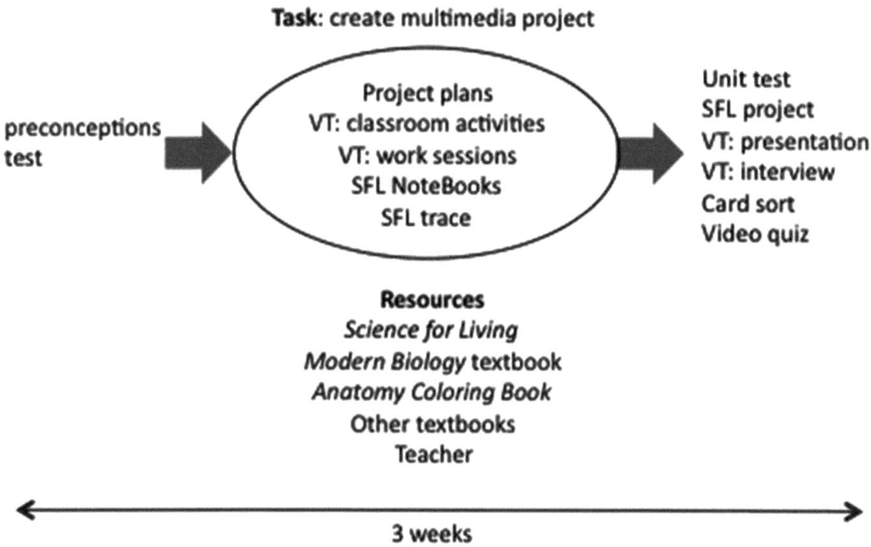

Fig. 14.3 Data collection during classroom use of *Science for Living* (*VT* = videotape data and *SFL* = digital files)

interviews. The preconceptions test, the unit test, project plans, projects, presentations, and the teacher's evaluations of the presentations were collected from the entire class with additional data collected from students identified by the teacher as average- and better-performing students.

Because the preconceptions test (Arnaudin & Mintzes, 1985) indicated that half of the students did not believe that the circulatory and digestive systems were connected, students were tasked with creating a multimedia presentation describing how food absorbed from the digestive system was transported to all the cells of the body. All but one student worked in groups of three or four. During semi-structured interviews, a representative sample of students was asked to perform a card sort of circulatory system components and to interpret an illustration of an experimental treatment to increase blood flow to heart muscle cells.

Data analysis involved reviewing and coding 160 h of videotape, computer logs, project plans, and other student-produced artifacts, semi-structured interviews, and pretest and posttests. In order to compare the ways in which students used SFL and how their understandings evolved over the 3-week period, I created numerous visual representations (Tufte, 1983) to capture both qualitative and quantitative aspects of the data. In the semi-structured interviews the key performance that discriminated between model builders and non-model builders was the ability to reason about a newspaper illustration that depicted an experimental treatment to increase blood flow to heart muscle cells after a heart attack. Just one student, Joanne, in the class of 28, was able to use her knowledge to describe and explain the normal functioning of the circulatory system, understand the illustration of the experimental alternative to heart bypass surgery, and reason beyond knowledge about which she felt confident. Other students, even very able students (as identified by the teacher), had to be reminded that heart muscle contracts, which was essential to understanding how this technique might work despite having viewed video of hearts beating in open chests.

In this case of model-based learning in situ, I described how Joanne planned to create a project that explained the structure, parts, and purpose of the circulatory system; how they work; and how the circulatory system works with the organs of the digestive system. In producing her project she sought information about the structure, function, behavior, and mechanisms of the circulatory system; integrated them into working mental models of the circulatory system; and used her evolving models in a variety of learning tasks to reason about the information she encountered in textbooks, in SFL, and in discussions with the teacher. Representations supported the formation and revision of her model of the heart by providing representations of not only the structures and functions of the circulatory system but also of the behaviors of the heart, blood vessels, and blood with close links between images and parallel text that further described behavior and causal mechanisms. Joanne emerged at the end of the study with integrated and useful knowledge of the circulatory system that included its structure, dynamic behavior, and some of the mechanisms producing that behavior (Buckley, 2000).

While access to the representations in SFL may have been an important enabling mechanism for Joanne's learning, it was not sufficient to ensure model building by the other students in the classroom. The list below contrasts the learning activities of Joanne and another student and highlights several other factors:

- Working as an individual to accumulate all the information needed versus working in a group that sought the information in a piecemeal way
- Interpreting the task as explaining how the organs work together versus describing the flow of food and blood without explanation of how that happens
- Engaging with the content of SFL versus using SFL as a presentation tool and source of illustrations
- Teacher encouragement to *find out for yourself* versus mini-lectures using analogies

What emerged from this study was a theory of model-based learning (MBL) grounded in rich descriptions (Buckley, 2000; Gobert & Buckley, 2000). Beginning from the premise that understanding requires the development of mental models (Johnson-Laird, 1983), model-based learning refers to the formation and subsequent development of mental models by a learner. Most often used in the context of dynamic phenomena, mental models organize information about how dynamic phenomena emerge from the interactions of its component parts. Mental models arise from the demands of some task that requires integration of multiple aspects and/or multiple levels of a system or situation (see Fig. 14.4). Model formation integrates prior knowledge and new information about the instance into a mental model of the situation. When the mental model is used to accomplish the task, it is evaluated for its utility in performing the task. If the mental model is deemed useful, it is reinforced and may become routinized with repeated use. If the mental model is deemed inadequate, it may be rejected and another model formed,

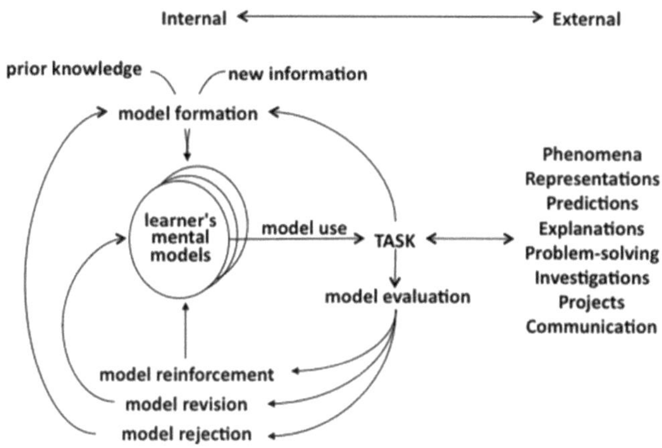

Fig. 14.4 Model-based learning

or it may be revised and then used to try again. Revisions may involve making changes to an element of the model or it may take the form of elaboration—adding elements to the model in order to better accomplish the task. Elements may also be dynamic systems. Ideally, model-based learning results in rich, multilevel, interconnected mental models that are extensible and useful for understanding the world.

Model-based learning begins with a task, whether explicit or tacit. That task is likely to be trying to understand or produce some phenomenon or representation thereof. External representations (text, diagrams, animations, gestures, physical or computer models) are generated from an individual or group's mental models. They may be categorized as either expressed or consensus models. Expressed models are representations of various types generated for a particular purpose. Consensus models, on the other hand, are models developed, agreed upon, and used by a group with some degree of permanence, such as the students in a class or the scientists and scholars of a domain (Gilbert & Boulter, 2000). This study demonstrated that while representations are important supports for model-based learning, they are not sufficient to ensure that it occurs.

BioLogica: Model-Based Genetics Learning with Dragons (2000–2006)

A similar conclusion was reached by the *GenScope* project, which created a computer-based manipulative (Horwitz & Christie, 2000) for helping high school students learn genetics. Learning gains associated with its use were initially quite

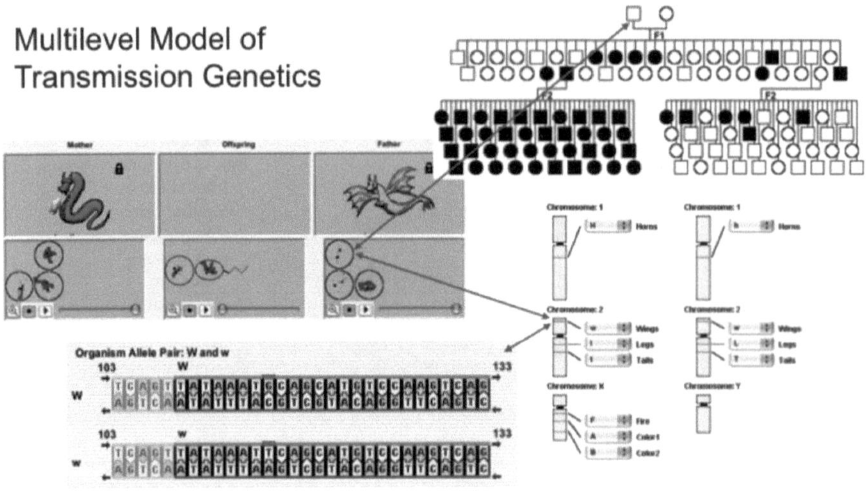

Fig. 14.5 *BioLogica* hypermodel

disappointing and improved only after the creation of curriculum materials for students and professional development for teachers (Hickey, Kindfield, Horwitz, & Christie, 2003; Horwitz & Christie, 1999). Building on this work, Horwitz and colleagues at the Concord Consortium created *BioLogica*, a hypermodel that guided students during computer-based investigations of genetics phenomena.

The *BioLogica* hypermodel employed representations linked via a multilevel simulation of transmission genetics. As shown in Fig. 14.5, *BioLogica* linked representations of base pairs in a molecular model of DNA to specific alleles in chromosomes. Chromosomes with specific alleles participated in models of meiosis and fertilization, which in turn produced organisms with specific traits. Organisms, in this case, dragons, could be bred, producing 40 offspring with varied traits. The frequency of traits in the resulting pedigree depended on the specific alleles of the parents and a probabilistic simulation of their distribution.

The representations in *BioLogica* were designed with investigations in mind. Students could change base pairs in the DNA model and see the impact on alleles. They could change the alleles in one dragon and see the change in its traits (horns, wings, color, legs, tails), breed a male and female, and see the impact on the inheritance of those traits over multiple generations of offspring. Using a progressive model-building approach (White & Frederiksen, 1998), the project developed 12 *BioLogica* activities that guided students through investigations of basic models of meiosis and fertilization and increasingly elaborate models of inheritance (monohybrid, dihybrid, sex-linked, and polygenic).

Scaffolding provided feedback and coaching to students. Two types of scaffolds were implemented—general scaffolds based on a large research base in educational psychology and model-based scaffolding elements to support the knowledge acquisition and reasoning required for progressive model building (Gobert, Buckley, &

Clarke, 2004). Scaffolding of each type was implemented in the form of questions, assigned tasks, or explanations that focused on a phase of model-based learning, followed by feedback. The nature of the feedback varied according to the pedagogical purpose of the scaffolding and was tailored to students' actions or answers. Within each activity the scaffolding faded as students progressed through the learning activity.

The tasks presented in the activities required reasoning from cause to effect (prediction) and from effect to cause (explanation). In some tasks students had to reason in both directions. For example, students were asked to demonstrate how two dragons with horns (a dominant trait) could produce offspring without horns (a recessive trait). Students had to reason from effect (hornless offspring) to cause (both parents had to possess a recessive allele). They had to change the alleles of the parents accordingly, then, after meiosis, select the appropriate gametes for fertilization, and observe the results (cause to effect).

Methods

As students manipulated the hypermodel, *BioLogica* captured very fine-grained, time-stamped data that included students' answers to embedded questions and their actions. These data were used to describe students' learning experiences as well as their understandings and to describe classroom implementations—which activities were used, when they were used, and when students took the pretest and posttests. Combined with teacher surveys this allowed us to create a description of *BioLogica* use in 54 classrooms with over 1,000 students (Buckley, Gobert, Horwitz, & O'Dwyer, 2010; Horwitz, Gobert, Buckley, & O'Dwyer, 2010).

In addition to the data collected while students interacted with *BioLogica*, pretest and posttests based on the paper and pencil instrument developed for the *GenScope* project (Hickey et al., 2003) were administered. The instrument was converted into an online version and validated through standard psychometric methods. Learning gains were determined through differences in pretest and posttest scores for each student (Buckley et al., 2004).

We learned that we could analyze the fine-grained data captured during student use of *BioLogica* to assess students' performance on tasks of varying complexity. In the Monohybrid activity, for instance, we analyzed student actions during a series of three tasks that increased in complexity as scaffolding decreased. We scored these tasks based on successfully completing the task, the number of attempts required to do so, and whether students were systematic or haphazard in their attempts. As the tasks increased in complexity and decreased in scaffolding, fewer students succeeded on the first attempt. Learning gains and performance on later tasks were correlated both with students' ability to use Punnett squares and to systematically conduct breeding experiments (Buckley et al., 2010; Horwitz et al., 2010).

For the purposes of this chapter, the important findings were related to our ability to gauge students' ability to reason with models of meiosis, fertilization, and inheritance as they solved problems and conducted investigations.

The *Calipers* Projects: Simulation-Based Assessments (2004–2012)

The *Calipers I* project extended the *BioLogica* work to simulation-based assessments. The project involved collaboration among Quellmalz and colleagues at SRI, as well as Horwitz and colleagues at the Concord Consortium. Together we designed and developed simulation-based summative assessments for ecosystems and force and motion for middle school classrooms (with students aged 11–13 years). The project used evidence-centered design (ECD) (Behrens, Mislevy, Bauer, Williamson, & Levy, 2004), a process that involves specifying (1) the knowledge and skills to be tested (student model), (2) the data that provide evidence that the student has demonstrated knowledge and skills at various levels of proficiency (evidence model), and (3) tasks requiring use of the targeted knowledge and skills, which would elicit this evidence (task model).

The *Calipers II* project at WestEd extended the use of simulations to develop curriculum-embedded formative assessment modules intended to promote and assess model-based learning in existing middle school science curricula (see www.simscientists.org). The Ecosystem suite is composed of two curriculum-embedded, simulation-based formative assessments (food web and populations), which the teacher inserts into a unit at key points to promote model-based learning and tailor instruction. A summative simulation-based benchmark assessment is used at the end of the unit to gauge student proficiency (Quellmalz et al., 2011).

The *Calipers II* project coupled model-based learning with evidence-centered design (ECD) as shown in Fig. 14.6 (Behrens et al., 2004; Buckley, 2012). Evidence-centered design guided specification of student models, tasks, and data collection and analysis.

We interpreted the student model in terms of a multilevel systems model, like that shown for ecosystems in Fig. 14.7. We then specified the tasks that would require use of the learner's mental model. The evidence model was then specified in terms of the data to be collected during model use and how the data would be analyzed.

A multilevel systems model is a representation that captures the systems view that phenomena are behaviors or properties that emerge from the interactions of system components (Wilensky & Reisman, 2006). The first two columns describe the generic system model levels—components, interactions, and emergent behavior. The third column describes the model levels and content targets for ecosystems.

The systems model guided development of the simulation-based representations to be used in the assessments. The Ecosystems suite represents the components as

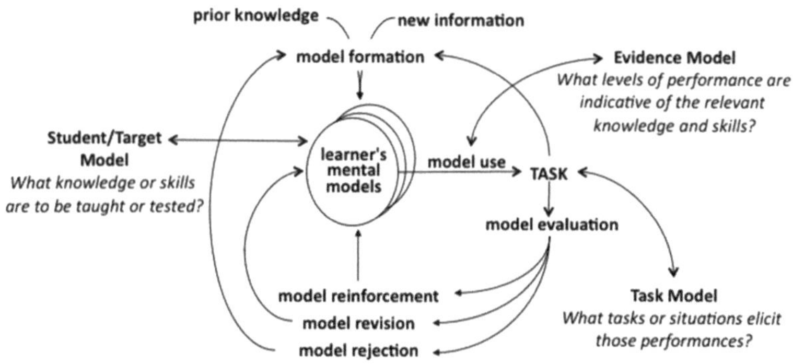

Fig. 14.6 Relationship of model-based learning and evidence-centered design

MODEL LEVELS	MODEL LEVEL DESCRIPTIONS	CONTENT TARGETS (CORE IDEAS)
COMPONENTS	What are the components of the system and their rules of behavior?	Every ecosystem has a similar pattern of organization with respect to the roles (producers, consumers, and decomposers) that organisms play in the movement of energy and matter through the system.
INTERACTIONS	How do the individual components interact?	Matter and energy flow through the ecosystem as individual organisms participate in feeding relationships within an ecosystem.
EMERGENT BEHAVIORS	What is the overall behavior or property of the system that results from many interactions following specific rules?	Interactions among organisms and the ecosystem's nonliving features cause the populations of the different organisms to change over time.

Fig. 14.7 Ecosystem target model

organisms that participate in feeding relationships, which over time produce the population dynamics of the emergent level. The primary representations in the *Calipers II Ecosystems* suite are the food web and the population model. The food web is represented as an animation of the interactions of organisms eating other

Fig. 14.8 Freeze frame of animation used to simulate interactions in the ecosystem

organisms (see Fig. 14.8). As students move the cursor over organisms, the organism and its name are highlighted in the legend, and vice versa. The population model shown in Fig. 14.9 simulates the fluctuation of populations over a 20-year period using the same calculations ecologists use to model population dynamics. It also displays an animation (middle box) of the interactions among the organisms, representing the simulation output with icons. Students set the starting values for the organisms, run the simulation that generates the graphs and the data table, and interpret the data. Students can use the data inspector (the vertical line with flags) to examine the data at various time periods.

The combination of content and inquiry targets guides specification of the task model, which is intended to elicit student performances that demonstrate the ability to use the targeted science knowledge and inquiry practices. The task model specifies how learners interact with ecosystem representations. We employed inquiry practices identified in the NAEP 2009 Science Framework (National Assessment Governing Board & U.S. Department of Education, 2008) to guide task development. Using the food web representation, students are asked to identify the producers and consumers based on their prior knowledge about the role of producers and consumers in ecosystems and their observations during the animations. They are then asked to construct a food web by drawing arrows that diagram the flow of matter and energy in the ecosystem. Students are able to replay the animation during this task. With the population model, students are able to make predictions, design and conduct experiments to test their predictions, and interpret the data to evaluate those predictions.

When these representations (set in different ecosystems) are used in embedded assessments, students are given immediate feedback through the reaction of the simulation to their inputs, multiple levels of coaching based on the type of errors or

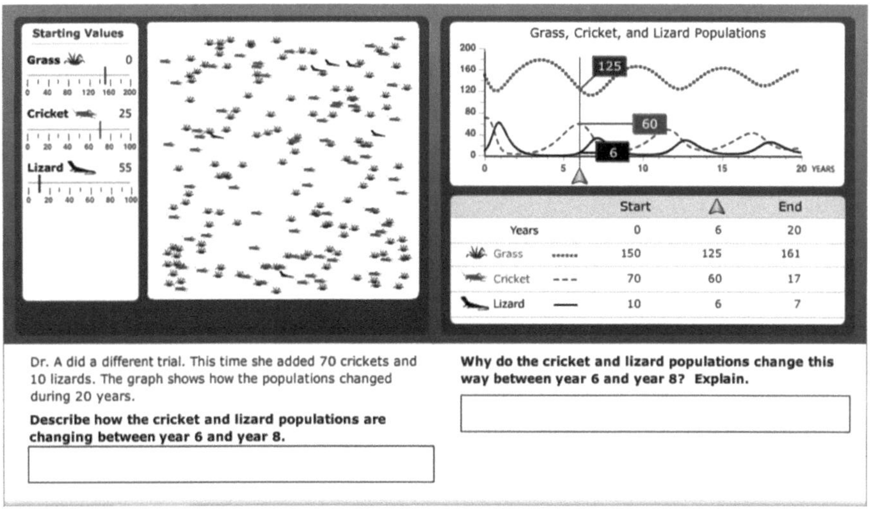

Fig. 14.9 Population simulation used in *Calipers II Ecosystems*

misconceptions demonstrated, and a progress report that identifies the amount of help they needed for each content or inquiry target. These interactions are examples of feedback and revision that can improve students' inquiry practices, content knowledge, and metacognitive skills (Clement & Rea-Ramirez, 2008; Pashler et al., 2007; White & Frederiksen, 1998).

As students work their way through the assessments, the learning management system (LMS) captures data generated by the assessment, as specified by the evidence model, which include answers selected, starting values, arrows drawn, and constructed verbal responses. These data are analyzed in real time and used to guide coaching and produce progress reports for the student and teacher. In the summative benchmark assessments, the data are used to calculate diagnostic variables that are combined with the teachers' scoring of students' constructed verbal responses. Both are passed along to a Bayes net that constructs estimates of each student's proficiency for both content and inquiry targets, which are reported to the student and teacher.

Methods

The *Calipers I* project demonstrated that simulation-based summative benchmark assessments for middle school topics on force and motion and ecosystems met psychometric standards for technical quality and could be effectively used to measure two different dimensions of science learning (content knowledge and inquiry practices). This approach yielded a more accurate measure of student ability than treating the science content knowledge as a single dimension (Quellmalz, Timms, & Buckley, 2010).

The *Calipers II Ecosystem* suite has been tested in cognitive labs, small-scale classroom feasibility tests, and a large-scale field test. The field test, conducted in 40 schools, 29 districts, and three states (Nevada, Utah, and North Carolina), demonstrated the technical quality (validity and reliability), feasibility, and utility of the *Calipers II* assessment suites. After professional development workshops, teachers inserted the formative embedded assessments into their regular curriculum. When students had completed the embedded assessments and other instructional activities, the teacher administered the summative benchmark assessment, which was set in yet another ecosystem and did not provide coaching. Students also took a 30-item, multiple-choice posttest of conventional items drawn from the AAAS item bank (now available online at http://assessment.p2061.org/). Other data included cognitive laboratory activities (think-aloud protocols), classroom observations, teacher surveys, and interviews.

The ecosystems benchmark assessment had a reliability of .76. Analysis showed statistically significant, although moderate, correlations of the student performance on the science content and inquiry measures from the benchmark assessment with their performances on the independent AAAS multiple-choice posttest. Correlations for inquiry (.57) were lower than the correlations for content (.64), supporting the interpretation that while the benchmark and posttest measure similar science content and practices, the measures were not exactly the same. This was expected as the simulation-based assessments were designed to measure content knowledge and inquiry practices that cannot be assessed fully with conventional items. The simulation benchmark assessment also distinguished student performance on inquiry practices more effectively than the multiple-choice posttest. While still performing below the level of the general population, English language learners (ELLs) and students with disabilities (SWDs) did better on the simulation-based benchmark assessment than they did on the posttest, thus narrowing the gap (Quellmalz, Timms, Silberglitt & Buckley, 2012).

To measure the impact of the curriculum-embedded simulation-based formative assessments on learning, a small randomized controlled trial (RCT) was conducted during spring 2011 in 24 classes, with five teachers and 763 students. Each teacher's classes were randomly assigned to a treatment or a control group; 464 students were in the treatment group classes, and 301 students were in the control group classes. Both treatment and control groups were given identical pretests and posttests containing 30 multiple-choice items and a summative, simulation-based benchmark assessment containing 40 measures. Preliminary analysis provides support for the potential benefits of curriculum-embedded, simulation-based, formative assessments in improving learning, when compared to business as usual. Effect sizes for this small RCT ranged from 0.28 to 0.52 after controlling for prior knowledge and teacher effects.

Discussion

The problem that had led researchers to launch this series of projects was the fragmented, inert, and inaccurate understanding that emerged from *taking biology* in school (Rosen, 1989). Whereas many saw the promise of hypermedia and interactive multimedia for addressing this problem, standardized multiple-choice tests were inadequate for measuring the rich, integrated, and extensible knowledge that were the learning goals of such hypermedia. These problems remain as evidenced by the findings of AAAS Project 2061 and the National Research Council (National Research Council, 2002; Stern & Roseman, 2004).

Since accountability often drives instruction, at least in the United States, we need *tests worth teaching to*. In the United States, the frameworks and standards for science education crafted by the National Research Council, the College Board, the National Assessment of Educational Progress, and the American Association for the Advancement of Science call for an increased emphasis on complex systems, the role of representations, models and modeling, and the science practices actually used by scientists (AAAS, 1993; College Board, 2009; National Assessment Governing Board & U.S. Department of Education, 2008; National Research Council, 2011). Individual states and consortia are working on assessments, including simulations, which address these standards. The SimScientists program (www.simscientists.org) at WestEd has been contributing to this effort through research and development efforts like that described for the *Calipers II* project. Much remains to be done.

Vidal (2009) argues that understanding biology requires an understanding of systems biology, the gene, the cell, the role of chemistry in biological processes, and evolution by natural selection. In order for students to develop the richly connected knowledge structures and reasoning skills associated with expertise in biology, these understandings must be woven together to create the linked multilevel mental models of living systems that can serve as a foundation for future learning, reasoning, and research. For genetics, the components (genes and chromosomes) interact in the processes of meiosis and fertilization, resulting in different models of inheritance (emergent behaviors), which produce genetic variation, a key component of evolution. Genetic variation enables some organisms to compete for resources more effectively than others. In *Calipers* assessments this competition, that is, interactions among organisms, is represented in animations of feeding behaviors. These interactions, in turn, produce the population changes of the emergent level over the shorter time spans of ecosystems and over the longer time spans associated with evolution. Such understanding does not often emerge from traditional instruction (Assaraf & Orion, 2010; Penner, 2000) and is not readily measured by traditional assessments (Quellmalz et al., 2011).

Our understanding of how to accomplish such a transformation of instruction and assessment is, like the biology knowledge of many students, fragmented. We need a systems view of biology learning to guide our research and development. It must encompass levels that range from policy research to fine-grained studies of students' interactions with the multiple representations that are the focus of this volume. Below I describe these levels and how they interact.

Policy Level

The international, national, and state frameworks and standards for science education often drive instruction and assessment. High-stakes testing that addresses subsets of the goals set forth in the frameworks may distort curricula in undesirable ways and often do not assess important reasoning, problem-solving, or inquiry goals (Darling-Hammond, 2010; Quellmalz, DeBarger, Haertel, & Kreikemeier, 2005). As policy makers try to develop assessment systems that measure worthwhile goals, research is needed that investigates both the intended and unintended consequences of these changes on science learning.

Standards and Assessments

Policy-level decisions are implemented by setting standards and designing assessments. In the *Calipers* projects the initial step in designing assessments for classrooms involves analyzing and aligning the standards from each of the relevant standards setting organizations (i.e., NRC, AAAS, College Board, NAEP) and crafting our target model. We specify the target of instruction or assessment not as a list of discrete learning objectives but as a complex system. It may not be sufficient to view complex systems as crosscutting elements (NRC, 2011); it may be necessary to reconceptualize educational goals at the policy level in terms of complex systems rather than discrete learning objectives (Goldstone, 2006).

Classroom Realities

In order to prepare students for the high-stakes tests each year, teachers carefully allot class time to the concepts to be tested. In primary school, science often takes a backseat to reading and mathematics. Formative assessments, although recognized as beneficial to learning, are implemented with varying degrees of fidelity (Black & Wiliam, 2009; Yue et al., 2008). Embedded formative assessments such as those developed and tested in the *Calipers II* project offer promise but must contend with the challenges presented by access to computers for such testing and competition for class time (Quellmalz et al., 2012).

Student Learning

This is the focus of the work described in this chapter. In order to develop useful and extensible mental models of phenomena, learners must interact with multiple aspects of phenomena either directly or indirectly via representations. In addition, learners often need considerable help in situating phenomena in the larger and smaller systems in which they interact. These projects have demonstrated some of the affordances for learning offered by linked multiple representations in general and for simulation-based representations in particular. Computer-based simulations and animations enable us to simultaneously represent multiple levels (components, interactions, and emergent behavior) and multiple aspects (spatial, dynamic, causal) of phenomena. Such representations of spatial, temporal, and causal phenomena support schema formation and mental model construction (Norman, 1993). Making the connections among system levels explicit benefits students' understanding (Ioannidou et al., 2010; Slotta & Chi, 2006; Vattam et al., 2011).

The ability to conduct active investigations of phenomena at both interaction and emergent levels is a major affordance of using simulation-based representations. Indeed, scientists express their mental models as conceptual models, computer models, and simulations in order to articulate their theories for testing, communicating, and consensus building (Giere, 1990; Gilbert & Boulter, 2000; Nersessian, 2008). Clement and Rea-Ramirez (2008) organized instruction that fosters students' model-based learning through the same cycles of generating hypotheses, experimentation, and modification engaged in by scientists. Engaging students in active investigations results not only in better inquiry practices but also increases in conceptual understanding and the understanding of nature of science (cf. Kolodner et al., 2003; Metz, 2004).

The *BioLogica* and *Calipers* projects have demonstrated the affordances of simulation-based representations for collecting fine-grained data suitable for formative and summative assessment. For *BioLogica* we created the evidence model post hoc by examining the data and drawing inferences across large numbers of students to create algorithms for interpreting the data (Buckley et al., 2010). The *Calipers II* project designed the evidence model ab initio, which enabled us to analyze the data in real time and use it for formative purposes. The *Calipers II* embedded assessments used these real-time analyses to provide (1) immediate feedback and coaching of both content and inquiry skills, based on students' answers and actions; (2) progress reports to students and teachers, which described student performances by system level and inquiry practices; and (3) follow-up classroom reflection activities including differentiated tasks that focused on the concepts or practices students found most difficult and provided opportunities to engage in scientific discourse as students combined their work into a larger group product (Quellmalz et al., 2012). These are examples of feedback and revision that can improve students' inquiry practices, content knowledge, and metacognitive skills (Clement & Rea-Ramirez, 2008; White & Frederiksen, 1998).

Student Interactions with Representations

Designing simulation-based representations to support model-based learning in biology requires multiple balancing acts. We wish to represent multiple levels of the target system and multiple aspects of phenomena (spatial, dynamic, causal), but we also want to avoid cognitive overload as the learner tries to make sense of what they are seeing. Considerable research conducted in the fields of multimedia learning (cf. Mayer, 2005b) and the learning sciences (cf. Sawyer, 2006) provides some general guidelines. We have found that the application of those guidelines to specific screens is often a blend of art and science involving deliberations that relate the specific screen back to the target model and science practices.

All three projects employed what Mayer (2005a) termed *spatial and temporal contiguity*, not only in terms of displaying graphics and text in close proximity, but also in terms of the levels of the systems represented. In *Science for Living* we linked textual descriptions with highlighted portions of in vivo images such as hearts beating in open chests (Buckley & Boulter, 2000). In ecosystems we displayed both how the organisms were interacting and the effect of that on the population levels over time. In *BioLogica* this took the form of always having two levels of the genetics model displayed so that, for example, the learner could select two gametes for fertilization and see the results in the traits of the new dragon on the same screen.

All three projects gave learners control over animations, playback of video, and inputs to simulations. Such interactivity can help the learner to avoid cognitive overload by controlling the flow of information for better processing (Mayer & Chandler, 2001). The particular advantage of simulations in this regard is that learners are also able to set inputs and see the effect of those inputs on the emergent behavior, particularly when learners are asked to predict what will happen before running the simulation. In all of the projects, we did our best to avoid unnecessary visual details that distract learners and increase extraneous processing (Mayer, 2005a).

The work described in this chapter documents some of the affordances of simulation-based representations for supporting model-based learning and reasoning in biology. It illustrates the potential of multilevel, linked, simulation-based representations for not only fostering model-based learning but also for transforming why, how, and when we assess important science learning.

Acknowledgments The author gratefully acknowledges the colleagues and funding that have made this work possible. The *Science for Living* project was supported by the Apple Classroom of Tomorrow and the Carnegie Corporation. The *BioLogica* project was funded by grants from the National Science Foundation and IERI. The *Calipers* projects are funded by grants from the National Science Foundation, whereas the large-scale field test was supported by the US Department of Education. Any opinions, findings, and conclusions or recommendations expressed in this chapter are those of the author and do not necessarily reflect the views of the funders.

References

AAAS. (1993). *Benchmarks for science literacy*. New York: Oxford University Press.

AAAS. (2011). *AAAS project 2061 science assessment website*, from http://assessment.aaas.org/

Arnaudin, M. W., & Mintzes, J. J. (1985). Students' alternative conceptions of the human circulatory system: A cross-age study. *Science Education, 69*(5), 721–733.

Assaraf, O. B.-Z., & Orion, N. (2010). System thinking skills at the elementary school level. *Journal of Research in Science Teaching, 47*(5), 540–563. doi:10.1002/tea.20351.

Behrens, J. T., Mislevy, R. J., Bauer, M., Williamson, D. W., & Levy, R. (2004). Introduction to evidence centered design and lessons learned from its application in a global e-learning program. *International Journal of Testing, 4*, 295–301.

Black, P., & Wiliam, D. (2009). Developing the theory of formative assessment. *Educational Assessment, Evaluation and Accountability, 21*(1), 5–31.

Buckley, B. C. (2000). Interactive multimedia and model-based learning in biology. *International Journal of Science Education, 22*(9), 895–935.

Buckley, B. C. (2012). Model-based learning. In N. M. Seel (Ed.), *Encyclopedia of the sciences of learning* (pp. 2300–2303). Drodrecht, The Netherlands: Springer.

Buckley, B. C., & Boulter, C. J. (2000). Investigating the role of representations and expressed models in building mental models. In J. K. Gilbert & C. J. Boulter (Eds.), *Developing models in science education* (pp. 105–122). Dordrecht, the Netherlands: Kluwer.

Buckley, B. C., Gobert, J. D., Horwitz, P., & O'Dwyer, L. (2010). Looking inside the black box: Assessing model-based learning and inquiry in BioLogica. *International Journal of Learning Technology, 5*(2), 166–190.

Buckley, B. C., Gobert, J., Kindfield, A. C. H., Horwitz, P., Tinker, B., Gerlits, B., et al. (2004). Model-based teaching and learning with hypermodels: What do they learn? How do they learn? How do we know? *Journal of Science Education and Technology, 13*(1), 23–41.

Clement, J. J., & Rea-Ramirez, M. A. (Eds.). (2008). *Model based learning and instruction in science*. London: Springer.

College Board. (2009). *Science: Standards for college success*. New York: The College Board.

Cronbach, L. J. (1985). *Essentials of psychological testing* (4th ed.). New York: Harper & Row.

Darling-Hammond, L. (2010). *Performance counts: Assessment systems that support high quality learning*. Paper presented at the The Council of Chief State School Officers (CCSSO), Washington, DC.

Duschl, R. A., Schweingruber, H. A., & Shouse, A. W. (2007). *Taking science to school: Learning and teaching science in grades k–8*. Washington, DC: The National Academies Press.

Ericsson, K. A., & Simon, H. A. (1993). *Protocol analysis*. Cambridge, MA: MIT Press.

Feltovich, P. J., Spiro, R. J., & Coulson, R. L. (1988). The nature of conceptual understanding in biomedicine: The deep structure of complex ideas and the development of misconceptions. In D. A. Evans & V. L. Patel (Eds.), *Cognitive science in medicine* (pp. 113–172). Cambridge, MA: MIT Press.

Giere, R. N. (1990). *Explaining science*. Chicago: University of Chicago Press.

Gilbert, J. K., & Boulter, C. J. (Eds.). (2000). *Developing models in science education*. Dordrecht, the Netherlands: Kluwer.

Gobert, J. D., & Buckley, B. C. (2000). Introduction to model-based teaching and learning in science education. *International Journal of Science Education, 22*(9), 891–894.

Gobert, J. D., Buckley, B., & Clarke, J. E. (2004). *Scaffolding model-based reasoning: Representations, cognitive affordances, and learning outcomes*. Paper presented at the American Educational Research Association, San Diego, CA.

Goldsmith, E. (1984). *Research into illustration*. Cambridge, UK: Cambridge University Press.

Goldstone, R. L. (2006). The complex systems see-change in education. *The Journal of the Learning Sciences, 15*, 35–43.

Hickey, D. T., Kindfield, A. C. H., Horwitz, P., & Christie, M. A. T. (2003). Integrating curriculum, instruction, assessment, and evaluation in a technology-supported genetics learning environment. *American Educational Research Journal, 40*(2), 495–538. doi:10.3102/00028312040002495.

Hmelo-Silver, C. E., & Pfeffer, M. G. (2004). Comparing expert and novice understanding of a complex system from the perspective of structures, behaviors, and functions. *Cognitive Science, 28*, 127–138.

Horwitz, P., & Christie, M. (1999). Hypermodels: Embedding curriculum and assessment in computer-based manipulatives. *Journal of Education, 181*(2), 1–23.

Horwitz, P., & Christie, M. A. (2000). Computer-based manipulatives for teaching scientific reasoning: An example. In M. J. Jacobson & R. B. Kozma (Eds.), *Innovations in science and mathematics education: Advanced design for technologies of learning* (pp. 163–191). Hillsdale, NJ: Lawrence Erlbaum Associates.

Horwitz, P., Gobert, J. D., Buckley, B. C., & O'Dwyer, L. M. (2010). Learning genetics with dragons: From computer-based manipulatives to hypermodels. In M. J. Jacobson & P. Reimann (Eds.), *Designs for learning environments of the future: International perspectives from the learning sciences* (pp. 61–87). New York: Springer.

Ioannidou, A., Repenning, A., Webb, D., Keyser, D., Luhn, L., & Daetwyler, C. (2010). Mr. Vetro: A collective simulation for teaching health science. *International Journal of Computer-Supported Collaborative Learning, 5*(2), 141–166.

Johnson-Laird, P. N. (1983). *Mental models*. Cambridge, MA: Harvard University Press.

Kolodner, J. L., Camp, P. J., Crismond, D., Fasse, B., Gray, J., Holbrook, J., et al. (2003). Problem-based learning meets case-based reasoning in the middle-school science classroom: Putting learning by design into practice. *The Journal of the Learning Sciences, 12*(4), 495–547.

Mayer, R. E. (2005a). Cognitive theory of multimedia learning. In R. E. Mayer (Ed.), *The Cambridge handbook of multimedia learning*. New York: Cambridge University Press.

Mayer, R. E. (Ed.). (2005b). *The Cambridge handbook of multimedia learning*. New York: Cambridge University Press.

Mayer, R. E., & Chandler, P. (2001). When learning is just a click away: Does simple interaction foster deeper understanding of multimedia message? *Journal of Educational Psychology, 93*, 390–397.

Merriam, S. B. (1991). *Case study research in education*. San Francisco: Jossey-Bass.

Metz, K. E. (2004). Children's understanding of scientific inquiry: Their conceptualization of uncertainty in investigations of their own design. *Cognition and Instruction, 22*(2), 219–290.

National Assessment Governing Board, & U.S. Department of Education. (2008). *Science framework for the 2009 national assessment of educational progress*. Washington, DC: National Assessment Governing Board, U.S. Department of Education.

National Research Council. (2002). *Knowing what students know: The science and design of educational assessment*. Washington, DC: National Academy Press.

National Research Council. (2011). *A framework for k-12 science education: Practices, crosscutting concepts, and core ideas*. Washington, DC: National Academies.

Nersessian, N. J. (2008). *Creating scientific concepts*. Cambridge, MA: MIT Press.

Norman, D. A. (1993). *Things that make us smart*. Reading, MA: Addison-Wesley.

Pashler, H., Bain, P., Bottge, B., Graesser, A., Koedinger, K., McDaniel, M., et al. (2007). *Organizing instruction and study to improve student learning*. Washington, DC: National Center for Education Research, Institute of Education Sciences, U.S. Department of Education.

Patel, V. L., Kaufman, D. R., & Magder, S. (1991). Causal explanation of complex physiological concepts by medical students. *International Journal of Science Education, 13*(2), 171–185.

Penner, D. E. (2000). Explaining systems: Investigating middle school students' understanding of emergent phenomena. *Journal of Research in Science Teaching, 37*(8), 784–806.

Quellmalz, E. S., DeBarger, A., Haertel, G., & Kreikemeier, P. (2005). *Validities of science inquiry assessments: Final report*. Menlo Park, CA: SRI International.

Quellmalz, E. S., Timms, M. J., & Buckley, B. C. (2010). The promise of simulation-based science assessment: The Calipers project. *International Journal of Learning Technology, 5*(3), 243–265.

Quellmalz, E. S., Timms, M. J., Buckley, B. C., Davenport, J., Loveland, M., & Silberglitt, M. D. (2011). 21st century dynamic assessment. In J. Clarke-Midura, M. Mayrath, & C. Dede (Eds.), *Technology-based assessments for 21st century skills: Theoretical and practical implications from modern research* (pp. 55–89). Charlotte, NC: Information Age.

Quellmalz, E. S., Timms, M. J., Silberglitt, M. D., & Buckley, B. C. (2012). Science assessments for all: Integrating science simulations into balanced state science assessment systems. *Journal of Research in Science Teaching, 49*(3), 363–393.

Rosen, W. G. (Ed.). (1989). *High-School Biology Today and Tomorrow*. Washington, DC: National Academy Press.
Sawyer, R. K. (Ed.). (2006). *The Cambridge handbook of the learning sciences*. New York: Cambridge University Press.
Shymansky, J. A., Yore, L. D., & Good, R. (1991). Elementary school teachers' beliefs about and perceptions of elementary school science, science reading, science textbooks, and supportive instructional factors. *Journal of Research in Science Teaching, 28*(5), 437–454.
Slotta, J. D., & Chi, M. T. H. (2006). Helping students understand challenging topics in science through ontology training. *Cognition and Instruction, 24*(2), 261–289.
Stern, L., & Roseman, J. E. (2004). Can middle-school science textbooks help students learn important ideas? Findings from project 2061's curriculum evaluation study: Life science. *Journal of Research in Science Teaching, 41*(6), 538–568.
Tufte, E. R. (1983). *The visual display of quantitative information*. Cheshire, CT: Graphics Press.
Vattam, S. S., Goel, A. K., Rugaber, S., Hmelo-Silver, C. E., Jordan, R., Gray, S., et al. (2011). Understanding complex natural systems by articulating structure-behavior- function models. *Journal of Educational Technology & Society, 14*(1), 66–81.
Vidal, M. (2009). A unifying view of 21st century systems biology. *FEBS Letters, 583*, 3891–3894.
Weiss, I. R., & Pasley, J. D. (2004). What is high-quality instruction? *Educational Leadership, 61* (5), 24–28.
White, B. Y., & Frederiksen, J. R. (1998). Inquiry, modeling, and metacognition: Making science accessible to all students. *Cognition and Instruction, 16*(1), 3–118.
Wilensky, U., & Reisman, K. (2006). Thinking like a wolf, a sheep, or a firefly: Learning biology through constructing and testing computational theories-an embodied modeling approach. *Cognition and Instruction, 24*(2), 171–209.
Yue, Y., Shavelson, R. J., Ayala, C. C., Ruiz-Primo, M., Brandon, P. R., Furtak, E., et al. (2008). On the impact of formative assessment on student motivation, achievement, and conceptual change. *Applied Measurement in Education, 21*(4), 335–359. doi:10.1080/08957340802347845.

Chapter 15
Secondary Students' Understanding of Genetics Using *BioLogica*: Two Case Studies

Chi-Yan Tsui and David F. Treagust

Introduction

In this chapter, we first discuss the theoretical aspects of learning genetics and learning with multiple external representations (MERs) by reviewing the literature relevant to our studies. Next, we reexamine our Australian study focusing on students' understanding in terms of gene conceptions and genetics reasoning from a cross-case analysis of data from three senior secondary schools in Perth (Tsui & Treagust, 2007, 2010). We also report on our recent Hong Kong study (Tsui, 2009), compare its results with those of the Australian study, and explore how students learned complex content in biology using MERs within different learning contexts, including the role of language in learning. Pseudonyms are used throughout this chapter to maintain anonymity of all participants in our studies.

Genetics is Conceptually and Linguistically Difficult

Over the past decades, research has shown that genetics not only is a conceptually difficult topic in school biology because that knowledge is organized at multiple levels but also is a linguistically difficult content area because of its large content-specific vocabulary (e.g., Bahar, Johnstone, & Hansell, 1999; Hackling & Treagust, 1984; Horwitz & Christie, 2000; Pearson & Hughes, 1988; Stewart, 1982; Venville & Treagust, 1998; Wood, 1996). Learning genetics requires multilevel thinking—phenotypes of an organism are at the macroscopic level, whereas cells and chromosomes are at the microscopic level, DNA is at submicroscopic level, and

C.-Y. Tsui • D.F. Treagust (✉)
Science and Mathematics Education Centre, Curtin University, GPO Box U1987, Perth, WA 6845, Australia
e-mail: C.Tsui@curtin.edu.au; D.Treagust@curtin.edu.au

D.F. Treagust and C.-Y. Tsui (eds.), *Multiple Representations in Biological Education*, Models and Modeling in Science Education 7, DOI 10.1007/978-94-007-4192-8_15,
© Springer Science+Business Media B.V. 2013

genotypes are at the symbolic level (e.g., Johnstone, 1991). Student understanding of genetics also depends on dealing with these concepts and processes simultaneously at several levels of organization, on connecting them as an interrelated whole (Marbach-Ad & Stavy, 2000), and on reasoning with concepts and processes across ontologically distinct levels (e.g., genes or DNA molecules are informational but the traits they control are physical) (Duncun & Reiser, 2007).

Science educators have recently called for improving the ways to teach the complexity of the gene concept and for using better approaches to address both the complex content of genetics and the inadequate current instructional methods and materials in schools (e.g., Duncan, Rogat, & Hmelo-Silver, 2009; Venville & Donovan, 2005). Genetics literacy—"being able to comprehend, use or respond to information about genetic phenomena and technologies" (Duncan et al., 2009, p. 657)—is needed for all citizens in order to better understand the emerging contemporary issues such as genetic modifications, genomics, or cloning and to make informed judgments and decisions.

Learning Genetics as Understanding: Gene Conceptions and Genetics Reasoning

In our studies, we considered student learning of genetics as conceptual understanding in terms of their gene conceptions and genetics reasoning. Theoretically, we drew on a multidimensional conceptual change framework (Tyson, Venville, Harrison, & Treagust, 1997) to address the acknowledged limitations of the traditional, largely epistemological conceptual change model of Posner, Strike, Hewson, and Gertzog (1982). In reexamining the results, we consider Vygotskian perspectives that emphasize the role of social and cultural contexts and that of language in learning (Vygotsky, 1968, 1978), as well as some perspectives about learning from psycholinguistic research (e.g., Kroll & Hermans, 2011; Lin, 2006).

The first focus in our Australian study was on student understanding as developing ontological conceptual change in conceptualizing the gene from being a particle to a sequence of instructions as in Venville and Treagust's (1998) study in which the grade 10 students developed their conceptions through a pathway indicating their progressively more sophisticated mental models of the gene (inactive particle gene → active particle gene → sequence of instructions gene → productive sequence of instructions gene). The second focus was on students' understanding in terms of reasoning that can be diagnosed by a two-tier instrument (Treagust, 1988) which we developed and used in three Perth schools for pre- and post-instructional evaluation of students' genetics reasoning (Tsui & Treagust, 2010). The two-tier diagnostic instrument was subsequently modified and used in our Hong Kong study. The two-tier test items evaluate students' genetics reasoning using Hickey and Kindfield's (1999) matrix of reasoning (Tsui & Treagust, 2003, 2010) (see Table 15.1).

Table 15.1 Six types of genetics reasoning adapted from Hickey and Kindfield (1999)

		Domain-general dimension of reasoning (novice ←→ expert)		
		Cause-to-effect reasoning	Effect-to-cause reasoning	Process reasoning
Domain- specific dimension of reasoning (simple ↕ complex)	Between generations	Monohybrid inheritance: mapping genotype to phenotype (Type II)	Monohybrid inheritance: mapping phenotype to genotype (Type IV)	Punnett squares (input/output reasoning): meiosis process event reasoning Mitosis process [a] (Type VI)
	Within generations	Mapping genotype to phenotype (Type I)	Mapping phenotype to genotype (Type III)	Mapping information in DNA base sequence (genotype) to amino acid sequence in protein synthesis (phenotype)[b] (Type V)

[a]Not included in Hickey and Kindfield's (1999) original types

[b]Not included in Hickey and Kindfield's (1999) original types but adapted from Venville and Treagust's (1998) sophisticated conception of the gene as being a productive sequence of instructions

As indicated by Table 15.1, genetics reasoning in our studies required students to use both logical reasoning (domain-general dimension) and information in their subject content (domain-specific dimension) for understanding. Novice reasoners often use mental representations of only one antecedent condition in reasoning tasks to arrive at the conclusion, whereas expert reasoners use two or more antecedent conditions in such reasoning processes and become more reflective and active in seeking alternatives and making inferences to draw conclusions (Lawson, 1992).

Therefore, we can explain how students reason in completing the tasks of genetics reasoning Types I–IV (see Table 15.1). For example, to solve pedigree problems that require Types III and IV reasoning, students need to reason by mapping given phenotypes to unknown genotypes of the parents (effect-to-cause), respectively, within and between generations. Mapping in Types III and IV is more difficult compared to that in Types I and II (cause-to-effect) because the former is not a one-to-one mapping, that is, more than one genotype may correspond to the same given phenotype. In solving human pedigree problems, Hackling and Lawrence (1988) also pointed out that the expert problem solvers are able to identify critical cues in the problems, test hypotheses with genotypes assigned to phenotypes, and use given evidence to support or falsify an alternative hypothesis before arriving at the answer. It was based on these six types of reasoning that we designed the interview reasoning tasks (Tsui & Treagust, 2003) and the two-tier diagnostic instrument in the Australian study (Tsui & Treagust, 2010).

Multiple Representations and BioLogica

Biology teachers have long been using different *external representations (ERs)* in classroom teaching to communicate ideas to students by voice, writing, drawings, diagrams, images, gestures, and so on. Students' conceptions can be regarded as their *internal representations* or mental models of an object or event (Duit & Glynn, 1996) constructed from the ERs of these entities. Models of scientific objects or processes can be considered as ERs for modeling in model-based learning which plays a central role in science education (Justi & Gilbert, 2002).

Visualizations, as ERs, have been important in learning since the advent of computer technology and are now being widely used for learning science and in the media to convey scientific information (Gobert, 2005). In our studies, we explored the pedagogical functions of using more than one form of external representations or *multiple external representations (MERs)* in learning (cf. van Someren, Reimann, Boshuizen, & de Jong, 1998). In particular, we utilized Ainsworth's (1999) functional taxonomy of MERs to argue that MERs can support learning in three ways: (1) by providing/supporting complementary information and/or cognitive processes, (2) by constraining interpretations or misinterpretations of phenomena, and (3) by promoting the construction of a deeper understanding of concepts through *abstraction*, such as detecting and extracting a subset of relevant elements from a representation; *extension* or extending knowledge learned in one representation to new situations with other representations; and *relations*, such as translating between two or more unfamiliar representations. However, learning with MERs may not always be useful because of the new costs and challenges (Ainsworth, Bibby, & Wood, 1997).

In this chapter, we argue that MERs appear to be a promising construct for improving learning of complex concepts in biology because biological knowledge is hierarchically organized (Marbach-Ad & Stavy, 2000) at ontologically distinct

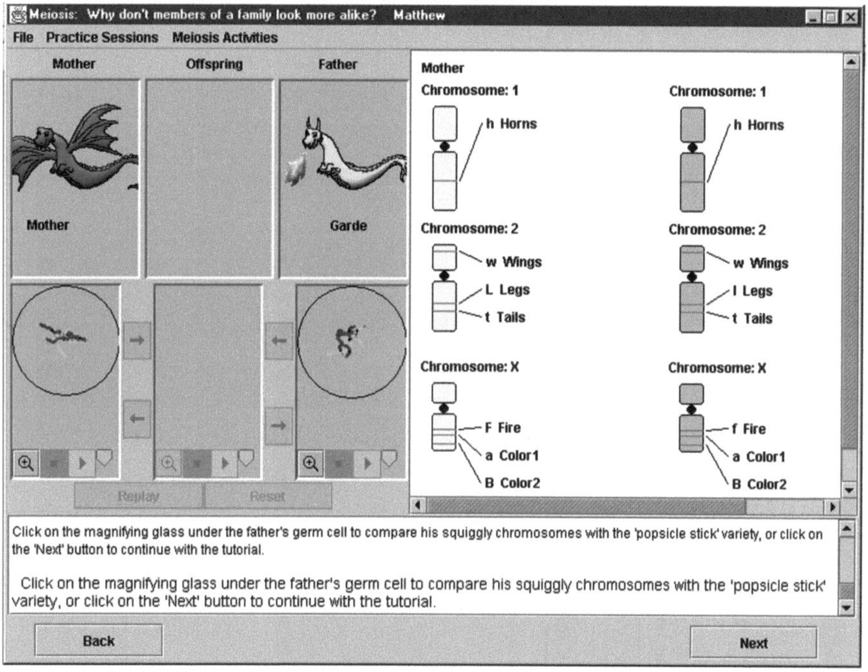

Fig. 15.1 A screenshot of *BioLogica* activity *Meiosis* showing organism level, cell level, and chromosome level of dragons (an imaginary species)

levels (Duncun & Reiser, 2007). The computer-based activities of *BioLogica* (Concord Consortium, 2002)—a *hypermodel* (Horwitz, 1995; Horwitz & Tinker, 2001) or an interactive, exploratory environment for learning genetics—were used in our studies. *BioLogica* features dynamically linked MERs of genetics that allow users to manipulate objects of genetics represented at different levels of biological organization—DNA, genes, chromosomes, gametes, cells, organisms, pedigrees, and populations—and observe the changes in their behavior as a result of manipulation in ways constrained by models based on transmission genetics and molecular/cellular mechanisms (Buckley et al., 2004; Gobert et al., 2011) (see Fig. 15.1).

BioLogica guides learners' interaction with the activities through a sequence of challenges, monitors their progress, and provides learners with feedback and helpful hints as they work through progressively more challenging activities. The interactions in these activities are controlled and implemented by a software component called *activity scripts* (Horwitz & Tinker, 2001) having different pedagogical functions—such as narratives, tasks and puzzles, representational assistance, reasoning models, explanations and feedback on actions and responses, embedded assessment questions, and reflective questions—that mediated the students' conceptual learning and reasoning (Buckley et al., 2004). The learning goals of eight *BioLogica* activities completed by most of the students in our studies are shown in Table 15.2.

Table 15.2 Learning goals of eight of the *BioLogica* activities (Concord Consortium, 2002)

Learning goals (genetic terms or concepts are underlined)	*BioLogica activities* (last one is a game)							
	Introduction	Rules	Meiosis	Inheritance	Monohybrid	Mutations	Mutation inheritance	Invisible dragons
1. Get to know the computer interface; change genes on chromosomes or genotype and observe phenotype changes	√							
2. Link representations with genetics, chromosomes, genes, and alleles	√							
3. Explore changing alleles and their combinations on phenotype		√						
4. Learn concepts of dominant/recessive genes, incomplete dominance, sex linkage, and homozygous/heterozygous gene combinations		√						
5. Understand monohybrid inheritance			√					
6. Under random shuffling of genes to gametes (sperms or eggs)			√					
7. Select gametes for fertilization to produce particular phenotype			√					
8. Show random distribution of traits through generations				√				
9. Take a look at probability				√				
10. Examine random selection effect on the distribution of traits					√			

11. Explore the use of Punnett squares to understand gene distribution combinatorics	√			
12. Familiarize with the molecular level—DNA view		√		
13. See how DNA changes result in genotype changes in the individual		√		
14. Generalize the notion of rules to include more than two alleles		√		
15. Become aware that mutated genes follow similar rules to non-mutated genes			√	
16. Practice using Punnett squares and solving probability problems			√	
17. Develop strategies for determining parents' genome				√

The data about the students' interactions (e.g., which screen a student used in answering a question or what graphic objects the student accessed and in what order) can be logged by the *BioLogica* program in the form of *log files* automatically generated and saved on an individual computer, a school network, or a remote server of a research center so that teachers and researchers can analyze the logged data. As Horwitz and Tinker (2001) explained, the log files that track student interactions in *BioLogica* are "a promising research tool that allows us to obtain at a distance detailed information about student thinking, knowledge, and problem-solving strategies" (p. 13).

A large-scale study on model-based learning using *BioLogica* in schools across the United States indicated that the experimental groups outperformed the control groups in understanding genetics (Buckley et al., 2004). Further analyses showed that those students with better understanding of models as multiple representations learned significantly better about the content of genetics in *BioLogica* activities than did those with less understanding of models as such (Gobert et al., 2011).

Research Questions

In this chapter we attempt to focus on two research questions about the understanding of genetics in terms of their gene conceptions and genetics reasoning by discussing and comparing the results of our Australian and Hong Kong studies: (1) What are the students' pre-instructional and post-instructional gene conceptions? (2) In what ways and how do the MERs of *BioLogica* promote students' genetics reasoning?

Method

Research Approach

In our Australian study, we adopted an interpretive research approach (Erickson, 1998) involving largely qualitative case study methods (Merriam, 1998). In particular, we drew on Merriam's three major features that characterize a case study—*particularistic*, *descriptive*, and heuristic—and the research strategies (e.g., *prolonged engagement, persistent observation, member checks,* and *triangulation*) for increasing the research rigor of a qualitative case study (cf. Denzin & Lincoln, 1994; Erickson, 1998).

School Context

The three case schools in our Australian study were School F (a state co-ed school), School O (an independent girls' school), and School U (a state co-ed school) in Perth, Western Australia. The study was first conducted in a 10th-grade class in

School F when genetics was taught and then in Schools O and U in the following year with similar methods. The four participating biology teachers had teaching experiences ranging from 9 to 27 years, and their participating students (72 girls and 17 boys), aged from 14 to 18 years, in three 10th-grade classes (Schools F and O) and two 12th-grade classes (School U), were mostly Australian-born and native speakers of English. Research ethics (e.g., voluntary participation, informed consent, use of pseudonyms for participants) was strictly followed (Tsui & Treagust, 2007, 2010).

The second case study involved action research in a government-subsidized girls' secondary school in Hong Kong, a special administrative region of China since 1997 (Tsui, 2009). This study was a learning project—a collaboration between the first author and a biology teacher with support from the school—for improving students' scientific reasoning and writing biology in English (Tsui, 2009). The 20 participating students who volunteered to take part were 10th-grade girls of average age of 15.6 years; they were all Hong Kong Chinese with English as their second language and Chinese as their first language (their native dialect is Cantonese and written language is Modern Standard Chinese). Before the study, the 10th graders in this Chinese Medium of Instruction (CMI) school had completed their first 3 years (grades 7–9) of secondary education in CMI. Since the beginning of the first semester of their 10th-grade year, these English-as-a-second-language (ESL) or English language learner (ELL) students had used English as the medium of instruction (EMI) for learning some subjects, including biology. This change in the medium of instruction is common in many CMI schools in Hong Kong. They had not learned genetics before this study because genetics was part of their 11th-grade biology curriculum.

Over 8 weeks in the second semester, these 10th graders learned genetics in weekly after-school computer sessions using *BioLogica* activities. Their biology teacher, Ms Chan, who had 15 years of teaching experience, collaborated with the first author to provide scaffolding and support in all the weekly computer learning sessions. Both the teacher and the first author are bilingual speakers of English and Cantonese.

Data Collection and Analysis

Australian Study

Although the participating teachers in the three schools all included *BioLogica* activities for their student learning, they also used other teaching aids and learning resources. The data from multiple sources—before, during, and after teaching with *BioLogica* activities—were collected: transcripts of semi-structured student interviews, online results of the two-tier pretests/posttests and open-ended questionnaires (delivered by *WebCT*, Curtin University's then e-learning system), computer log files on students' usage of *BioLogica*, classroom observation field notes and audio recordings transcripts, the first author's reflective journals, and teachers' handouts and other school documents.

To evaluate Australian students' understanding in terms of gene conceptions, we analyzed their open-ended questionnaire responses, interview and lesson transcripts, and other qualitative data. We interviewed 26 target students in the three Perth schools, selected from each class on the basis of their scores in the online pretests on genetics reasoning to include students from high and low groups. The interview protocols used were the same in the three schools except that for School U no reasoning tasks were included (Tsui & Treagust, 2007, 2010). We used the two-tier posttest to diagnose students' understanding of genetics in terms of reasoning and analysis of some target students' log files. Both the two-tier tests and interview reasoning tasks were designed to evaluate students' six types of genetics reasoning.

Hong Kong Study

In this study, only five sources of data used in the Australian study were collected—interviews of students, open-ended questionnaire (gene conception) and two-tier posttest (genetics reasoning), *BioLogica* log files (tracking student interactions with MERs), and teacher's handouts and other documents. We also analyzed students' written answers to the parallel open-ended questionnaire in the paper-and-pencil pretest and posttest *What do you know about a gene?* for identifying their gene conceptions using the framework of Venville and Treagust (1998). We interviewed four target students, from the high- and low-ability group based on their school examination results, before and after instruction. Unlike the Australian study, we used the two-tier posttest only to diagnose students' understanding of genetics in terms of reasoning to respect the biology teacher's suggestion. We also conducted analyses of the log files and correlation analyses to explore the relations between students' genetics reasoning and other variables.

Results

Identifying Common Gene Conceptions

In a cross-case analysis of the Australian students' gene conceptions before and after instruction—based on their responses to an open-ended questionnaire *What do you know about a gene?* in the online pretest and posttest—we identified five common gene conceptions of the 10th graders in a way similar to the findings of Venville and Treagust (1998). A student could hold more than one gene conception. As shown in Table 15.3, the most common gene conception was: "A gene is from parents/grandparents."

In our Hong Kong study, the analysis of students' written answers to the same parallel open-ended questionnaire in the paper-and-pencil pretest and posttest (*What do you know about a gene?*) indicated that their gene conceptions could be

Table 15.3 Gene conceptions of Australian grade 10 students

Gene conception[a]	Quotes from online *WebCT* questionnaire and interview transcripts	Number of conceptions (%) Pretest ($n = 63$)	Posttest ($n = 60$)
A gene is from parents/ grandparents	. . .genes are inherited from our family. It could be from generations ago. You can get a mixture of your families genes so you might have your dads hair and your mums eyes (Laurie, School F; pretest)	36 (57.1)	30(50.0)
A gene is/part of a chromosome	Genes have something to do with chromosomes which you receive from your parents and ancestors (Nelly, School F; pretest)	3(4.8)	25(41.7)
A gene is/part of DNA	Information about your characteristics that are passed on to you from your parents through your DNA (Andrea, School O, pretest)	16(25.4)	15(25.0)
A gene determines a trait/ characteristic	Genes are the determining factors in the development and purpose of cells of an organism (Luke, School F; pretest)	27(42.9)	37(61.7)
A gene is information for controlling development/ making proteins	Um. Well, genes . . . made up of the genetic code in the DNA, which tells the body to make proteins, and um, they just carry the information which tells the body how it should work and stuff and how it should develop (Andrea, School O, post-instructional interview)	3(4.8)	4(6.7)

[a]Based on Venville and Treagust (1998)

categorized into four gene conceptions along a pathway of progressively more sophisticated conceptions of the gene as reported by Venville and Treagust (1998) (see Table 15.4). These results suggest that the Hong Kong students improved their understanding of the gene in terms of developing progressively sophisticated conceptions of the gene after their learning with eight *BioLogica* activities they had done weekly over 2 months.

Learning to Write Genetics with Confidence: Some Examples

Although only some Hong Kong students could fully express their understanding in writing about genetics, most of them improved their confidence in writing English despite their grammatical and other errors. Bilingual support (e.g., bilingual glossary of genetic terms in English and Modern Standard Chinese), on-site scaffolding, weekly feedback of the first author to the students by returning to them their log

Table 15.4 Change in students' gene conceptions in the Hong Kong study

Gene conception[a]	Quotes from pretest[b] or posttest	Number of conceptions (%) Pretest ($N = 20$)	Posttest ($N = 20$)
C1: a gene as a passive particle from the parents	A gene is a factor that has passed from our parents to us; everyone has got different genes[c] (S16[d], pretest)	15 (75.0)	3 (15.0)
C2: a gene as an active particle that determines a trait	A gene will affect our appearance, for example different nose, mouth, eyes and ears... (S4, posttest)	4 (20.0)	11 (55.0)
C3: a gene as sequence of instructions or information	Genes record about the growth, function of cells/tissues/organs. As they are in the nucleus, they can give out messages to "order the cell" (S1, pretest)	2 (10.0)	6 (30.0)
C4: a gene as productive sequence of instructions for proteins or information for proteins	Gene is a length of DNA which contains information about one protein..., which allows us to do many things and it also control us in our lives (S13, posttest)	0 (0.0)	1 (5.0)

[a]Based on Venville and Treagust (1998)
[b]Students were allowed to answer the open-ended questionnaire either in English or Chinese in the pretest but must write in English in the posttest
[c]Translated from the student's written Modern Standard Chinese
[d]Student number

files with feedback comments, and collaborative classroom discourse with mixed-code (English and Cantonese) discussions—all appeared to be conducive to their learning (Tsui, 2009). For example, some students like Mei-ling (see Figs. 15.2 and 15.3) who used Chinese and diagrams to represent their gene conceptions at the pretest became more confident at the posttest to write in English; some used mixed-code in their representations.

The pretest and posttest open-ended questionnaires were both *What do you know about a gene?* and at the pretest students were allowed to write in either Chinese or English and use diagrams to illustrate their answers, but at the posttest they were asked to write in English (for some examples of students' answers, see Figs. 15.2, 15.3, 15.4, and 15.5).

Comparing Genetics Reasoning of Students from Hong Kong and Perth

In our Australian study, for the students in all three schools, a paired *t* test indicated that their genetics reasoning posttest scores were significantly higher than their pretest scores at $p < 0.01$ (see Table 15.5).

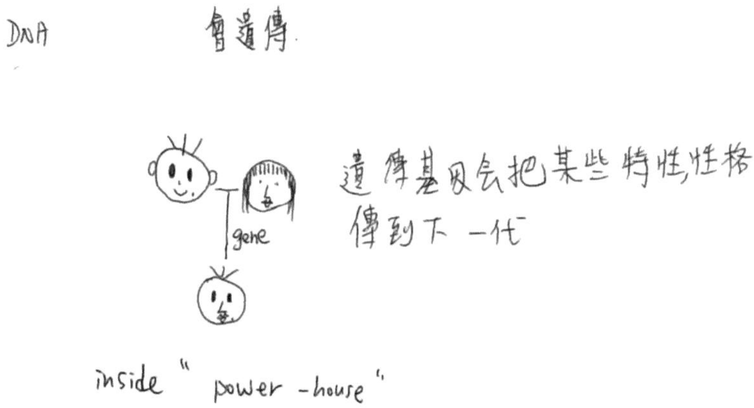

"[DNA] is inheritable" (top) and "Genes can pass certain traits, characters to the next generation." (on the right of the pedigree diagram with the word "gene")
She also mentioned that DNA is "inside 'power house' " or a mitochondrion of a cell.

Fig. 15.2 Pretest answers of Mei-ling (16 years old) who used Chinese to describe her gene conception. Her pre-instructional gene conception was categorized as C2 (see Table 15.4)

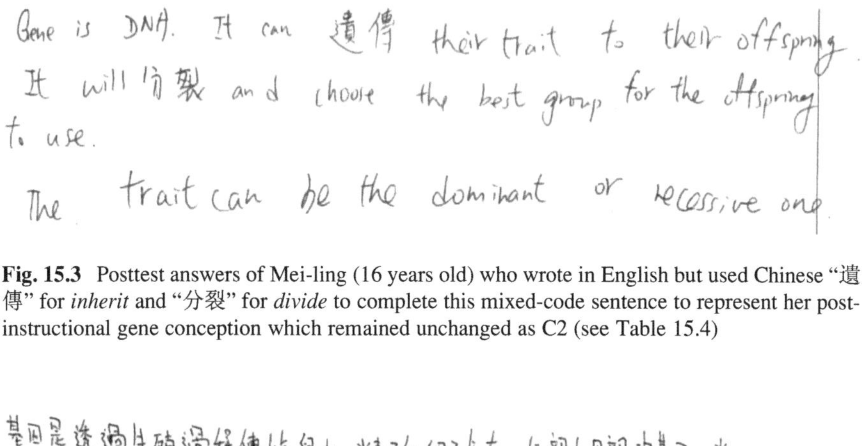

Fig. 15.3 Posttest answers of Mei-ling (16 years old) who wrote in English but used Chinese "遺傳" for *inherit* and "分裂" for *divide* to complete this mixed-code sentence to represent her post-instructional gene conception which remained unchanged as C2 (see Table 15.4)

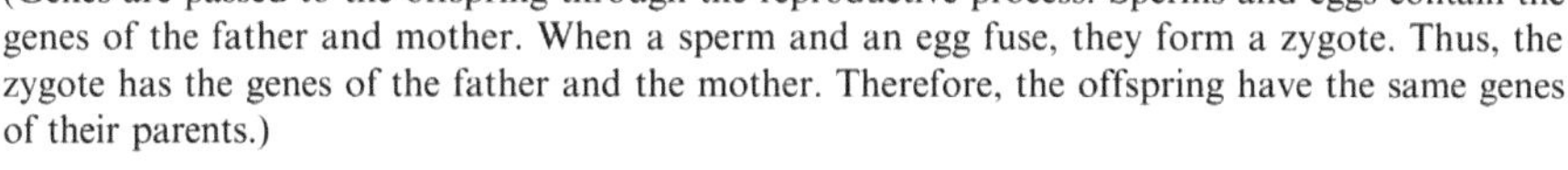

(Genes are passed to the offspring through the reproductive process. Sperms and eggs contain the genes of the father and mother. When a sperm and an egg fuse, they form a zygote. Thus, the zygote has the genes of the father and the mother. Therefore, the offspring have the same genes of their parents.)

Fig. 15.4 Pretest answers of Lai-ming (16 years old) whose answers were in Chinese. Her pre-instructional gene conception was categorized as C1 (see Table 15.4)

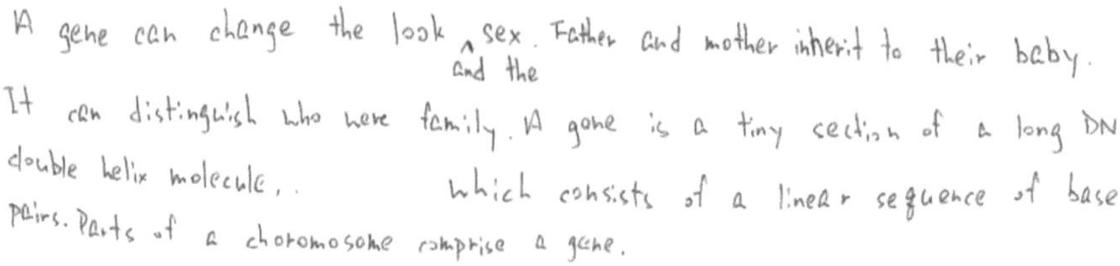
A gene can change the look and the sex. Father and mother inherit to their baby.
It can distinguish who were family. A gene is a tiny section of a long DNA
double helix molecule, . which consists of a linear sequence of base
pairs. Parts of a choromosome comprise a gene.

Fig. 15.5 Posttest answers of Lai-ming (16 years old) whose gene conception had progressed from C1 to C3 after instruction (see Table 15.4)

Table 15.5 Comparison of genetics reasoning pretest and posttest scores in three Australian schools

	Pretest			Posttest				
School	*n*	*M*	*SD*	*N*	*M*	*SD*	*t*	*p*
F (grade 10)	24	13.89	18.17	24	54.86	24.81	5.66	.000**
O (grade 10)	31	12.46	12.65	31	49.76	20.80	9.86	.000**
U (grade 12)	13	46.15	25.12	13	65.68	18.48	3.61	.004*

$*p < .01$; $**p < .001$

One limitation in our Hong Kong study is that there was no genetics reasoning pretest for a pre-post comparison; however, an analysis of descriptive statistics of the posttest scores ($M = 31.25$, $SD = 16.42$) of the Hong Kong students ($N = 20$) showed that their posttest reasoning patterns by types were as predicted according to the difficulty level of the six types of reasoning (Tsui & Treagust, 2003, 2010). Analyses also showed that they had achieved the similar patterns at the posttest comparable to those of their Australian counterparts as indicated in Figs. 15.6 and 15.7; nevertheless their mean scores were much lower because they had not previously studied genetics in school.

Analysis of other results suggest that student performance in genetics reasoning in our Hong Kong study depended on their prior knowledge of biology and English language proficiency as indicated by Pearson correlation analyses using the students' school examination scores before the study—prior knowledge of biology ($r = .512$; $p = .021$, two-tailed; $N = 20$) and English language proficiency ($r = .57$; $p = .008$, two-tailed; $N = 20$) were significantly correlated with the genetics reasoning scores in the posttest of the study ($p < .05$).

Analyses of Students' Log Files

The computer log files that tracked students' interaction with *BioLogica* were subsequently analyzed to explore how students learned during the computer activities. Log file specifications (e.g., time in screen, interaction time, inputs to model, answers, typing time) and their analysis can be useful for understanding how

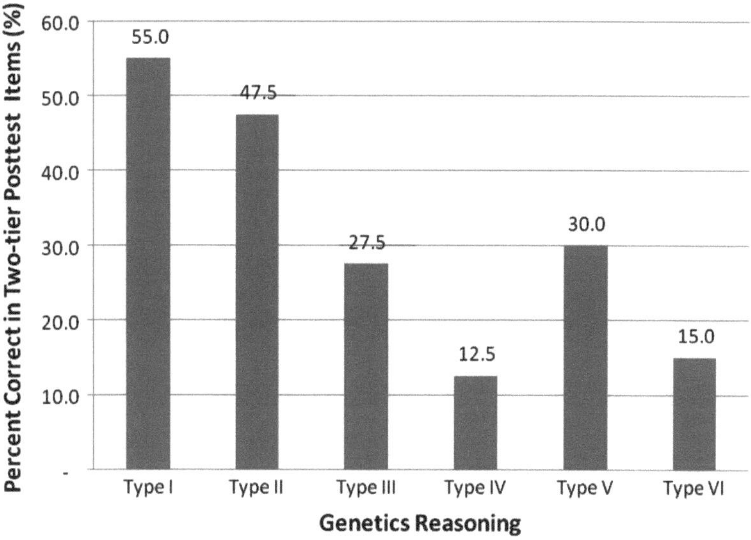

Fig. 15.6 Genetics reasoning by types in two-tier posttest of Hong Kong 10th-grade students ($N = 20$)

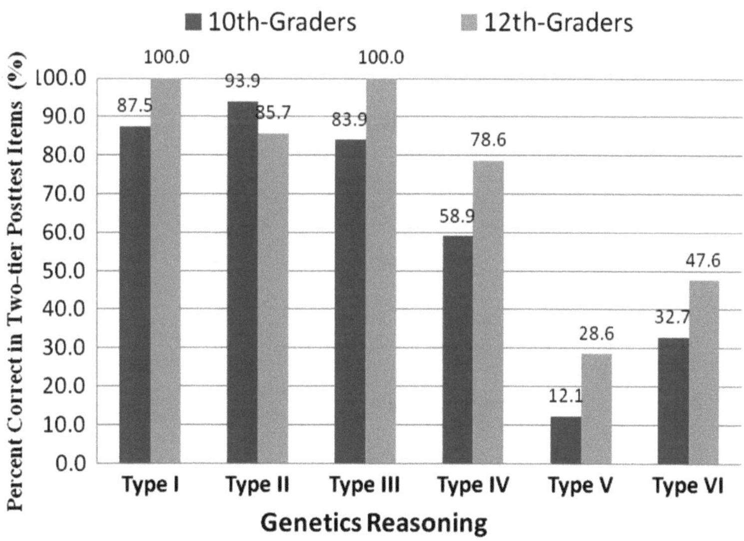

Fig. 15.7 Genetics reasoning by types in two-tier posttest of Australian 10th graders ($n = 56$; $n = 33$ for Types II and V items) and 12th graders ($n = 14$)

students interact with the MERs in terms of their model-based learning and their modeling skills (Buckley et al., 2004; Gobert, 2005).

There was also a limitation in our Australian study. We did not have a complete set of log files so that we could only analyze some case studies of students' log files to examine how they interacted with the MERs of *BioLogica* and with each other (from

Table 15.6 Dialogic interactions between Helena and May during the BioLogica activity Monohybrid

Time	Line	Transcript of dialogue from audio recordings	Helena's *Monohybrid* log file segments (from 16:13:45 to 16:14:10)
16:13:45	1	May: If you use the same two dragons again	. . .
	2	do you think. . .	<date>2002.08.06.16.13.45 08/06/02 | 16:13:45 </date>
	3	Helena: Mine is different to yours.	Got a plain-tailed dragon in 2 tries. Next cross will have 30 offspring.
	4	May: You'll get a fancy tailed baby. Oh there	</action>
	5	you go. After three tries	<action>
	6	you get a fancy. . .	<date>
	7	Helena: What do you do? Mine's different to	2002.08.06.16.14.10 08/06/02 | 16:14:10
	8	yours.	</date >
	9	May: What have you done? Okay, click off.	Created a total of 30 offspring, of which 16 have plain tails and 14 have fancy tails.
	10	Now do the same thing as you did to get	</action>
	11	the first one. Go from the circle. The	. . .
	12	black circle.	
	13	Helena: Whoops.	
	14	May: The little black circle, and go to that	
	15	white square. There you go.	
	16	Helena: Mm hm. You do the same thing?.	
	17	May: But you got it [a plain-tailed] after two	
	18	tries. (Reading from screen) "A	
	19	question for you. If you made say 30	
	20	more babies how many do you I think	
	21	will have fancy tails?. . . what did you	
	22	do?. . ."	

classroom audio recordings) during the activities. For example, a dyad of 12th graders, Helena and May of School U in Perth, had the following episode in which they had dialogic interactions while working on a task of the *Monohybrid* activity that could be interpreted by juxtaposing Helena's log file with a reconstructed screenshot of the *BioLogica* program she was using (see Table 15.6 and Fig. 15.8).

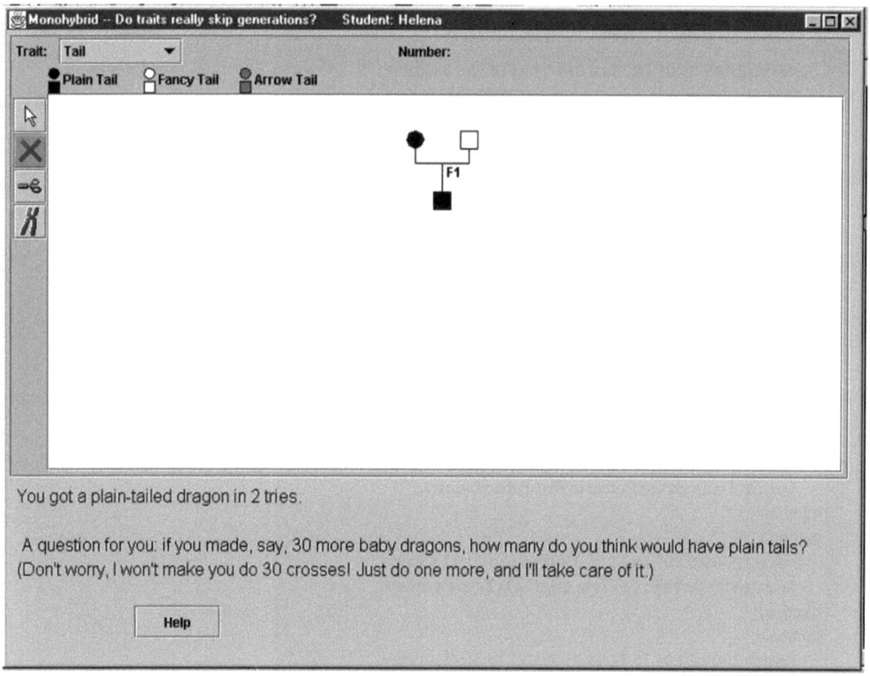

Fig. 15.8 A screenshot of the *BioLogica* activity *Monohybrid* reconstructed based on the information of the corresponding log file of Helena (from 16:13:45 to 16:14:10)

Verbatim transcription of the audio recordings of the two students' dialogic interactions in Table 15.6 indicates that Helena was being encouraged (e.g., lines 4–6 in Table 15.6) and scaffolded (lines 9–12; 14–15) by May, her more confident peer, during this same predict-observe-explain (POE) task on which they were working at their own computers next to each other. After May had read out the question on the screen (lines 18–22 in Table 15.6), Helena did not answer the question to *predict* what would happen as indicated in the log file (no text was logged between markup tags "</action>" and "<action>" before the time 16:14:10), but she went on with the POE task to breed 30 baby dragons to *observe* what happened next. The last part of the log file in Table 15.6 summarizes the results of Helena's action. Then, in the next part the computer would ask the users to explain (not shown in Table 15.6).

In the Hong Kong study, we had collected a complete set of the log files of all students and we analyzed in detail the log files of four target students and the students' errors in the activities. The following episode illustrates Mei-yee's (one of the four target students) interactions with the computer during the *Monohybrid* activity. She was working on a task of Type II reasoning (see Table 15.1) as illustrated by a snippet of the log file tracking her interactions with *BioLogica* that corresponded to the reconstructed screenshot at 16:31:48 (see Figs. 15.9 and 15.10).

```
<action>
    <date> 2009.03.05.16.28.10 03/05/09 | 16:28:10 </date>
    Selected the wrong zygotes in the tt X Tt Punnett square.
</action>
<action>
    <date> 2009.03.05.16.29.46 03/05/09 | 16:29:46 </date>
    Looked Dad's chromosomes in first Punnett square.
</action>
<action>
    <date> 2009.03.05.16.29.49 03/05/09 | 16:29:49 </date>
    Looked Mom's chromosomes in first Punnett square.
</action>
<action>
    <date> 2009.03.05.16.30.11 03/05/09 | 16:30:11 </date>
    Looked Mom's chromosomes in first Punnett square.
</action>
<action>
    <date> 2009.03.05.16.30.19 03/05/09 | 16:30:19 </date>
    Looked Dad's chromosomes in first Punnett square.
</action>
<action>
    <date> 2009.03.05.16.31.23 03/05/09 | 16:31:23 </date>
    Selected the wrong zygotes in the tt X Tt Punnett square.
</action>
<action>
    <date> 2009.03.05.16.31.48 03/05/09 | 16:31:48 </date>
    Selected the right zygotes in the tt X Tt Punnett square.
</action>
```


Fig. 15.9 A snippet of Mei-yee's log file corresponding to the computer-user interactions that followed the screenshot in Fig. 15.10

As the log file in Fig. 15.9 indicates, at 16:28:10, Mei-yee had just successfully completed her first task to use the Punnett square to work out the possible combinations of alleles in a monohybrid cross between two dragon parents (a fancy-tailed dad and plain-tailed mom, i.e., Tt × tt). Then, she was asked to select all the zygotes in the Punnett square that would develop into plain-tailed baby dragons in order to work out the proportion of plain-tailed baby dragons in the offspring.

The log file in Fig. 15.9 continues to indicate that after Mei-yee's first attempt to select the right zygotes failed, she repeatedly viewed the chromosomes of the parents' to check out their genotypes (by clicking on the dragon icons). However, she made another wrong attempt before she finally selected the right zygotes at 16:31:48. That is, the two zygotes with genotype tt or two of the four possible cases in the cross (tt × Tt) that would develop into plain-tailed baby dragons. In so doing, she had achieved the two learning goals of the *Monohybrid* activity for using a Punnett square in solving problems in Mendelian genetics (see Table 15.2).

We next analyzed the log files of three selected activities—*Meiosis, Monohybrid*, and *Mutations*—by counting the number of students' errors in using the *BioLogica* activities, including their wrong answers to questions and unsuccessful attempts to

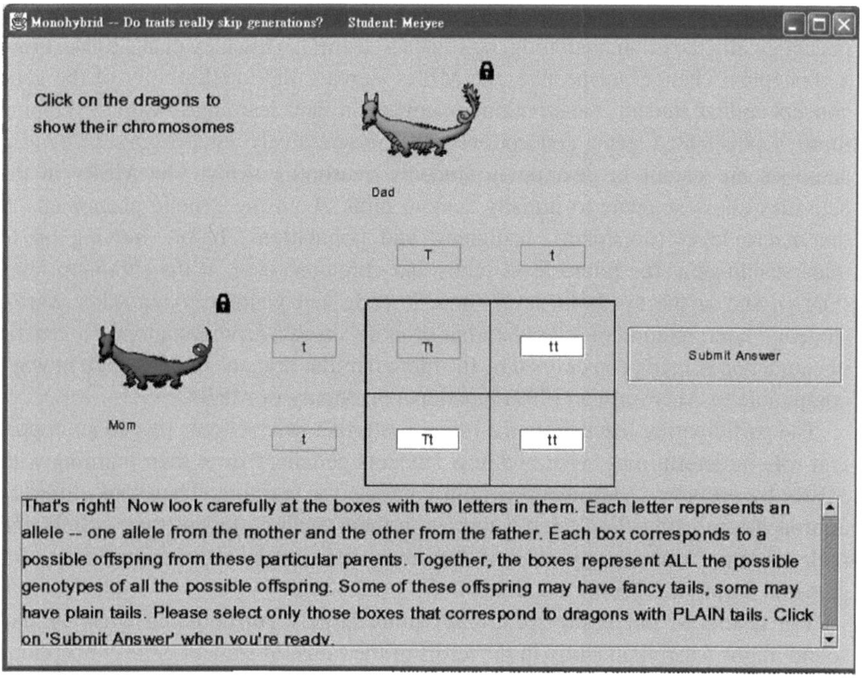

Fig. 15.10 A reconstructed screenshot of Mei-yee's interactions with *BioLogica Monohybrid* activity as tracked by the log file with the snippet shown in Fig. 15.9

solve problems (e.g., Mei-yee made two errors during part of the *Monohybrid* activity as indicated by the log file in Fig. 15.9). We wanted to find out the relation between these students' errors and their genetics reasoning as indicated by the two-tier posttest results. The results of an SPSS correlation analysis indicated that students' genetics reasoning skills ($r = -.428$; $p = 0.034$, one-tailed; $N = 19$) had significant negative correlation ($p < .05$, one-tailed) with their errors in using these three *BioLogica* activities. These results suggest that the tasks, puzzles, and embedded assessment questions of the *BioLogica* activities can be used to evaluate students' understanding of genetics in terms of the six types of reasoning as they work through the progressively challenging activities.

Discussion and Conclusions

The findings of our studies suggest that the MERs of *BioLogica* provided students with complementary information and processes about genetics across the dynamically linked levels of organization. These manipulable MERs, particularly the visual-graphical representations of the genetic phenomena, co-deployed simultaneously with scripts or texts—including narratives, tasks and puzzles, representational assistance, reasoning models, and explanations and feedback—are pedagogically useful in mediating the students' learning (Buckley et al., 2004). From a conceptual change perspective, the MERs increase the intelligibility of the gene concept so that students can continue to engage in their learning toward developing more sophisticated gene conceptions. The progressively challenging *BioLogica* activities are useful in developing students' reasoning skills. The MERs in the activities allow students to initially start to think about the genetic phenomena at the macro level (organisms, pedigrees, and populations) before moving on to understandings at the micro level (cells and chromosomes), at the submicro level (DNA), and at the symbolic level (genetic code and genotypes). In other words, students' interpretation of a less familiar or more abstract representation of a genetic phenomenon is being constrained by the more familiar dragons in *BioLogica* in ways compatible to Ainsworth's (1999) functional taxonomy of MERs.

The participating teachers in the three Australian case schools played an important role in determining what and how students benefited from their learning with MERs by providing various classroom contexts for learning. They took different approaches in using *BioLogica* activities in their teaching to suit their beliefs and their students' learning styles, thus providing different learning opportunities for students during the genetics course.

The results of our Hong Kong case study largely corroborated what we had found in our Australian study in the terms of the range of student gene conceptions and their reasoning skills. It is interesting that the reasoning pattern of the Hong Kong students was similar to that of the Australian counterparts in Perth notwithstanding the linguistically and culturally different learning contexts across the schools in the two cities. Visualization can play an important role in scaffolding knowledge construction and conceptual understanding for the Hong Kong students who are *English language learners* as shown by some studies (e.g., Dixon, 1995).

Most students in our Hong Kong study appeared to have learned some reasoning skills and improved in their confidence to write biology in English. Although the Hong Kong students had not learned genetics before the study, their prior knowledge of genetics in Chinese acquired from the media and the Internet appeared to have helped their understanding of genetics in English. Code-mixing in classroom discourse helped these ELL students to access and capitalize on their L1 (Cantonese) linguistic resources for learning concepts in science in their L2 (English) (e.g., Lin, 2006). This interpretation is compatible with the sociocultural perspectives of learning, especially in terms of the constructs of *zone of proximal development* (Vygotsky, 1978) and *verbal thought* (Vygotsky, 1968).

Psycholinguistic research has indicated that bilinguals are able to learn concepts by way of the developmental shift from lexical mediation between their L1 and L2 to direct word-concept access or conceptual links in their L1 and then L2 (Kroll & Hermans, 2011) as shown in Fig. 15.11.

Accordingly, such a shift depends on the bilingual learners' increasing ability to directly process the concepts in L2 without L1 mediation. For example, using mixed-coding and code-switching, a Hong Kong science teacher can embed key terms in L2 and concepts in a rich L1 semantic context and then illustrate L2

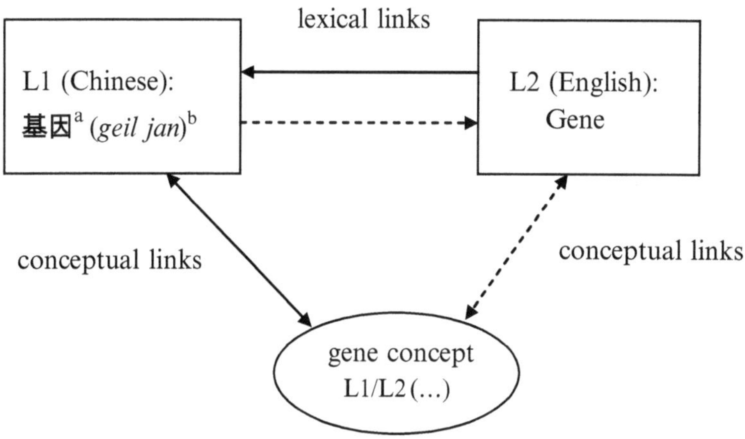

[a]The word "gene" translated into Modern Standard Chinese in written script; "基" means "fundamental" and "因" means "factor;" in Mandarin (Putonghua) the phonetics (*pīnyīn*) for the word "gene" is *jīyīn*; therefore, the Chinese translation of "gene" is semantically and phonemically similar to the English equivalent.

[b]Phonetics to indicate the pronunciation of "基因" in Cantonese (spoken Chinese dialect in Hong Kong) is from the online dictionary of the Chinese University of Hong Kong: http://humanum.arts.cuhk.edu.hk/Lexis/lexi-can/

Fig. 15.11 A proposed model adapted from Kroll and Hermans (2011, p. 18) to illustrate how Hong Kong ELL students might possibly learn the concept of the gene in a bilingual way. The links indicated by *dotted arrows* will become *solid* when learners have acquired better L2 skills

abstract scientific concepts with concrete L1 everyday life experiences and examples and so on (Lin, 2006). In terms of the second pedagogical function of MERs, the interpretation of biological concepts in a less familiar L2 representation can be constrained by its more familiar L1 representation for better understanding.

Unlike the Australian study, where the teachers taught genetics and used *BioLogica* activities to a lesser or greater extent in class to support student learning, the Hong Kong students learned genetics largely from *BioLogica* activities. The Hong Kong students completed all eight *BioLogica* activities and were seldom absent from the after-school program. Therefore, we have reason to believe that the causal relation between the usage of the interactive activities and student understanding should be stronger in the Hong Kong study. Just as in our Australian study, we found that mere engagement in the *BioLogica* activities interacting with the MERs may not be useful for developing deep understandings. Apart from the difference in the individual and classroom factors, interactions with the MERs in *BioLogica* need to be mindful and intrinsically motivated for students to benefit from such interactions in developing their understanding (Tsui & Treagust, 2004).

We summarize here the implications of our research discussed in this chapter. First, the MERs of *BioLogica* within different classroom contexts—Australian teachers' stories and games, web-based activities, and other approaches, as well as Hong Kong's bilingual and mixed-code classroom discourse—appeared to provide different learning opportunities for students to undergo conceptual change toward developing more sophisticated gene conceptions. Second, complementary and constraining functions of the MERs appeared to promote students' construction of deep understanding of the genetic phenomena. Thus, they were able to move on to coherently *relate* the hierarchically arranged objects and events of genetics, *abstract* the genetic phenomena (phenotypes and inheritance patterns) to symbols (genotypes and DNA code) for reasoning and problem solving, and *extend* such understandings (e.g., sex-linkage) to real-life human examples. Third, MERs of *BioLogica* appeared to provide scaffolding for ELL students with limited English language skills for developing scientific reasoning by way of visual-graphical representations dynamically linked to texts. Furthermore, bilingual representations and discourse also might have scaffolded these ELL students to develop better understanding.

To conclude, these two case studies have provided some detailed evidence and thick descriptions (Merriam, 1998) for the claims that learning with multiple representations can be pedagogically useful (cf. Ainsworth, 1999) within different learning contexts for students' conceptual understanding and reasoning in biology, particularly for students from diverse backgrounds. We believe this is important at a time when the latest trend of science education is directed toward globalization (cf. Chiu & Duit, 2011).

Acknowledgments Our thanks go to the participating teachers and students in Perth and Hong Kong schools. We are grateful to Paul Horwitz of the Concord Consortium of the USA for granting us permission for using *BioLogica* in our studies and for reviewing an earlier draft of this chapter. We also thank Barbara Buckley for her comments on the final draft of this chapter.

References

Ainsworth, S. (1999). The functions of multiple representations. *Computers in Education, 33*(2/3), 131–152.

Ainsworth, S., Bibby, P. A., & Wood, D. J. (1997). Information technology and multiple representations: New opportunities – new problems. *Journal of Information Technology for Teacher Education, 6*(1), 93–104.

Bahar, M., Johnstone, A. H., & Hansell, M. H. (1999). Revisiting learning difficulties in biology. *Journal of Biological Education, 33*(2), 84–86.

Buckley, B. C., Gobert, J. D., Kindfield, A. C. H., Horwitz, P., Tinker, R., Gerlits, B., et al. (2004). Model-based teaching and learning with BioLogica^{tm}: What do they learn? How do they learn? How do we know? *Journal of Science Education and Technology, 13*(1), 23–41.

Chiu, M.-H., & Duit, R. (2011). Globalization: Science education from an international perspective. *Journal of Research in Science Teaching, 48*(6), 553–566.

Concord Consortium. (2002). *BioLogica teachers' guide*. Retrieved June, 2002, from http://mac.concord.org/downloads/

Denzin, N., & Lincoln, Y. S. (Eds.). (1994). *Handbook of qualitative research*. Thousand Oaks, CA: Sage.

Dixon, J. K. (1995). Limited English proficiency and spatial visualization in middle school students' construction of the concepts of reflection and rotation. *Bilingual Research Journal, 19*(2), 221–247.

Duit, R., & Glynn, S. (1996). Mental modelling. In G. Welford, J. Osborne, & P. Scott (Eds.), *Research in science education in Europe: Current issues and themes* (pp. 166–176). London: The Falmer Press.

Duncun, R. G., & Reiser, B. J. (2007). Reasoning across ontologically distinct levels: Students' understanding of molecular genetics. *Journal of Research in Science Teaching, 44*(7), 938–959.

Duncan, R. G., Rogat, A. D., & Hmelo-Silver, C. E. (2009). A learning progression for deepening students' understandings of modern genetics across the 5th-10th grades. *Journal of Research in Science Teaching, 46*(6), 655–674.

Erickson, F. (1998). Qualitative research methods for science education. In B. J. Fraser & K. G. Tobin (Eds.), *International handbook of science education* (pp. 1115–1173). Dordrecht, The Netherlands: Kluwer.

Gobert, J. D. (2005). A model of molecular visualization. In J. K. Gilbert (Ed.), *Visualization in science education* (Vol. 1, pp. 61–72). Dordrecht, The Netherlands: Springer.

Gobert, J. D., O'Dwyerb, L., Horwitz, P., Buckley, B. C., Levy, S. T., & Wilensky, U. (2011). Examining the relationship between students' understanding of the nature of models and conceptual learning in biology, physics, and chemistry. *International Journal of Science Education, 33*(5), 653–684.

Hackling, M. W., & Lawrence, A. J. (1988). Expert and novice solutions of genetic pedigree problems. *Journal of Research in Science Teaching, 25*(7), 531–546.

Hackling, M. W., & Treagust, D. F. (1984). Research data necessary for meaningful review of grade ten high school genetics curricula. *Journal of Research in Science Teaching, 21*(1), 197–209.

Hickey, D. T., & Kindfield, A. C. H. (1999, April). *Assessment-oriented scaffolding of student and teacher performance in a technology-supported genetics environment*. Paper presented at the annual meeting of the American Educational Research Association, Montreal, Canada.

Horwitz, P. (1995). Linking model to data: Hypermodels for science education. *The High School Journal, 79*(2), 148–156.

Horwitz, P., & Christie, M. A. (2000). Computer-based manipulatives for teaching scientific reasoning: An example. In M. J. Jacobson & R. B. Kozma (Eds.), *Innovations in science and mathematics education: Advanced design for technologies of learning* (pp. 163–191). Hillsdale, NJ: Lawrence Erlbaum Associates.

Horwitz, P., & Tinker, R. (2001). Pedagogica to the rescue: A short history of hypermodels. *Concord Newsletter, 5*(1), 1–4.

Johnstone, A. H. (1991). Why is science difficult to learn? Things are seldom what they seem. *Journal of Computer Assisted Learning, 7*, 75–83.

Justi, R. S., & Gilbert, J. K. (2002). Science teachers' knowledge about and attitudes towards the use of models and modelling in learning science. *International Journal of Science Education, 24*(12), 1273–1292.

Kroll, J. F., & Hermans, D. (2011). Psycholinguistic perspectives on language processing in bilinguals. In M. S. Schmid & W. Lowie (Eds.), *Modeling bilingualism: From structures to chaos* (pp. 15–36). Amsterdam/Philadelphia: John Benjamins.

Lawson, A. E. (1992). What do tests of "formal" reasoning actually measure? *Journal of Research in Science Teaching, 29*(9), 965–983.

Lin, A. (2006). Beyond linguistic purism in language-in-education policy and practice: Exploring bilingual pedagogies in a Hong Kong science classroom. *Language and Education, 20*(4), 287–305.

Marbach-Ad, G., & Stavy, R. (2000). Students' cellular and molecular explanations of genetic phenomena. *Journal of Biological Education, 34*(4), 200–205.
Merriam, S. B. (1998). *Qualitative research and case study applications in education.* San Francisco: Jossey-Bass.
Pearson, J. T., & Hughes, W. J. (1988). Problems with the use of terminology in genetics education: 1 A literature review and classification scheme. *Journal of Biological Education, 22*(3), 178–182.
Posner, G. J., Strike, K. A., Hewson, P. W., & Gertzog, W. A. (1982). Accommodation of a scientific conception: Toward a theory of conceptual change. *Science Education, 66*(2), 211–227.
Stewart, J. H. (1982). Difficulties experienced by high school students when learning basic Mendelian genetics. *The American Biology Teacher, 44*(2), 80–84. 89.
Treagust, D. F. (1988). Development and use of diagnostic tests to evaluate students' misconception in science. *International Journal of Science Education, 10*(2), 159–169.
Tsui, C.-Y. (2009, October). *Developing scientific reasoning with multiple external representations: A case study in a Hong Kong secondary school.* Paper presented at the International Conference of the East-Asian Science Education (EASE), Taipei, Taiwan.
Tsui, C.-Y., & Treagust, D. F. (2003). Genetics reasoning with multiple external representations. *Research in Science Education, 33*(1), 111–135.
Tsui, C.-Y., & Treagust, D. F. (2004). Motivational aspects of learning genetics with interactive multimedia. *The American Biology Teacher, 66*(4), 277–286.
Tsui, C.-Y., & Treagust, D. F. (2007). Understanding genetics: Analysis of secondary students' conceptual status. *Journal of Research in Science Teaching, 44*(2), 205–235.
Tsui, C.-Y., & Treagust, D. F. (2010). Evaluating secondary students' scientific reasoning in genetics using a two-tier diagnostic instrument. *International Journal of Science Education, 32* (8), 1073–1098.
Tyson, L. M., Venville, G. J., Harrison, A. G., & Treagust, D. F. (1997). A multidimensional framework for interpreting conceptual change events in the classroom. *Science Education, 81*, 387–404.
van Someren, M. W., Reimann, P., Boshuizen, H. P. A., & de Jong, T. (Eds.). (1998). *Learning with multiple representations.* London: Pergamon.
Venville, G. J., & Donovan, J. (2005). Searching for clarity to teach the complexity of the gene concept. *Teaching Science, 51*(3), 20–24.
Venville, G. J., & Treagust, D. F. (1998). Exploring conceptual change in genetics using a multidimensional interpretive framework. *Journal of Research in Science Teaching, 35*, 1031–1055.
Vygotsky, L. S. (1968). *Thought and language.* Cambridge, MA: MIT Press.
Vygotsky, L. S. (1978). *Mind in society: The development of higher psychological processes.* Cambridge, MA: Harvard University Press.
Wood, R. (1996). *From my DNA to Darwin: Learning and teaching in genetics.* D.Phil thesis, University of Waikato, Hamilton, New Zealand.

Chapter 16
The Hidden Hand that Shapes Conceptual Understanding: Choosing Effective Representations for Teaching Cell Division and Climate Change

Kai Niebert, Tanja Riemeier, and Harald Gropengießer

An Episode from a Teaching Experiment on Cell Division

In this teaching experiment (Riemeier & Gropengießer, 2008), an onion, whose roots had grown into a water-filled jar, was shown to three girls aged 15 years. The girls were asked to describe what had happened with the onion. Initially, the students were surprised by this question. Roots and plants commonly grow—so what? When the researcher asked how onion roots grow, the girls enumerated conditions of growth like the need for water or nutrients. After a while, one girl remembered the cellular structure of the onion, whereupon all three students explained the growth of onion roots by multiplication of cells. The onion produces a lot of cells, and in the case of a sufficient number of cells, the human eye is able to see a root. This process of multiplying cells was named *cell division* by the students. The growing number of cells brings forth the growth of onion roots. One student outlined her conception in a drawing (see Fig. 16.1).

At this point, a learning activity was offered to the girls. A bar of chocolate was shown, and the students were asked to break it into squares and subsequently compare this process to cell division. In doing so, the girls recognized that despite the increased number of chocolate pieces, the pieces were smaller than the whole bar of chocolate. "But in case of a cell, it wouldn't yield anything; it would be the same size." The girl assumed the cell would not divide in the sense of getting smaller but rather ending up with two cells of similar size: "It is more like copying itself." The other students agreed upon this. After a while, one student pushed it a little bit further: "The cell divides itself in the middle and grows thereupon."

K. Niebert (✉) • T. Riemeier • H. Gropengießer
Institute for Science Education, Leibniz Universität Hannover, 30167 Hannover, Germany
e-mail: niebert@idn.uni-hannover.de; tanja.riemeier@googlemail.com;
gropengiesser@idn.uni-hannover.de

D.F. Treagust and C.-Y. Tsui (eds.), *Multiple Representations in Biological Education*, Models and Modeling in Science Education 7, DOI 10.1007/978-94-007-4192-8_16,
© Springer Science+Business Media B.V. 2013

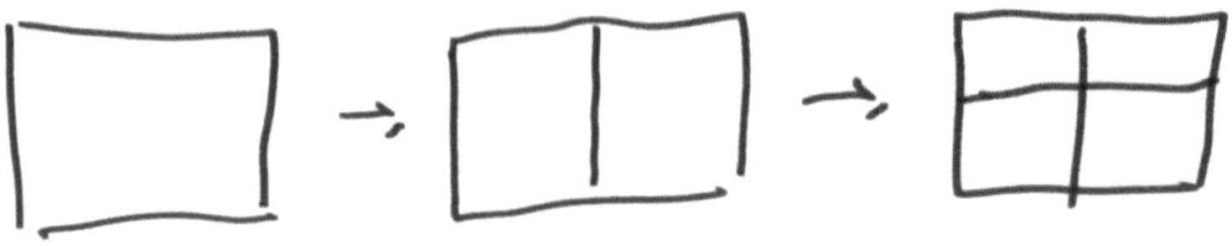

Fig. 16.1 Drawing of cell division by a student aged 15 years

What Happened? A Brief Analysis of the Learning Episode

At the outset, the students conceptualized "growth" as a normal process that happens perpetually. There was no need for them to explain this process. It just happens, if the conditions of growth are given. Later, however, *cells* came to mind, and the term *cell division* was used. Even though the students were using this scientific term, they adhered to its literal meaning. They thought of division and thus multiplication of cells exclusively. Thus, the students got the idea that more cells suffice to accomplish the growth of onion roots. The students were seduced by a chain of false reasoning: (1) More pieces are obtained by division, (2) more pieces of root cells result in more root, and (3) more root means growth of the root. To put it more briefly, a growing number of cells will lead to root growth.

Breaking a bar of chocolate shed doubt on the students' former certainty and stimulated analogical thinking (e.g., Aubusson, Harrison, & Ritchie, 2006) in connection with new hypotheses. Finally, this learning activity led the students to a conception of root growth—by cell division and cell growth—that is adjacent to scientific theory. Obviously, breaking chocolate helped students in the conceptual reconstruction of growth. We have evidence of learning progress in several teaching experiments, but how could we come up with such a remote idea as to have students break a bar of chocolate for the process of learning about the cellular basis of onion root growth? And still more challenging is the question: Why was this learning activity successful?

Striving for Effective Representations

The above analysis of the learning episode shows that students initially held conceptions on growth and cell division. These preconceptions we regard as points to start with in further learning. The goals and objectives of teaching and learning require consideration of scientists' conceptions of this particular science content. To this end, we carried out what we call scientific clarification, that is, the critical analysis of scientific terms, conceptions, and content structure from an educational point of view (Duit, Gropengießer, & Kattmann, 2005). In contrast to the students' preinstructional conceptions, cell division means to scientists that the division of cells is accompanied by the growth of cells. Otherwise, no growth would occur. The students' conceptions initially indicated no understanding of cell growth in connection with division. They focused exclusively on the multiplication of cells through division.

Having identified students' conceptions and perspectives on the one hand and clarified scientific theory on the other, the task turns to shaping a learning environment. Students' conceptions and the clarified scientific view on root growth are of equal importance for the design of a learning environment; however, they differ in function. Students' perspectives mark the precondition and source of teaching and learning, whereas the clarified scientific view constitutes the goals and objectives. The pathway of learning works through a designed learning environment, which should challenge students' ideas on growth and cell division and foster conceptual reconstruction (Duit & Treagust 2011; Kattmann, 2008). The learning environment primarily consists of effective representations that let students reflect on their initial ideas (cf. Chap. 15 by Tsui & Treagust, this volume; Chap. 6 by Yarden & Yarden, this volume).

Schemata that Shape Understanding of Cell Division

Breaking a bar of chocolate fosters students' conceptual development toward scientific understanding of cell division and growth, insofar as we have evidence that this is an effective representation. If this is the solution, what is the problem of understanding the idea of cell division?

Schemata Are Shaping Understanding

We answer the question of understanding based on the theory of experientialism (Gropengießer, 2003; Lakoff & Johnson, 1980, 1999). According to this theoretical framework, we hold true that thought is embodied, that is, our basic conceptions grow out of bodily experience. Our basic categories of thought and concepts arise out of perception, body movement, and experience with our physical and social environment. Experiences like *up-down*, *center-periphery*, *front-back*, or *inside-outside* are conceptualized in schemata. These schemata are conceptual systems arising from direct experience. For instance, our conception of *up-down* is organized as the verticality schema that is grown directly out of our experience with gaining an upright position. There are several other schemata like the *container schema* or the *source-path-goal schema* (see also Lakoff, 1990) which are conceptual structures grounded in bodily experience that is understood directly. These schemata shape our conceptual understanding.

In our episode, the girls were surprised to be questioned about the growth of onion roots. Growth is seen as a normal process that needs no explanation. We experience growth in everyday life. Consequently, growth of living things—like an onion root—is an embodied conception. Growth to us means *becoming bigger* because we experience that all living organisms (i.e., trees, humans, dogs) get bigger while growing. In contrast, the cellular structure of the onion root cannot be experienced directly in everyday life. We do not have any experiences with the process of cell division on the microscopic level in our normal course of life. However, each interviewee described some conception of how a cell is imagined and understood. Obviously, a concept of cell division is not embodied in the same way as the above-mentioned schemata. Instead, they are thought of in an imaginative way. Imaginative thinking is accomplished mainly by metaphors and analogies. Thus, guided by experientialism, we distinguish between embodied conceptions and imaginative conceptions (Lakoff 1990). The latter are not directly grounded in experience but draw on the structure of our experience. We use our embodied schemata to explain abstract phenomena. Thus, imagination can be seen as bridging the gap between our experience and abstract phenomena. We employ conceptions from a source domain (i.e., the *container schema*) and map them onto an abstract target domain (i.e., *cells*) to understand abstract phenomena. Thus, the use of imagination requires source-target mapping. The structure of a source domain is projected onto a target domain.

Experiential Basis of (Cell) Division

For instance, the process of cell division cannot be experienced directly in everyday life, although the students of our episode thought about growth through division of cells. For them, cells have to multiply for growing onion roots, and this multiplication could be done by dividing the cell into two halves. The students called this process cell division. Based on experientialism, we explored the conceptual schema of *division*. Grounded in our everyday experiences, two different meanings of *division* can be distinguished according to the outcomes of the process of parting (see Fig. 16.2). Division may be conceptualized as resulting in (a) *more* single parts or (b) *smaller* parts than the whole object.

Thinking about growth solely through division of cells follows a logic outlined as follows: Growth requires becoming more, whereas becoming smaller sounds contradictory. Viewed from this perspective, the girls' idea of cell division in the context of growth is traceable and to be expected. The girls found it obvious that division and the resulting multiplication of cells explain onion root growth. In the context of a classroom, teachers probably would judge the term cell division as denoting the underlying scientific idea.

To develop the scientific way of thinking, we decided to shed light on the division schema. We searched for representations likely to advance reflection about the different meanings of division. In our episode, dividing a bar of chocolate helped to bring the schema of division to students' critical attention. Nonetheless, it is also effective to divide a sheet of paper (Schneeweiss & Gropengießer, 2010). Thereupon, the students were able to recognize that cells have to divide and increase to a normal size. A representation of the conceptual schema of division induced students' reflection and fostered conceptual development toward scientific understanding.

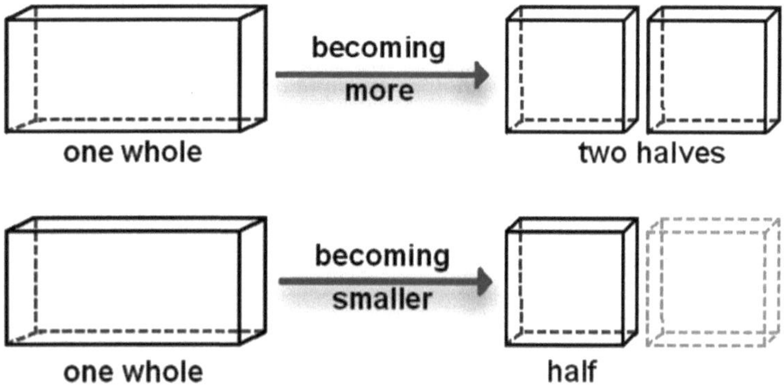

Fig. 16.2 Experience-based meanings of *division*

Reexperiencing and Reflecting on the Division Schema

Following the ideas of experientialism, we hold to the view that language and thought are based on the same conceptual structures. Language is thereby a window into students' conceptions. Generally, we distinguish between three domains—cognition, language, and reality in a broad sense (Gropengießer, 2003; Richards & Ogden, 1923). Conceptions belong to the cognitive domain; they are expressed on the linguistic level through different symbols of speech or drawing. Therefore, students' statements are representations of their conceptions. The latter refers to a specific referent, that is, to an object, phenomenon, or occurrence. In our episode, students' conceptions are related to the referent *cell*, and we investigated these conceptions by analyzing students' drawings and their spoken language in video data. Breaking the bar of chocolate let the students reexperience the act of division and furthermore inspired them to reason by analogy concerning cell division. The students used this representation of the division schema to reflect on its meaning. Other instances showed that different representations of the division schema, like cutting a sheet of paper, could also induce reflections on cell division. This substantiates our assertion that a representation of the underlying division schema fosters students' conceptual development.

So far, we analyzed a learning episode from one of our teaching experiment series to provide insight into our theoretical framework. We have shown that students' understanding of growth through cell division is based on a basic schema of division (see Fig. 16.2). This is grounded in everyday experience and is thereby embodied in this context. In an attempt to understand cell division (i.e., a field where students had little or no previous experience), imagination comes into play. The structure of the division schema that serves as a source domain is projected onto cell division that serves as a target domain. The vague occurrence of cell division in the microcosm is understood in terms of an embodied division schema grown out of everyday experience. By reexperiencing and reflecting on the conceptual schema of division, students are able to develop conceptions that are scientifically more adequate.

Schemata Employed in Understanding Climate Change

In the following section, we give another example of a source-to-target mapping in the instance of climate change. Research on students' conceptions of climate change has shown that their conceptions often differ from the scientific perspective (Andersson & Wallin, 2000; Schreiner, Henriksen, & Hansen, 2005) and that alternative conceptions are very resistant to conceptual change (Ekborg & Areskoug, 2006; Sterman & Booth-Sweeney, 2007).

Conceptions of Climate Change

In Table 16.1, we present the scientific view and alternative conceptions of the global carbon cycle expressed in interviews by students at the age of 18. The analysis of these conceptions guided by experientialism shows that students and scientists have different conceptions, but both refer to the same schemata (Niebert & Gropengießer, 2011). We then present learning activities that focus on the schemata that serve as source domains in understanding the global carbon cycle as the target domain. Finally, we present students' conceptual development by using these learning activities.

The Intergovernmental Panel for Climate Change (IPCC, 2007) report described global carbon flows between different carbon pools. In the diagram (see Table 16.1), the pools are indicated as boxes and the flows as arrows; the flow rates and pool sizes are indicated by figures. Carbon seesaws between the boxes and cannot be lost. Since carbon may return on its path to one of the boxes from which it came, this is understood as the cycling of carbon. The figure illustrates this conception; the text specifies change—as long as there are just natural carbon flows, there is an assumed balance. Actually, flow rates differ from zero by anthropogenic effect. This is seen as a disturbed balance that causes global warming.

The conception *man-made* CO_2 shows that some students do not take CO_2 to be a natural component of the atmosphere, whereas the conception *natural versus man-made* CO_2 implies that CO_2 emitted by burning has another structure than CO_2 emitted by respiration.

Schemata of Climate Change

Although on a content level these conceptions are very different, scientists and students refer to the same schemata in thinking of the global carbon cycle: the *container-* and the *source-path-goal schema:*

- The container schema, as shown in Fig. 16.3a, is based on the experience that our body is a container, which has a sharp border between inside and outside crossed by input and output (Johnson, 1987).

Table 16.1 Conceptions of the causes of global warming

<table>
<tr><td>Scientific conception</td><td>Anthropogenic imbalance
"Carbon cycles between the atmosphere, oceans and land biosphere. [. . .] These fluxes [. . .] are approximately in balance when averaged over longer time periods. These [. . .] fluxes have become significantly different from zero." In [the] figure, the natural or unperturbed exchanges among oceans, atmosphere and land are shown. . . (IPCC, 2007, 501 ff.)
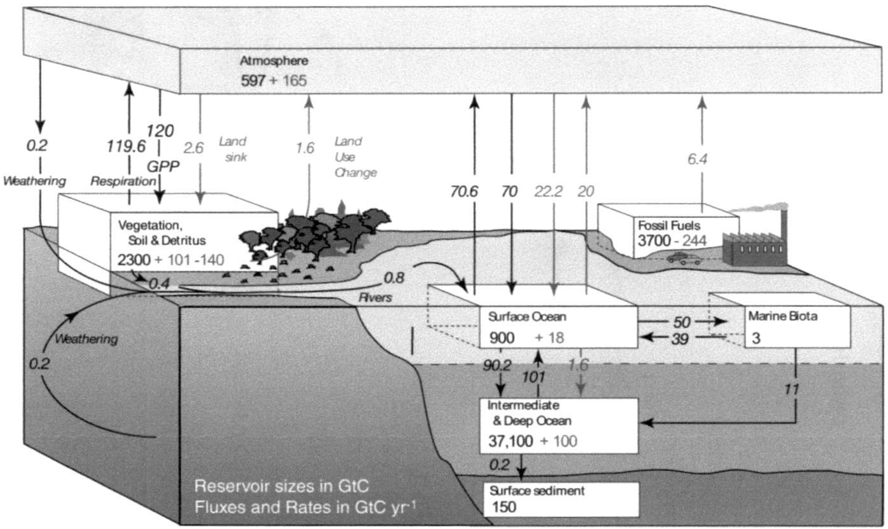
</td></tr>
<tr><td>Alternative conceptions</td><td>Man-made CO_2
"CO_2 is emitted into the atmosphere by burning coal and oil. Burning biofuel or wood does not emit CO_2. A normal atmosphere does not contain CO_2". (Detlef, 18 year)
Natural versus man-made CO_2
"Humans emit CO_2 by respiration. This CO_2 is captured by the plants. It is a fact, that the CO_2 emitted by burning has another structure than the CO_2 emitted by respiration. Thus, the CO_2 from burning cannot be captured again by photosynthesis." (Gustav, 18 year)</td></tr>
</table>

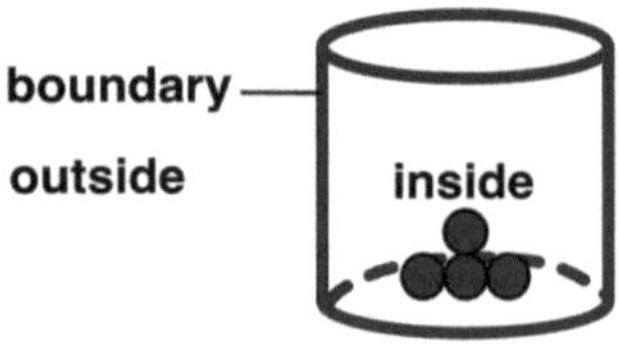

Fig. 16.3 The container schema

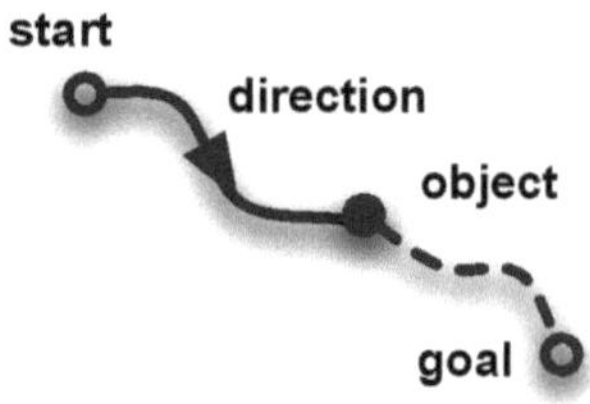

Fig. 16.4 The source-path-goal schema

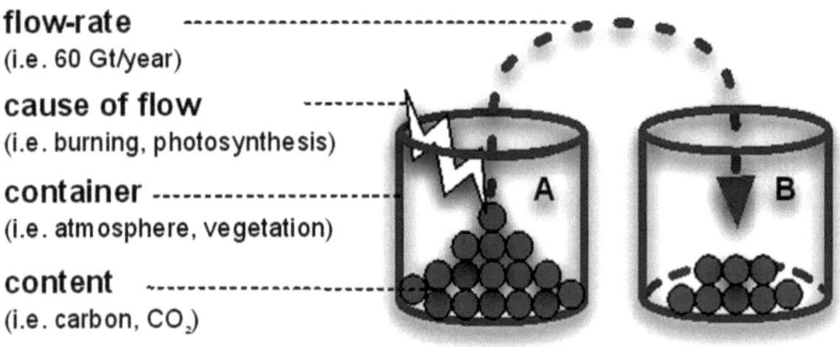

Fig. 16.5 The container-flow schema

- The source-path-goal schema, as shown in Fig. 16.4, is based on our locomotive experience of moving from A to B. An object (i.e., a person) moves from a starting point to a goal. The moving direction is defined by start and goal (Lakoff & Johnson, 1999).

As for the carbon, these two schemata are combined into a more complex *container-flow schema*. This container-flow schema, as shown in Fig. 16.5, is used to think of the atmosphere, ocean, and vegetation as containers enclosing carbon, which flows from one container to another (i.e., from fossil carbon to the atmosphere) by different causes (i.e., burning, respiration).

Terms like *into*, *flow between*, *flux*, *cycle between*, and *emission* indicate the use of the container-flow schema. The figure taken from the IPCC report combines several containers and flows (see Figs. 16.3, 16.4) into a complex container-flow schema (16.5) resulting in a quite typical model of the carbon cycle. Earth scientists often name containers as reservoirs, while they describe their content as the carbon pool. Scientists and students differ in the use of the container-flow schema. Scientists ascribe climate change to unbalanced flow rates of carbon into the atmosphere and thus an increasing amount of content (CO_2) in the atmosphere. Students use the container-flow schema differently to understand climate change. They either attribute climate change to the mere existence of part of the content (man-made CO_2) or to the existence of a different part of the content (man-made vs. natural CO_2) in the atmosphere (c.f. Table 16.1). Scientists source climate change in the carbon flow, whereas students ascribe it to the existence of a specific part in the content.

In understanding climate change, the container-flow schema is accompanied by two more conceptions: the distinction between natural versus man-made and the balance schema. Metaphor analysis shows the conceptions *man-made CO_2* and *natural versus man-made CO_2* have emerged from the judgment *natural is good, whereas man-made is bad*. This resembles the fallacy appeal to nature (Moore, 1996). Based on this judgment, the man-made CO_2 is attributed with devastating and detrimental properties, whereas an atmosphere without CO_2 or only with natural CO_2 is in an undisturbed, healthy state. The balance schema shapes our conceptual system with our first attempts of walking instead of crawling. This schema comprises logic, where each change is followed by a counterchange (Lakoff, 1990). Whereas scientists mainly use the balance schema to denote the causes of climate change (from balanced to unbalanced carbon flows), students distinguish between natural and man-made carbon content in the atmosphere as natural and man-made kinds of CO_2.

Learning with Representations of Schemata

Based on the analyzed comprehension of students and scientists, we defined the learning demand as (1) learning environments on the global carbon cycle should encourage students to reflect on their content-specific use of the container-flow schema and (2) reflecting on the distinction between natural and man-made should aim at reconstructing the causality of climate change: from *man-made matter* to *man-made cause*.

Design of the Learning Environments

Thus, we developed learning environments with representations of the schemata that students and scientists used to understand the carbon cycle (Niebert, 2009). The students were asked to transfer the scientific model of the carbon cycle to a materialized model of the container-flow schema. First, they interpreted and discussed a scientific representation of the carbon cycle that makes the container-flow schema explicit as shown in Fig. 16.6a. After doing so, they transferred their conceptions discussed to a materialized model of the container-flow schema with labeled boxes as containers and balls as carbon particles as shown in Fig. 16.6b. Thus, students could bring their conception of the carbon cycle into a material representation.

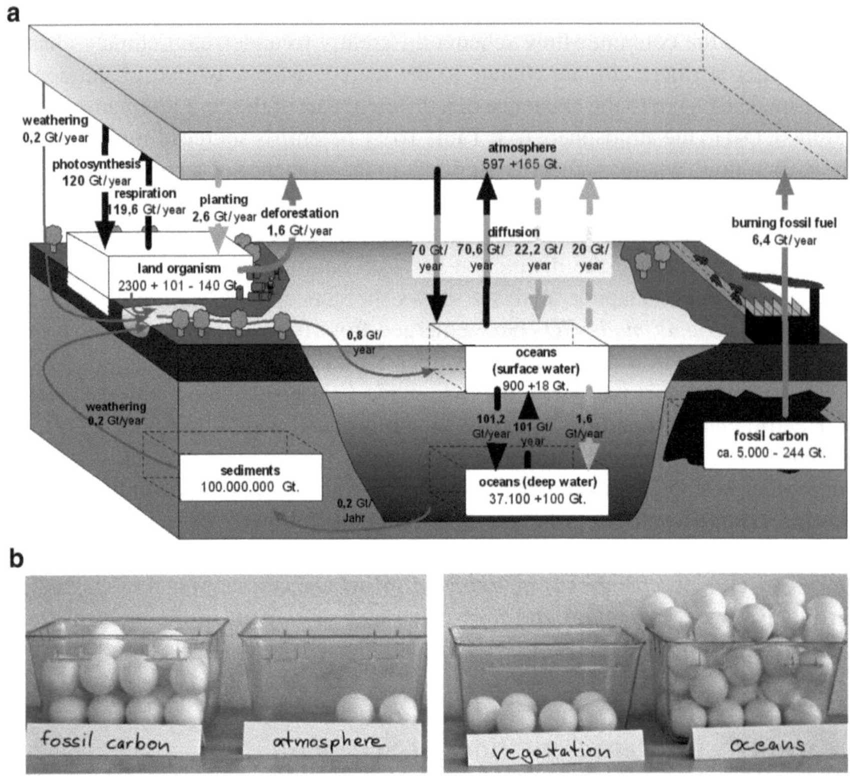

Fig. 16.6 Two representations of the carbon cycle

While working with the model, students were asked to explain the causes of climate change by using their model. We expected explanations based on unbalanced carbon flows into the atmosphere (i.e., burning and deforestation). The natural carbon flows between oceans/atmosphere and vegetation/atmosphere should be recognized as balanced carbon flows. It should become clear that it is not the amount but the balance of the carbon flows that matters.

Additionally, students who used the distinction between *natural* and *man-made* were asked to reflect it upon the model. This was intended to show CO_2 as a naturally occurring compound of the atmosphere and that CO_2 from burning fossil carbon and respiration is, in this respect, identical in structure and quality.

Understanding the Carbon Cycle by Modeling Conceptions

As shown in the following transcript of a teaching experiment (Niebert, 2009), a student reconstructed her conceptions of the causes of climate change. Initially, she argued, based on the distinction natural versus man-made, that the mere existence

Table 16.2 Container schema in the carbon cycle

	alternative conceptions		scientific conception
statement	**man-made CO_2** *Burning fossil fuel emits detrimental CO_2, burning wood does not emit CO_2.*	**man-made vs. natural CO_2** *Burning fossil fuel emits man-made CO_2, while respiration emits natural CO_2.*	**imbalance in carbon cycle** *Burning fossil carbon and cutting trees emits more CO_2 than trees can capture.*
container-schema	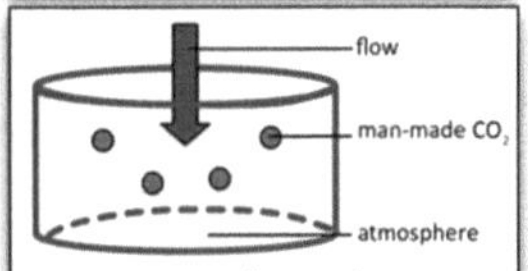 **appearance of a certain content**	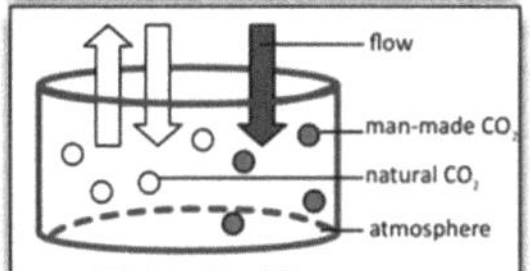 **addition of a different content**	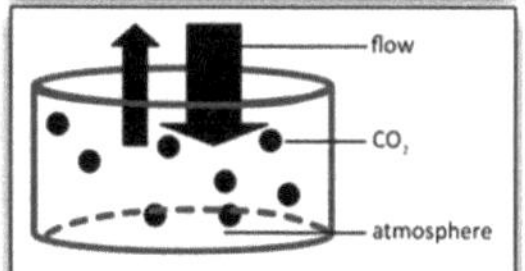 **increasing content**

of CO_2 is the cause of global warming (man-made CO_2). While at the end of the teaching experiment, her argumentation was based on a balance schema with too much CO_2 (*unbalanced carbon cycle*).

Interview at the Beginning of the Teaching Experiment

Interviewer: You said global warming is caused by CO_2. Please tell me where CO_2 comes from.

Tina: CO_2 is emitted by the industry, and it is not possible to reduce the CO_2 concentration to zero because of industrialization. The only way it would be possible is when we use nothing but renewable energy.

As shown in the above transcript at the beginning of the teaching experiment, Tina referred to the conception "man-made CO_2," where for her, carbon dioxide is produced solely by industrialization, that is, by burning fossil carbon (cf. Tables 16.1 and 16.2). Her conception implied that using renewable energy exclusively would reduce CO_2 emissions to zero.

Modeling the Carbon Cycle

Tina: Carbon enters the atmosphere from the organisms by respiration, and photosynthesis captures it again. Carbon from the oceans enters the atmosphere, but the same amount goes back into the oceans. There is a natural, a balanced cycling. [. . .] By deforestation, more CO_2 enters the atmosphere, and deforestation decreases photosynthesis because there are fewer trees. The carbon from deforestation stays in the atmosphere, because it cannot get down again. With the carbon from coal and oil, it is the same. It stays in the atmosphere, because not all CO_2 can be captured again; there is too much.

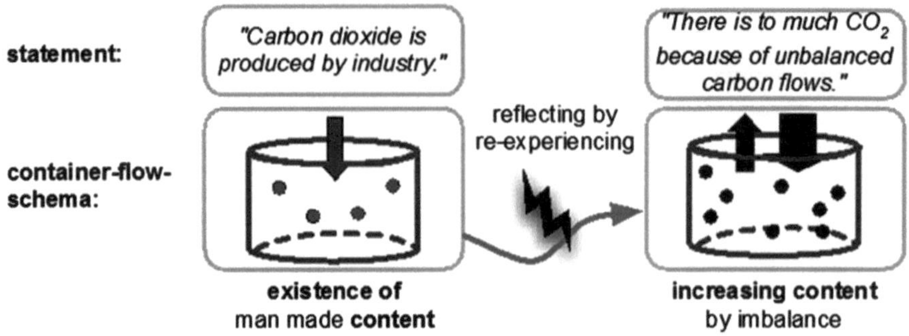

Fig. 16.7 Tina's conceptual development from man-made content to imbalance

While modeling the carbon cycle, Tina worked out the idea of a combination of balanced and unbalanced carbon flows and the cause of climate change. On the level of the employed conceptions, Tina reconstructed her conceptions from *man-made* CO_2 to *anthropogenic imbalance*. Modeling the carbon cycle with containers and balls seemed to help Tina to reconstruct her conceptions. In her argumentation at the end of the teaching experiment, she traced global warming not from the existence of CO_2 but from too much CO_2. For Tina, there is too much carbon emitted into the atmosphere to be captured by photosynthesis (see Fig. 16.7).

Modeling Schemata as a Way of Reconstructing Conceptions

Students who adhered to the conception *natural versus man-made* CO_2 were asked, in addition, to read a narrative—adapted by Niebert (2009)—from Levi's (1975) *The Periodic Table*. Levi described the carbon cycle by the cycling of a virtual carbon particle:

> Our character, a little carbon atom lies for hundreds of millions of years, bound to three atoms of oxygen and one of calcium, in the form of a limestone. In 1840 a man's pickaxe sent it on his way into the world of change. The carbon atom we are speaking of, accompanied by its two oxygen-satellites was therefore borne by the wind along a row of vines. It had the good fortune to brush against a leaf, penetrate it, and be nailed there by a ray of the sun... (p. 231)

The following example shows a student who reflected on his application of the judgment *natural versus man-made* on which he based his argumentation on climate change.

Interview at the Beginning of the Teaching Experiment

Gustav: It is a fact that the CO_2 emitted by burning has another structure than the CO_2 emitted by respiration. Thus, the CO_2 from burning cannot be captured again by photosynthesis.

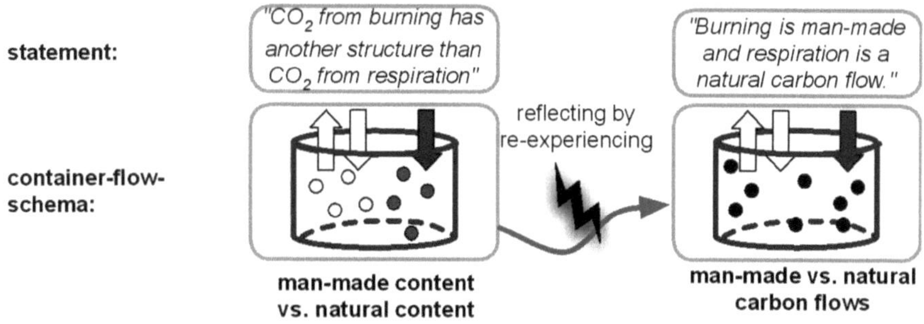

Fig. 16.8 Conceptual development in the use of the natural versus man-made schema

At the beginning of the teaching experiment, Gustav reassured himself about the two kinds of CO_2 he employed in his argumentation on climate change. He argued with the distinction between natural and man-made CO_2.

While Modeling the Story

Gustav: My idea with the natural and the man-made CO_2 was humbug, because in the story, the carbon, which was burned, is captured again by photosynthesis. And if the tale is right, the idea of a natural and a man-made CO_2 with different properties must be wrong. It is not the matter. The cause of emitting CO_2—the burning—is man-made. The emission of CO_2 by respiration is natural.

After modeling the carbon cycle, Gustav rejected the distinction between natural and man-made CO_2. The reason for this conceptual development (see Fig. 16.8) is the idea that *CO_2 is CO_2* which is mediated by the story, where CO_2 emitted—from fossil carbon as well as CO_2 emitted by respiration—is fixed again by photosynthesis. But after all, the distinction between natural and man-made played an important role in Gustav's argumentation. After modeling the carbon cycle in a container-flow schema, he assigned the distinction between natural and man-made no longer to the matter (CO_2) but to the cause for the carbon flow (burning and respiration).

Uncovering the Hidden Hand

So far, we have given evidence of the schemata that students employed in their efforts to understand cell division and climate change, and their understanding was shaped by schemata that were based on their everyday experiences. We now go one step further and uncover the "hidden hand" (Lakoff & Johnson 1999, p. 13) that shapes understanding.

Experiential Basis of Understanding

Viewed from everyday life, the growth of onion roots bears no question. As a matter of course, students perceive plants as growing. In the explanation of growth by cell division, the schema of division comes into play, which is structured by *becoming more* and *becoming smaller* (see Fig. 16.2). The girls drew from that aspect of the schema, which seems to fit: *Growth means more* cells. What was not considered is that division means cells *becoming smaller* too. They explained onion root growth by its production of more cells. At this point, the *division* schema with its aspect *becoming more* functions as the hidden hand that shapes conceptual understanding. Even though viewed from a scientific perspective, it forms an unsatisfactory idea. The hidden hand is like groping in the dark and getting hold of what seems to fit first.

Uncovering and highlighting the hidden hand draws attention to its effect on understanding. Breaking a bar of chocolate serves as a representation of the division schema. In some regards, this is like reexperiencing the schema of division for the sake of reflecting on it. This measure helped the girls to develop a conception toward scientific understanding. The schema of division conceived as *becoming more* and *smaller* is combined with the schema of *growth.*

In the case of understanding climate change, students as well as scientists use the *container-flow* schema to comprehend the causes of global warming. The *container-flow* schema combines the *container* schema with the *source-path-goal* schema, both simpler than the combined schema. Beyond that, scientists connect many containers with different flows in a quasi-circular manner in order to model biogeochemical cycles. This rather complex model of the carbon cycle poses two considerable learning difficulties to students, even though students have a basic understanding of the *container-flow* schema at hand.

First, students like Tina use fewer containers and flows than necessary, at least for understanding the causes of climate change, that is, omitting the ocean as a container. This is mainly a matter of mastering the required complexity. Second, students like Gustav think that the container's content (i.e., man-made CO_2) causes climate change. Scientists also use the distinction between natural and man-made CO_2, but would assign this to the flows of CO_2 that are man-made. Scientists would not talk about two kinds of CO_2. In this case, it is easier to think of man-made CO_2 than of man-made flows of CO_2. Working with and reflecting upon a representation of a container-flow-model (see Fig. 16.6) in addition to information about the carbon cycle (presented in a science-like and a narrative context) helps students to develop more scientific conceptions: from *man-made* CO_2 to *man-made cause of carbon flow*.

Mesocosm—the World We Live In

We have shown how the hidden hand of schemata shapes our understanding. Highlighting the hidden hand helps students to reconstruct their conceptions. Given the relevance and the crucial role of the schemata employed for understanding scientific phenomena, we now examine the world of the origin of the hidden hand. It is the normal world that Vollmer (1984) called the *mesocosm*. It is "that section of the real world we cope with in perceiving and acting, sensually and motorially [. . .]" (p. 87). The mesocosm is a *world of medium dimensions* that reaches from a blink to a lifetime, from *light as a feather* to *heavy as an elephant*, from a *hair's breadth* to the *horizon*, and so forth. These dimensions explicitly refer to a human's sensory abilities and are perceivable and tangible. In contrast, macrocosmic structures like the biosphere, our solar system, or the mass of the moon are not part of the mesocosm, because our cognitive apparatus was evolutionarily adapted to medium dimensions. The same holds for microcosmic entities such as cells or structures like molecules. In the macrocosm and the microcosm, we encounter imperceptible entities, at least not in our everyday life experience. Whereas it is reasonable to differentiate between meso-, macro-, and microcosm in an educational view, biologists actually study life on many more levels of biological understanding. This is different in the fields of chemistry education, where it is common to focus on micro-macro thinking (e.g., Bucat & Mocerino, 2009; van Berkel, Pilot, & Bulte, 2009) and establish correspondences between macroscopic and submicroscopic levels of chemical operation (e.g., Bucat & Mocerino).

The distinction among meso-, macro-, and microcosm is of prognostic value for the degree of students' difficulties in understanding scientific phenomena. The way we are as human beings restricts us to medium dimensions in interacting with our environment. Thus, our basic concepts and schemata are of mesocosmic origin. We are confined to comprehend microcosmic as well as macrocosmic phenomena in terms of mesocosmic concepts and schemata. This is one of the reasons science is hard to grasp.

Scientific understanding depends to a large degree on technologically extended perception. Biologists, for example, use microscopes and chromatographs to experience microcosmic phenomena and develop scientific conceptions. Scientific understanding depends to a great extent on imagination too. Imagination is employed in understanding the cellular processes of root growth or climate change. Schemata acquired in the mesocosm are used to comprehend phenomena in the microcosm as well as in the macrocosm. Thus, scientific understanding can be traced back to experience in the mesocosm.

This insight bears important consequences for instructional interventions. First, experiences necessary for scientific understanding—especially those that originate from the microcosm and macrocosm—have to be provided. This is in accordance with Vosniadou and Ioannides's (1998) demand to provide *meaningful experiences*. Second, the scientific view has to be explained and outlined carefully. This is what teachers usually concentrate on. Third, *metaconceptual awareness* has to be facilitated (Vosniadou & Ioannides). In particular, we advocate reflection on schemata employed in understanding. It has not escaped our attention that we left the fundamentally social nature of the use of representations out of consideration (e.g., Kozma, 2003; Mortimer & Scott, 2003).

Three Representations for Effective Learning Environments

It is common knowledge in science education that multiple representations foster scientific understanding. However, educational practice shows that some representations are more effective than others. There is evidence that the majority of representations like metaphors, analogies, and models that teachers use to explain scientific topics are not adopted by their students (Harrison & Jong, 2005) nor are they understood in the anticipated way (Harrison & Treagust, 2006). However, we have found that some representations are even more crucial for conceptual reconstruction. To begin with, we distinguish, from an experiential point of view, among three different kinds of these representations.

Representations Afford Experiences

In the episode on root growth, a real onion was presented. To be sure, a photograph of an onion with a root or a realistic drawing of an onion with a root would have sufficed but that would be a representation serving as a second-hand experience. There are multiple representations that afford experiences—such as photomicrographs or electromicrographs, a chromatogram, recordings of action potentials, and a view of a DNA sequencing gel. These representations, whether of first- or second-hand origin, prepare the ground for the development of conceptions. Empirical methods in science are often means for students to experience beforehand imperceptible entities with the help of technical devices, for example, a microscope or a chromatograph. Whereas the core of our conceptual system is grounded in everyday experience, many scientific conceptions are grounded just as thoroughly in scientific inquiry.

Representations Denote Conceptions

The figure on the global carbon cycle (see Fig. 16.6a), for instance, represents a scientific conception of the movement of carbon on earth. Every depiction of a mental model falls into this category of representations—including the cell cycle, the citric acid cycle, the structure of a DNA molecule, the equation of photosynthesis, and the Punnett square. There are many well-known representations of this kind—such as figures, models, symbolic systems, or scientific terms—that represent the scientific way of thinking. We found that learning environments based solely on this kind of representation often pose problems to students. For a scientist, these representations might be adequate and understandable because they refer to common scientific experience. The challenge for students is to relate these representations obtained by scientific experience to the scientific phenomena that they are meant to represent.

Representations Depict Schemata

Breaking a bar of chocolate and moving balls from one labeled box to another are both materialized representations of cognitive schemata employed in understanding cell division and the carbon cycle. By working with these representations, students reexperience the inherent structure of the schema and reflect on how they employ it in their effort to understand the phenomenon. This kind of representation throws light on the hidden hand that shapes students' conceptual understanding.

In our teaching experiments on cell division and climate change, we chose learning environments with representations that depict schemata. Reexperiencing and reflecting on the hidden hand helps students to understand complex and abstract phenomena. To this end, students need to work with representations that throw light on the schema they employed in their endeavor to understand. Awareness of the schemata can be deliberately deployed to understand the scientific conception of the phenomena. A comparison of the different kinds of representations is needed as well. For example, students develop an adequate conception of the global carbon cycle by reexperiencing the containers and transferring these conceptions to the representations taken from other sources, for example, the IPCC. As for learning cell division, students reflect on the schema of division and use this knowledge to understand the scientific conception. Beyond that, the students are encouraged to critically think about limits and possibilities of the representations they used so far. Awareness of the hidden hand that shapes our conceptual understanding enables teachers to choose effective representations and design learning environments that foster learning.

References

Andersson, B., & Wallin, A. (2000). Students' understanding of the greenhouse effect, the societal consequences of reducing CO_2 emissions and the problem of ozone layer depletion. *Journal of Research in Science Teaching, 37*(10), 1096–1111.

Aubusson, P. J., Harrison, A., & Ritchie, S. M. (2006). Metaphor and analogy. In P. J. Aubusson, A. Harrison, & S. M. Ritchie (Eds.), *Metaphor and analogy in science education* (pp. 1–9). Dordrecht, The Netherlands: Springer.

Bucat, B., & Mocerino, M. (2009). Learning at the sub-micro level: Structural representations. In J. K. Gilbert & D. Treagust (Eds.), *Multiple representations in chemical education* (pp. 11–29). Dordrecht, The Netherlands: Springer.

Duit, R., Gropengiesser, H., & Kattmann, U. (2005). Towards science education research that is relevant for improving practice: The model of educational reconstruction. In H. Fischer (Ed.), *Developing standards in research on science education* (pp. 1–9). London: Taylor & Francis.

Duit, R., & Treagust, D. F. (2011). Conceptual change – Still a powerful framework for improving the practice of science instruction. In B. F. Fraser, K. Tobin, & C. McRobbie (Eds.), *Second international handbook of science education* (pp. 107–118). Dordrecht, The Netherlands: Springer.

Ekborg, M., & Areskoug, M. (2006). How student teacher's understanding of the greenhouse effect develops during a teacher education programme. *NorDiNa, 5*, 17–29.

Gropengiesser, H. (2003). *Lebenswelten, Denkwelten, Sprechwelten. Wie man Vorstellungen der Lerner verstehen kann* [Worlds of Living, thinking, and talking. How to understand student's conceptions]. Oldenburg, Germany: Didaktisches Zentrum.
Harrison, A. G., & Jong, O. D. (2005). Exploring the use of multiple analogical models when teaching and learning chemical equilibrium. *Journal of Research in Science Teaching, 42*(10), 1135–1115.
Harrison, A. G., & Treagust, D. F. (2006). *Teaching and learning with analogies. Metaphor and analogy in science education* (pp. 11–24). Dordrecht, The Netherlands: Springer.
Intergovernmental Panel for Climate Change. (2007). *Climate change 2007: The physical science basis*. Geneva, Switzerland: Cambridge University Press.
Johnson, M. (1987). *The body in the mind. The bodily basis of meaning, imagination, and reason*. Chicago, IL: The University of Chicago Press.
Kattmann, U. (2008). Learning biology by means of anthropomorphic conceptions. In M. Hammann, M. Reiss, C. Boulter, & S. D. Tunnicliffe (Eds.). *Biology in Context Learning and teaching for the 21st century. A selection of papers presented at the VIth Conference of ERIDOB* (pp. 7–17). London.
Kozma, R. (2003). The material features of multiple representations and their cognitive and social affordances for science understanding. *Learning and Instruction, 13*, 205–226.
Lakoff, G. (1990). *Women, fire, and dangerous things: What categories reveal about the mind*. Chicago, IL: The University of Chicago Press.
Lakoff, G., & Johnson, M. (1980). *Metaphors we live by*. Chicago IL: The University of Chicago Press.
Lakoff, G., & Johnson, M. (1999). *Philosophy in the flesh: The embodied mind and its challenge to western thought*. New York: Basic Books.
Levi, P. (1975). *The periodic table*. London: Penguin.
Moore, G. E. (1996). *Principia Ethica*. Dover, MA: Dover Philosophical Classics.
Mortimer, E., & Scott, P. (2003). *Meaning making in secondary science classrooms*. Maidenhead, PA: Open University Press.
Niebert, K. (2009). Der Kohlenstoffkreislauf im Klimawandel. [The carbon cycle in climate change]. *Unterricht Biologie, 349*, 34–40.
Niebert, K., & Gropengiesser, H. (2011). CO_2 causes a hole in the atmosphere« Using laypeople's conceptions as a starting point to communicate climate change. In W. Leal (Ed.), *The economic, social and political elements of climate change* (pp. 603–622). Berlin, The Netherlands: Springer.
Richards, I. A., & Ogden, C. K. (1923). *The meaning of meaning*. Orlando, FL: Harcourt Brace Jovanovich.
Riemeier, T., & Gropengiesser, H. (2008). On the roots of difficulties in learning about cell division. *International Journal of Science Education, 30*(7), 923–939.
Schneeweiss, H., & Gropengiesser, H. (2010). Lernervorstellungen und Vorstellungsentwicklungen zu Bakterienkolonien [Students' conceptions and their conceptual development on microbes]. In U. Harms & I. Mackensen-Friedrichs (Eds.), *Lehr- und Lernforschung in der Biologiedidaktik. Band 4* (pp. 85–98). Innsbruck: Studienverlag.
Schreiner, C., Henriksen, E. K., & Hansen, P. J. K. (2005). Climate education: Empowering today's youth to meet tomorrow's challenges. *Studies in Science Education, 41*, 3–50.
Sterman, J., & Booth-Sweeney, L. (2007). Understanding public complacency about climate change: Adults' mental models of climate change violate conservation of matter. *Climatic Change, 80*(3), 213–238.
van Berkel, B., Pilot, A., & Bulte, A. (2009). Micro-macro thinking in chemical education: Why and how to escape. In J. K. Gilbert & D. Treagust (Eds.), *Multiple representations in chemical education* (pp. 31–54). Dordrecht, The Netherlands: Springer.
Vollmer, G. (1984). Mesocosm and objective knowledge. In F. M. Wuketits (Ed.), *Concepts and approaches in evolutionary epistemology* (pp. 69–121). Dordrecht, The Netherlands: D. Reidel Publishing Company.
Vosniadou, S., & Ioannides, C. (1998). From conceptual development of science education: A psychological point of view. *International Journal of Science Education, 20*(10), 1213–1230.

Chapter 17
Analogy and Gesture for Mental Visualization of DNA Structure

Anveshna Srivastava and Jayashree Ramadas

Introduction

The birth of molecular biology was significantly marked by the discovery of the double-helical structure of the DNA molecule by Watson and Crick (1953a). The general correctness of this structure was gradually proven in the subsequent years by substantial research on the structural as well as functional aspects of the DNA molecule. The structure of DNA had immediate functional implications: "It follows that in a long molecule many different permutations are possible, and it therefore seems likely that the precise sequence of the bases is the code which carries the genetical information" (Watson & Crick, 1953b, p. 965).

Conceptual understanding in molecular biology involves integration of the macro (genetic traits), micro (cell), and molecular (gene) levels. Building up of the DNA molecular structure and its location at the cellular level leads to an understanding of its biological significance, for example, in genetic expression. Marbach-Ad and Stavy (2000) remarked that the difficulty in understanding and linking these different organizational levels is "because sometimes one level (e.g., the macro level) belongs to one discipline (e.g., biology), and the other level (e.g., the molecular level) belongs to different discipline (e.g., chemistry)" (p. 201). In fact, the integration occurs in multiple ways—one that includes concepts from various disciplines, another that involves the macro, micro, and the molecular levels, and finally, the structure-function linkages within and across these levels.

Structural-functional linkages have been identified as a problem area in elementary genetics (Lewis, 2004; Marbach-Ad, 2001). In a study of major problem areas in biological sciences as identified by students, Bahar, Johnstone, and Hansell (1999) reported that the structure and function of the DNA and RNA molecule

A. Srivastava (✉) • J. Ramadas
Homi Bhabha Centre for Science Education, Tata Institute of Fundamental Research, Mankhurd, 400088 Mumbai, India
e-mail: anveshna.sriv@gmail.com; jr@hbcse.tifr.res.in

D.F. Treagust and C.-Y. Tsui (eds.), *Multiple Representations in Biological Education*, Models and Modeling in Science Education 7, DOI 10.1007/978-94-007-4192-8_17,

© Springer Science+Business Media B.V. 2013

was considered as a topic of relatively low difficulty. However, we make a case here that students do have a problem in understanding the basic 3-D structure of the DNA molecule.

Structure of the DNA Molecule

The double-helical structure of the DNA molecule can be visualized as two right-handed helices coiled around a central axis. Each helix is composed of a sugar-phosphate backbone, and each (deoxyribose) sugar molecule in this backbone is attached with a nitrogenous base through a glycosidic bond to form a nucleoside unit. The nitrogenous bases—purines (adenine or guanine) or pyrimidines (thymine or cytosine)—are paired in a complementary fashion where adenine (A) forms two hydrogen bonds with thymine (T) and guanine (G) forms three hydrogen bonds with cytosine (C). The nitrogen bases of the DNA molecule are planar ring structures of equal length which are perpendicular to the central DNA axis and also to their attached sugar molecules. Orientation of the nitrogenous base pairs and the specific hydrogen bonding between the complementary base pairs give rise to a basic ladder shape, which is coiled into a right-handed helix of specific dimensions.

Textbook Representations of DNA Structure

In Indian schools, the chemical prerequisites for learning the DNA molecule in biology are built up from middle school till the higher secondary level (for students aged 17) as part of the chemistry curriculum. The higher secondary biology textbook used in our study, published by Maharashtra State Board of Secondary and Higher Secondary Education (MSBSHSE, 2009), introduces the DNA molecule by describing the components of nucleotides, the pentose sugar, phosphate group, and the nitrogenous bases, with their chemical formulas. The analogy of a *twisted ladder* is usually followed by two kinds of diagrammatic representations. The first is a schematic representation of the *DNA double helix*, depicting two crisscrossing wavy ribbonlike strands, in which there are (sugar-phosphate) links labeled "S-P-S-P" in the backbone (see diagram a in Fig. 17.1). Connecting the backbone are the skeletal structures of the nitrogenous base pairs with the respective number of hydrogen bonds with dimensional details indicated in Angstroms (Å). The accompanying text mentions the angle between successive base pairs and also that each "spiral turn" contains 10 pairs of nucleotides (p. 15).

The second diagram (see diagram b in Fig. 17.1) is the "detailed molecular structure," which is a ladder structure containing skeletal outlines of the pentagonal sugar molecules—connected with the phosphate groups—and labels of the 3′ and 5′ ends of the two strands. The sugar molecules are shown attached with purines (two joined circles) or pyrimidines (one circle). The hydrogen bonds between the

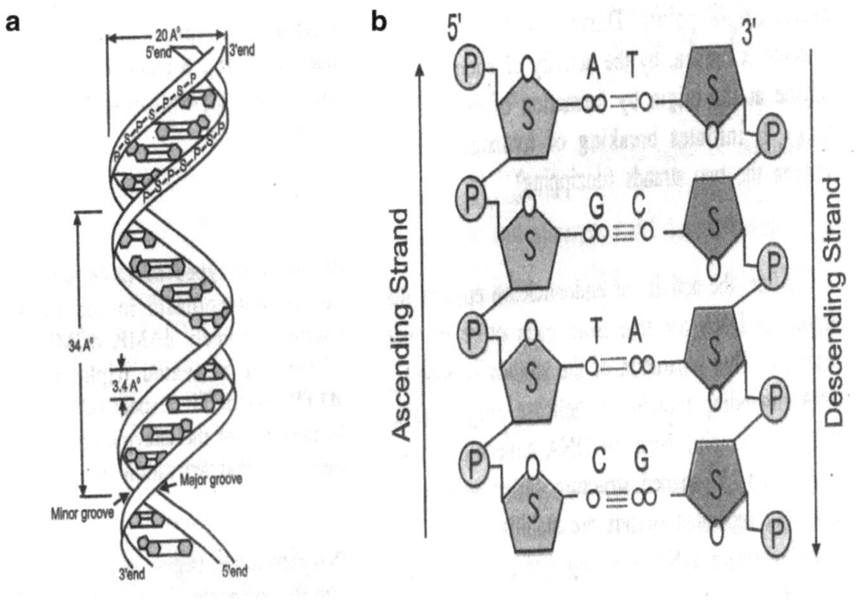

Fig. 17.1 Textbook diagrams (MSBSHSE, 2009, p. 15): (**a**) double helix and (**b**) ladder structure (Reprinted with permission)

complementary bases are represented through either two (for A–T) or three (for G–C) dotted lines. Thus, by the end of high school, students are introduced to standard diagrams of the DNA molecule. The *twisted ladder* is an analogy for DNA structure which has considerable potential to help students mentally visualize the structure of DNA. Our interest in this study was in seeing whether students are able to sustain the analogy in order to form a mental image of the 3-D molecular structure of DNA.

Role of Multiple Representations in Learning

Multiple external representations (MERs) are believed to support learning by providing complementary information or processes and by constraining the learner's interpretation of a new representation using a familiar representation to help learners to understand the information carried by this new representation; MERs also support construction of deeper understanding through abstraction, extension, and relations among representations (Ainsworth, 1999). Tsui and Treagust (2003) applied Ainsworth's MER functions to their study on genetics reasoning in Australian classrooms, which was conducted in the context of computer-aided learning with detailed analysis of students' learning and reasoning in genetics as they used the MERs.

The question of how MERs could connect with internal mental representations is one important area of interest in science pedagogy. Recent research on the embodied view of cognition suggests that our reasoning is enabled significantly by our ability to participate in actions in the world and that our internal representations are not amodal (propositional) but linked to our sensorimotor perceptions and actions (Clark, 1997; Barsalaou, 1999). One direct implication of the embodied view is that MERs connect to internal representations through the learner's perceptions and actions.

Drawing on the embodied view of cognition, we suggest that a possible pedagogical route from external to internal (mental) representations might be taken through the use of gesture. Goldin-Meadow and Beilock (2010) argued that gestures affect thinking by grounding it in action and may even have a more powerful influence on thoughts than action itself resulting in a rich internal representation that incorporates the sensorimotor properties required to act out in the world. This insight from cognitive science was used by Padalkar and Ramadas (2011) in proposing a pedagogical purpose for deliberately designing gestures in science. Padalkar and Ramadas argued that gestures might be used to internalize a natural phenomenon, a scientific model, or properties of space. It is important that the gestures in Padalkar and Ramadas' study served not only to link external representations with internal mental ones, but they were also designed to link two types of external representations (concrete models and diagrams).

Mathai and Ramadas (2009) proposed tasks calling for changing the viewpoint of an observer to encourage mental visualization of body systems. This parallels Goldin-Meadow and Beilock's (2010) hierarchies of gestures and actions—*character viewpoint gestures* reflect actual movements, *observer viewpoint gestures* capture the goal object or its trajectory, and *metaphorical gestures* represent abstractions—as well as the suggestion that character and observer viewpoint gestures, if used in sequence, could provide a bridge between concrete actions and more abstract representations. We therefore suggest that (a) gestures could be used to link external and internal representations, (b) gestures could be used to link together different MERs into an integrated internal representation, (c) real or imagined manipulations or transformations of structure and imagining a change in one's viewpoint could enable mental visualization of the structure, and (d) character viewpoint gestures or actions could help in making a molecular structure (e.g., DNA) more comprehensible to students.

A complementary approach to building internal mental representations, particularly visual ones, is that of analogy. Gentner (1989) defined analogy as a mapping from a base (familiar) domain to a target (unfamiliar) one. Previous research has shown that analogy is useful in visualization, model-based reasoning, knowledge construction, and understanding (e.g., Duit, 1991; Justi & Gilbert, 2006; Harrison & Treagust, 2006). Like gesture, analogy has potential to help students construct mental visual models from multiple external representations. Therefore, in this study, we used a combination of gesture and analogy of the twisted ladder using a character viewpoint simulation for encouraging visualization of the 3-D structure of the DNA molecule.

This Study

In this study, we examined students' reasoning processes in understanding the 3-D nature of the DNA molecule through the integration of prerequisite concepts from physics and chemistry, supported by appropriate simple and low-cost multiple external representations (MERs) of DNA structure in terms of the following research questions:

1. Are students able to link the ladder analogy with common 2-D diagrams of DNA structure to form a mental model of the 3-D structure of the molecule?
2. Can we use gesture to link the 2-D representations and the ladder analogy with the 3-D concrete models of DNA structure?
3. Can we use mental simulation of changing observer viewpoint to link the 2-D representations and the ladder analogy with the 3-D concrete models of DNA structure?

Through a screening test, we selected five students aged 17–19 years, who were enrolled in the first year of a 3-year bachelor degree course in biological sciences at a university in India (see Table 17.1).

We used a microgenetic research design (Flynn & Siegler, 2007; Siegler, 2006; Siegler & Crowley, 1991; van der Aalsvoort et al., 2009) which is appropriate for situations that involve rapid transitions in learning by tracing the processes of the students' learning under dynamic, in vivo conditions. The three important attributes of a microgenetic research design developed in Siegler and Crowley (1991) and modified in Siegler (2006) are:

- Observations span the period of rapidly changing competence.
- Within this period, the density of observations is high, relative to the rate of change.
- Observations are analyzed intensively, with the goal of inferring the representations and processes that gave rise to them (p. 469).

The students are observed very closely during the period of learning, and then these observations are revisited again and again for a finer understanding of the patterns that depict "change in real time as how it occurs" (van der Aalsvoort, van Geert, & Steenbeek, 2009, p. 9).

In our study, observations were carried out during six individual sessions held over 9 days. Each session involved a clinical interview-cum-teaching sequence for 1–1.5 h for each student per day. The language of the interview was English except for some occasions when Marathi and, occasionally, Hindi were used for two of the interviewees: Nitin and Aakriti. The prerequisites for the sessions lay within the syllabus for secondary and higher secondary schools recommended by the State Board. The sessions on days 1 through 4 focused on initial assessment and recall of prerequisite concepts in biology and chemistry. Brief sequences of direct instruction were included in order to bridge some inevitable gaps in understanding. The issue of three-dimensionality of DNA structure was addressed on days 4 through 6, and these data were analyzed microgenetically.

Table 17.1 Demographic information of participants in this study

Name of the student[a]	Age (in years)	Gender	Mother tongue	Degree pursuing (bachelors)	Courses taken in the current semester[b]
Anuja	18	F	Marathi	Microbiology	MPC
Sharada	18	F	Oriya	Biotechnology	BMC
Nitin	19	M	Marathi	Microbiology	MPC
Sandhya	17	F	Telugu	Biotechnology	BMC
Aakriti	18	F	Hindi	Microbiology	MPC

[a]Names are changed to preserve anonymity
[b]*MPC* Microbiology, Physics, Chemistry, *BMC* Biotechnology, Microbiology, Chemistry

Multiple Representations of the DNA Backbone and the Nitrogenous Base Pairs

The students were asked to draw the textbook diagrams (the ladder and helical structures in diagrams (a) and (b) in Fig. 17.1) and recall the well-known ladder analogy for DNA structure. The DNA backbone was represented by five simple models (M1 to M5 in Table 17.2). M1 was comprised of a sheet of paper laid on a table, and the students were asked to consider its long edges to represent the two DNA backbones. M2 consisted of two antiparallel pencils laid on the table and considered as the two DNA backbones. M3 was a variant of M2 where the two antiparallel pencils (the backbones) were made to stand erect on the table. M4 was a cutout model depicting the two backbones, each consisting of two phosphate groups attached with one sugar molecule at its $3'$ and $5'$ positions, fixed on a cardboard base. M4 showed the molecular details of the two sugar-phosphate backbones.

M5, or the *clothespin model*, was adapted from Venville (2008). The students were provided with two plastic tubes along which they could string interlocking clothespins of four different colors (green, yellow, blue, and pink) to represent the complementary DNA bases. The students were asked to construct the M5 model to depict first the ladder structure and then the helical representation of the DNA molecule.

In combination with models representing the DNA backbone, two types of representations of the nitrogenous base pairs were introduced. The first representation consisted of cardboard cutouts of the different nitrogenous bases (see Fig. 17.2) which was first suggested by James Watson's own account of his discovery of base pairing as recounted in an online video program (Cold Spring Harbor Laboratory's DNA Learning Center, n.d.). The students were to use these cutouts against the M4 model to depict the orientation of the base pairs in the molecular model while indicating the position of attachment of the bases with the sugar molecules in the DNA backbone.

The second base pair representation comprised the palm *gesture,* in which the palm represents a nitrogenous base (purine or pyrimidine) and the straightened

Table 17.2 Multiple representations of the DNA backbone

Model No.	Backbone representation	
M1	Long edges of a sheet of paper (laid on the table)	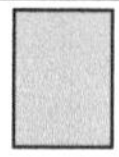
M2	Two (anti)parallel pencils (laid on table)	
M3	Two (anti)parallel pencils (held to stand erect on table)	
M4	Cardboard cutout of a sugar molecule attached with two phosphate molecules (two sets) standing on a cardboard base	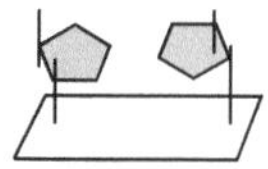
M5	Clothespin model (ladder representation of DNA which can be assembled on a table and then twisted to form a helix)	

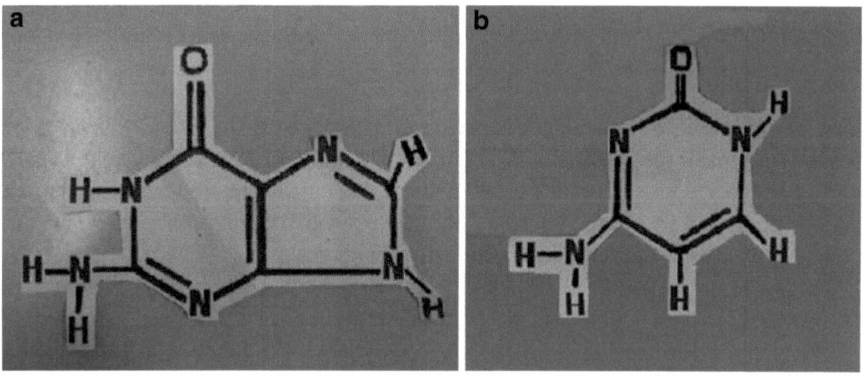

Fig. 17.2 Cutouts of molecules of nitrogenous bases: (**a**) purine base and (**b**) pyrimidine base

fingers represent the complementary nitrogenous base (pyrimidine or purine) (see Fig. 17.3). The students used this gesture to imitate the orientation of the base pairs in the ladder against the models M1–M5, as appropriate.

The last type of representation was the ladder analogy, in which the backbone and the base pair representations were combined. The students were asked to visualize first a straight ladder and then a twisted ladder. The ladder analogy was used as a reminder to the students about the DNA structure while they attempted to show the base pair orientation with the help of the palm gesture or the cutouts. If the analogy by itself did not work, then the students were instructed to mentally simulate the action of walking up the straight ladder and, in that situation, consider how the steps of the ladder would be oriented. The gesture and mental simulation

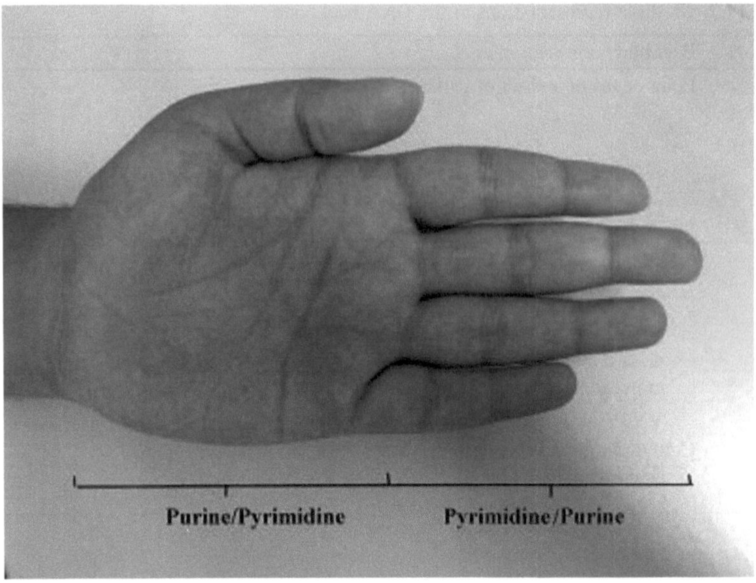

Fig. 17.3 Palm gesture with palm and straightened fingers representing a complementary base pair

were also used for the helical ladder structure in model M5. The mental visualization (of the straight or the twisted ladder) and the simulation (of walking up the ladder) correspond, respectively, to the observer viewpoint and character viewpoint gestures/ actions discussed by Goldin-Meadow and Beilock (2010). Here, the actions are, of course, not actually carried out but mentally simulated.

Preparing the Background (Days 1, 2, and 3)

On day 1, we examined students' understanding of the concept of DNA as the genetic material. We probed their familiarity with terms such as *genetic material*, *gene*, *heredity,* and so on. The students were asked about *cells*, the location of genetic material, and DNA as genetic material. All the students except Nitin had problems in understanding the relationship between a gene and DNA, for example, Anuja said, "I am confused that it (gene) is inside the DNA or the DNA is inside the gene." Also, all the students did not understand Hershey and Chase's classical experiment which proved that DNA is the genetic material. Each day, from day 2 till day 6, began with students' diagrammatic representations of the DNA ladder and the double helix as some approximation of the two familiar textbook diagrams (see Fig. 17.1). On day 2, we focused on recapitulating elementary background related to the chemistry of the DNA molecule and reintroduced to them the idea of nitrogenous bases (purines and pyrimidines) with the electronegative nitrogen atoms which can form hydrogen bonds with another nitrogenous base. On day 3, the students explored different pairing possibilities between the bases using cardboard cutout models of the bases (see Fig. 17.2). They eventually used the cutouts to form the A–T double hydrogen bond and G–C triple hydrogen bonds to demonstrate that the base pairs were planar and of identical lengths.

Introduction to the Nucleoside (Day 4)

At the start of day 4, the students were introduced to the palm gesture (see Fig. 17.3), and were then asked to imagine its correspondence with the planar base pairs, and to use the gesture against the M1 and/or M2 model. All the students began with an incorrect gesture, that is, they showed the base pairs in the plane of the straightened parallel backbones (Episode I to be discussed in the following sections). Day 4 then continued with questions and tasks which required revisiting of the concepts such as chemical bonds and the valencies of atoms depicted in the cutouts of the nitrogenous bases and the sugar molecule. The students were shown the M4 model of the sugar-phosphate backbone and were then asked to depict base pair orientation against it through the palm gesture as well as through the cutouts of the bases. The day also involved instruction regarding heterocyclic molecules, functional groups, and IUPAC numbering conventions for bases and sugars. This line of discussion was significant to help students understand the structure of the nucleotide unit and the antiparallel nature of the two DNA strands.

Sharada and Aakriti were unclear about concepts—such as atomic structure, valency, electronegativity, and bonding of atoms comprising the bases—and hence, the whole of day 4 session was directed toward building of their chemistry background pertaining to atomic structure and bond formation and they were introduced to M4 only on day 5. The purpose of days 2, 3, and 4 was to familiarize the students with the planar structures formed through the bonding of the purines and pyrimidines and the chemistry involved in the formation of individual DNA units along with the introduction of gesture and analogy as tools for visualizing the orientation of the nitrogenous base pairs. Student interactions on days 5 and 6 then dealt largely with the three-dimensionality of DNA structure, which was analyzed microgenetically.

Data Analysis

Video recordings of all the 6 days sessions were the major data source along with journal notes and students' written data. The video data from day 4 to day 6 were subjected to a time-sequence analysis with microgenetic method. The video recordings of the five students, of a time interval from between 189 and 235 min, were scanned for *episodes* that consisted of continuous stretches of time during

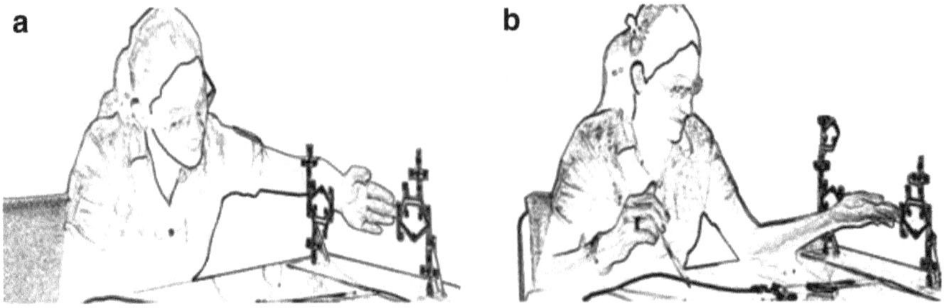

Fig. 17.4 Palm gesture used with M4 model: (**a**) incorrect (−) gesture and (**b**) correct (+) gesture

which the students engaged themselves with the three-dimensionality of the DNA molecule. An episode had either one or more *events* where the learner made a guided or a spontaneous attempt to depict base pair orientation or twisting of the M5 backbone. The base pair orientation was indicated by their palm gesture, that is, placing of the palm against the DNA backbones (M1–M5) or through similar placing of the cutouts of the base pairs (against M4 only) (see Fig. 17.4). The backbone models (M1–M5) in use during that episode were noted, along with the correctness ("+" event) or the incorrectness ("−" event) of placing of the base pairs. The time interval was counted from the start of day 4 as "$t = 0$."

We discuss here the detailed analysis of the sequence of correct (+) and incorrect (−) events for two of the five students, Anuja and Sandhya, as examples and the specific backbone models (M1–M5) referred to in each event of the episode (see Tables 17.3 and 17.4). A summary of the "+ve" transitions, that is, from incorrect (−) to correct (+) events, for all the five students is given in Table 17.5. The unshaded events in Tables 17.3 and 17.4 indicate that the straight ladder structure is under discussion. Models M1–M4 are always straight ladder structures. If model M5 is being used, or if the gesture is being made in air (i.e., without support of one of the backbone models), then the ladder structure under discussion could be straight (an unshaded event) or helical (a shaded event).

Students' Understanding of the Ladder Structure

At the beginning of day 4, it was clear to us that all the students were visualizing the "steps" of the DNA ladder to be flat. The first event on day 4 for every student was a "−" event, referring to a straight ladder structure where the students depicted the base pair orientation in the plane of the DNA backbones. This turned out to be a strongly held misconception, probably reinforced by diagrams (see diagram b in Fig. 17.1) which are common in textbooks.

The initial incorrect palm gesture in Episode I on day 4 was followed up by between 30 and 55 min of questions-cum-instruction related to the formation of the

Table 17.3 Microgenetic analysis of episodes related to three-dimensionality of DNA Structure for Anuja

Day	Day 4											Day 5							Day 6					
Start time[a]	7.5 min		37.1 min								55.5 min	74.09 min		122.3 min	125.6 min		134.4 min		164.2 min					
Episode no. (duration)	**I** (0.3 min)		**II** (5.6 min)								**III**	**IV** (0.4 min)		**V**	**VI** (1.1 min)		**VII** (3.0 min)		**VIII** (2.2 min)					
Event[b] +							M3	M2	M4	M4 (c)	M4 (c)	Air	Air	M5	M5	M5	M5	M5			M4	M4 (c)		M4 (c)
												z	z	x			z	z						
Event[c] –	M1	M1	M4	M1	M2	M2													M4 (c)	M4			M4 (c)	

M5 ladder construction (start time: 75.0 min)
M5 helix formation (start time: 119.3 min)

M4 (c) indicates that the cutouts of the N-bases were being used to show orientation. In all other cases, the palm gesture was being used
The shaded events depict palm gesture in reference to the helical model, in M5 or in air
0, none of the base pairs twisting; x, only two base pairs twisting; y, partial or nonuniform twisting; z, uniform twisting
[a]Start time: The start time denotes the beginning of the episode with day 4 starting at $t = 0$
[b]Event +: palm gesture or cutout orientation (c) perpendicular to DNA axis (correct)
[c]Event – : palm gesture or cutout orientation (c) parallel to DNA axis (incorrect)

Table 17.4 *Microgenetic Analysis of Episodes Related to Three-Dimensionality of DNA Structure for Sandhya*

Day	Day 4															
[1]**Start time**	4.4 min		36.2 min			42.6 min		46.6 min				52.4 min	57.3 min			
Episode No. (Duration)	**I** (0.8 min)		**II** (2.3 min)			**III** (0.3 min)		**IV** (2.2 min)				**V**	**VI** (2.0 min)			
[2]**Event +**					M4		M4 (c)	M4 (c)		M4 (c)	M4 (c)	M4 (c)	M4 (c)	M4 (c)		M4 (c)
[3]**Event -**	M1	M2	M4	M4		M4 (c)			M4 (c)						M4 (c)	

Day 5															Day 6					
71.1 min	121.3 min														151.4 min					156.4 min
VII	**VIII** (4.3 min)														**IX** (3.0 min)					**X**
Air					M1	M5		M5	M1	M2			Air	M5 x		M5 y	Air	Air y	Air z	Air z
	M5	Air	M3	M1			M5				Air 0	Air 0			M5 y					

M5 ladder construction (Start time – 71.2 min)
M5 helix formation (Start time – 106.5 min)

[1]Start Time : The start time denotes the beginning of the episode with Day 4 starting at t=0
[2]Event + : Palm gesture or cutout orientation (c) perpendicular to DNA axis (correct)
[3]Event - : Palm gesture or cutout orientation (c) parallel to DNA axis (incorrect)
M4 (c) indicates that the cutouts of the N-bases were being used to show orientation. In all other cases, the palm gesture was being used.
The shaded events depict palm gesture in reference to the helical model, in M5 or in Air.
0: none of the base pairs twisting; x: Only two base pairs twisting; y: Partial or non-uniform twisting; z: uniform twisting

Table 17.5 Summary of number of "+ve" transitions and their contexts

Name of the student	No. of "+ve" transitions	Context of the transitions
Anuja	3	1. **Ladder**[a] **analogy with mental simulation**, 2. reminder about gesture against M1, 3. reminder about orientation
Sharada	2	1. **Ladder analogy**, 2. palm gesture
Nitin	7	1. **Ladder analogy with mental simulation**, 2. palm gesture, 3. palm gesture, 4. reminder of earlier orientation, 5. reminder of earlier orientation; 6. ladder analogy with mental simulation, 7. ladder analogy with mental simulation
Sandhya	8	1. **Ladder analogy with mental simulation**, 2. **ladder analogy**, 3. reminder about base positioning, 4. reminder about earlier gesture, 5. palm gesture, 6. ladder analogy with mental simulation, 7. ladder analogy, 8. reminder about the base placement
Aakriti	4	1. Ladder analogy, 2. ladder analogy, 3. **ladder analogy with mental simulation**, 4. **ladder analogy**
Total	**24**	Ladder analogy (6), ladder analogy with mental simulation (7), palm gesture (4), reminders (7)

[a]All contexts which had direct bearing on the "Aha!" moment of the student are given in *bold* font

nucleoside and bonding of the DNA base pairs; after which, the students were asked to repeat the palm gesture (Episode II). Although all the students began with the incorrect *in-the-plane-of-the-backbone gesture*, Tables 17.3 and 17.4 show that they quickly changed to the correct gesture (in Episode II or Episode III). We refer to this as a "+ve" transition, indicating a realization of the three-dimensionality of the ladder structure. It was striking, however, that the correct response was not stable in any of the students. As the interviews proceeded, all the students showed a series of "−ve" and "+ve" transitions, that is, they kept switching between the correct and incorrect responses. This was notwithstanding the fact that the correct response was often accompanied by an "Aha!" moment (to be described later) and positive encouraging feedback (e.g., a broad shared smile and "good!" or "very good!") from the interviewer. The type of model being used during the episode was one factor which may have determined their responses.

For Anuja, the first "+ve" transition happened with her use of M3, that is, when she picked up the parallel pencils (representing the DNA backbone) lying on the table and held them to stand vertically (Episode II). She sustained the correct orientation through day 4 and even day 5, when she worked with M5, the clothespin model. But on day 6, when she returned to the M4 (cutout) model, she reverted to a series of incorrect and correct orientations (Episode VIII) (see Table 17.3).

For Sandhya, the first "+ve" transition happened on day 4, using the palm gesture with M4. However, when in the next episode, four minutes later, she had to place the base pair cutouts against the M4 model, she reverted to the incorrect orientation. Over a total interval of 16.7 min on the same day (Episodes III–VI), using the M4 (c) base cutouts, Sandhya showed a series of three "−ve" and three "+ve" transitions. In Episodes VIII and IX, too, Sandhya showed four "−ve" and four "+ve" transitions while working with the straight and then helical M5 model (see Table 17.4).

Students' Understanding of the Helical Structure

The palm gesture was used with models M1–M4 to represent the fact that the base pairs of DNA were planar (of equal lengths), parallel to each other, and perpendicular to the two backbones, just like the steps of a ladder. The next task for the students was to depict the base pairs orientation in a helical ladder. In this task, they had to maintain the base pairs locally perpendicular to the two backbones and to the axis of the helix but show that each base pair was twisted (by 36°) with respect to its adjacent base pair. This could be indicated by the students positioning their two palms in parallel planes but angularly displaced with respect to each other, either in air or against the M5 (clothespin) model.

In Tables 17.3 and 17.4, the shaded events indicate that the students were showing the base pair orientation in the helical structure. A "+" or "−" event indicates that the base pair is shown perpendicular (correct) or parallel (incorrect) to the axis of the helix. The twisting of the base pairs is shown by a "0," "x," "y," or "z" in the shaded boxes, with 0 for no twisting of the bases, x for relative twisting of two base pairs only, y for nonuniform or partial twisting of some base pairs, and z for uniform or continuous twisting of all base pairs such that the first pair is aligned with the eleventh one (correct response).

Before the M5 model was constructed, the students were asked whether the base pair orientation would change if the straight ladder was twisted to form a helical one. It was interesting that only Anuja and Sharada stated that the base pair orientation would change in the helix, while the other three students stated that the bases would remain parallel, exactly as in the straight ladder structure. Anuja and Sharada indicated a continuous twisting in air with the base pairs perpendicular to the DNA axis (Anuja, Episode IV) (see Table 17.3).

The construction of the M5 model is indicated by two arrows below Tables 17.3 and 17.4, a hollow arrow for the straight ladder and a shaded one for the twisted ladder. The straight ladder construction involved attaching the clothespins (bases) to the plastic tubing (backbone) and pairing the A–T and G–C bases. With some help, Anuja, Sandhya, and Sharada placed the bases equidistant along the backbone. However, when it came to twisting the ladder, something unexpected happened. Anuja and Sandhya crossed the two backbones and, instead of making a helix, pressed the backbones and the bases flat on to the table, so that the ladder looked like a textbook diagram (see diagram a in Fig. 17.1). Nitin did the same, even before he was asked to form the helix.

All the five students except Nitin remembered that there were 10 base pairs in one helical turn, and there was a 36° angle involved somewhere, but none guessed that 36° was the constant angle between the base pairs. Even as she handled the M5 helical model, Anuja still thought that only the two base pairs at the "center" were turning (Episode V). This was in contradiction to the correct gestures in air that she had shown in Episode IV (see Table 17.3). Notwithstanding their problems with the M5 model, all the students except Nitin had some idea of a helical shape as in a telephone cord, spiral-bound notebook, or a spiral staircase. Nitin, however, was misled by the Marathi term *sarpil* for helix, meaning "snake-like," which he illustrated with a wavy 2-D shape made from stiff wire. When shown a wire wound around a pencil, he said in Marathi, "It is like a snake wound around a tree."

Next, there was a pedagogical intervention to remind the students about "10 base pairs in a helical turn," "one turn is 360°," and "10*36° = 360°." In all the students, this led to an "Aha!" moment, that is, sudden realization or acceptance of the fact of uniform turning of the base pairs, indicated verbally or through a convincing facial expression. The intervention took place in or after the final gesture episode for all the students except Anuja, for whom the intervention happened in Episode VII (see Table 17.3). We cannot tell about the stability of this learning, since it happened at the very end of the sessions. The "Aha!" moments were more prominent in the contexts of the "+ve" transitions (parallel to perpendicular orientation of the base pairs) which are analyzed next.

Context of the "+ve" Transitions

Throughout days 4–6 when students were questioned about the orientation of the base pairs, they frequently switched between a "−" (incorrect) response (base pairs locally in the plane of the backbone) and a "+" (correct) one (base pairs locally perpendicular to the plane of the backbone). The "−ve" ("+" to "−") transitions were all unconscious ones, whereas the "+ve" ("−" to "+") transitions were usually the result of an interjection or a hint by the interviewer. Of the 19 "−ve" transitions for all the students, 12 took place when the students used the cutouts with the M4 model. Here, they had to simultaneously grapple with the chemical bonding between the bases and the sugar molecule and the orientation of the base pairs with respect to the backbones. They had to recall that the bases were to be bonded with the carbon atom at the "first (prime)" position of the sugar molecule and that it was the nitrogen atom at the first and the ninth position of a purine and a pyrimidine, which bonded, respectively, with the sugar molecule. For Sandhya, several negative transitions happened while using the M5 model where she had the twin task to consider the perpendicular orientation of the bases to the backbone or axis, as well as the angular turn of base pairs (see Table 17.4).

The "+ve" transitions were interesting because they represented a learning episode. Hence, we asked: What were the types of intervention that led to "+ve" transitions? Table 17.5 summarizes the number of "+ve" transitions for each student and the context of each transition. The first "+ve" transition for each student occurred after they were given the ladder analogy: "Have you seen a ladder?" Initially, for Anuja, Nitin, and Sandhya, the ladder analogy by itself did not help. So the interviewer followed it up with instruction to the student to (mentally): "Try to climb the ladder. Where will you step? How will you place your foot?" This instruction to mentally simulate walking up the ladder immediately led to an "Aha!" moment and a quick correction of the gesture or the cutout orientation. Anuja, Sharada, Sandhya, and Aakriti spontaneously laughed out aloud. Sharada asked incredulously, "The real ladder?!" She then proceeded to correct her orientation without further instruction for mental simulation. Nitin was generally more reserved in his expression, but he, too, gave a hint of a smile with vigorous shaking of his head, showing that he had realized something.

Out of the total of 24 "+ve" transitions for the five students, 13 transitions came about when the interviewer gave the ladder analogy by itself or accompanied by instruction to mentally simulate walking up the ladder. Sandhya and Aakriti had a second "Aha!" moment with just the ladder analogy, after the instruction to simulate had been given in a previous episode or event. Possibly mental simulation recurred in those events, spontaneously, without students being cued explicitly by the interviewer.

After the initial "Aha!" moment, seven of the subsequent "+ve" transitions occurred simply with a reminder to the students about their previous gesture or orientation. Four of these transitions occurred when the students spontaneously corrected their gesture. Of these self-corrections, two occurred while gesturing with the M1 model. The other two occurred with the M4 model, when the students were asked to use the palm gesture. Thus, after the "Aha!" moment, a simple reminder about the use of the palm gesture was sufficient to bring about a "+ve" transition.

Visualizing the 3-D Structure of DNA

The results of this study were striking and surprising to us. We anticipated that biology students might have some problem in visualizing the precise 3-D structure of the DNA molecule. We were not too surprised when all the students in our sample initially thought that the DNA base pairs (the "steps" of the ladder) were in the plane of the backbone. This was a misconception from the common textbook diagrams (e.g., diagram b in Fig. 17.1), and we found the same misconception in senior biologists.

What surprised us then was the difficulty that students had in correcting their apparently simple misconception. All of them had one or more "Aha!" moments when they realized that the base pairs were "really" like the steps of a ladder, that is, planar and perpendicular to the backbone. But, while dealing with the molecular model M4 (which required students' demonstration of palm gesture only against M4) and M4 (c) (which required students to place base cutouts against M4) or the helical models (M5), they rapidly and repeatedly forgot this simple fact. The difficulty here probably lay in a limitation of the working memory to simultaneously hold in their mind the molecular structure as well as orientation of the base pairs.

The second surprise came when Anuja, Nitin, and Sandhya on day 5 constructed the DNA helix as two crisscrossing backbones with base pairs between them, forcibly flattening them to lie flat on the table! Undergraduate science students in urban India are exposed to the image of the DNA helix not only in their classrooms but also in the media and the Internet. All the students in our sample had attended tutorial classes, in which they had been exposed to clear and more detailed diagrams about DNA structure compared to those in their regular textbooks. Despite this considerable exposure, they had not realized the essential three-dimensionality of DNA structure. The palm gesture with analogy and mental simulation helped convert the 2-D representations to 3-D ones. Pozzer-Ardenghi and Roth's (2005) work made salient the role of gestures and body orientations as semiotic resources which are usually unavailable in textbooks. We have proposed further that analogy and mental simulation can crucially enhance the effectiveness of gestures.

Palm Gesture as an Instructional and a Diagnostic Tool

The palm gesture could be a basic, simple tool to convey the orientation of the base pairs in the ladder structure. We used this gesture as a means to connect the multiple models (M1–M5) of the DNA backbone. The palm gesture is powerful and flexible enough that it is not tied to any specific orientation of the backbone. Models M1 and M2 were laid flat on the table, M3 and M4 were standing up, and M5 could be rotated in any direction. Gestures in air could be done in any direction, as did the students sometimes during this study. The palm gesture served to abstract out the idea of base pair orientation, independent of the particular model that was being used. It was as a diagnostic tool for us to begin with, but as the interaction proceeded, it also became an instructional tool.

Use of Analogy for Visualization

The ladder analogy was crucial in correcting the students' base pair orientation. The planarity of the base pairs arises due to the hydrogen bonds between them, while their perpendicularity to the DNA backbone comes from glycosidic bonds between the bases and the sugar molecules. The helical ladder structure of DNA is formed due to the tendency of the bases to avoid contact with water and stack one above the other, an arrangement that is further stabilized by Van der Waals forces and polar interactions between the adjacent bases (Woski & Schmidt, 2002).

Structure-function linkages in biology help students make sense of what they learn and are thought to play a role in mental visualization in understanding the human body systems (Mathai & Ramadas, 2009). The structural peculiarity of the DNA molecule is directly consequential to Chargaff's rule, whereby the ratios of the adenine base to thymine base and that of guanine base to cytosine base are always very close to unity (Kauffman, 2003). Implications of the DNA physical structure are evident in the functions of DNA replication, transcription, and translation, whereby DNA copies itself to maintain genetic constancy, forms RNA and proteins contributing to phenotypic expression, and affords mutation and evolution. In the absence of this deep knowledge about functional features, the ladder analogy in this study helped students find a beautiful and pleasing consistency between a simple structure that they knew and DNA structure that they had to learn.

In the framework of Goldin-Meadow and Beilock (2010), the ladder analogy by itself is observer centric, and the palm gesture is an observer viewpoint gesture. We found that these were not sufficient in most cases to bring about learning. We then had to ask students to imagine that they were actually stepping on the ladder, that is, getting inside the model. This could be seen as the equivalent of character viewpoint gestures or actions, which might have provided for the students a bridge between an imagined concrete action and the abstract representation of base pair orientation. Our results showed that, though students did not spontaneously link the ladder analogy with their textbook diagrams, gesture could be used to link 2-D representations with multiple 3-D models of DNA structure; and mental simulation—involving changing the observer viewpoint, to one from inside the molecule—could effectively link the ladder analogy with the molecular structure of DNA.

References

Ainsworth, S. (1999). The functions of multiple representations. *Computers and Education, 33*, 131–152.

Bahar, M., Johnstone, A. H., & Hansell, M. H. (1999). Revisiting learning difficulties in biology. *Journal of Biological Education, 33*(2), 84–86.

Barsalaou, L. W. (1999). Perceptual symbol systems. *Behavioral and Brain Sciences, 22*, 577–560.

Clark, A. (1997). *Being there: Putting brain body and world together again.* Cambridge, MA: MIT Press.

Cold Spring Harbor Laboratory's DNA Learning Center. (n.d.). *Discovering the double helix structure of DNA, James Watson, video with 3D animation and narration.* Retrieved April 9, 2011, from http://www.dnalc.org/view/15492-Discovering-the-double-helix-structure-ofDNA-JamesWatson-video-with-3D-animation-and-narration.html

Duit, R. (1991). On the role of analogies and metaphors in learning science. *Science Education, 75*(6), 649–672.

Flynn, E., & Siegler, R. (2007). Measuring change: Current trends and future directions in microgenetic research. *Infant and Child Development, 16*, 135–149.

Gentner, D. (1989). The mechanisms of analogical learning. In S. Vosniadou & A. Ortony (Eds.), *Similarity and analogical reasoning* (pp. 199–241). Cambridge, MA: Cambridge University Press.

Goldin-Meadow, S., & Beilock, S. L. (2010). Action's influence on thought: The case of gesture. *Perspectives on Psychological Science, 5*(6), 664–674.

Harrison, A. G., & Treagust, D. F. (2006). Teaching and learning with analogies. In P. J. Aubusson, A. G. Harrison, & S. M. Ritchie (Eds.), *Metaphor and analogy in science education* (pp. 11–24). Dordrecht, The Netherlands: Springer.

Justi, R., & Gilbert, J. (2006). The role of analog models in the understanding of the nature of models in chemistry. In P. J. Aubusson, A. G. Harrison, & S. M. Ritchie (Eds.), *Metaphor and analogy in science education* (pp. 119–130). Dordrecht, The Netherlands: Springer.

Kauffman, G. B. (2003). DNA structure: Happy 50th birthday! *Chemistry Educator, 8*, 219–230.

Lewis, J. (2004). Traits, genes, particles and information: Re-visiting students' understandings of genetics. *International Journal of Science Education, 26*(2), 195–206.

Maharashtra State Board of Secondary and Higher Secondary Education. (2009). Chapter 2: Biotechnology. In *Standard XII 'Biology'* (pp. 13–33). Pune, India: MSBSHSE.

Marbach-Ad, G. (2001). Attempting to break the code in student comprehension of genetic concepts. *Journal of Biological Education, 35*(4), 183–189.

Marbach-Ad, G., & Stavy, R. (2000). Students' cellular and molecular explanations of genetic phenomena. *Journal of Biological Education, 34*(4), 200–205.

Mathai, S., & Ramadas, J. (2009). Visuals and visualization of human body systems. *International Journal of Science Education, 31*(3), 439–458.

Padalkar, S., & Ramadas, J. (2011). Designed and spontaneous gestures in elementary astronomy education. *International Journal of Science Education, 33*(12), 1703–1739.

Pozzer-Ardenghi, L., & Roth, W.-M. (2005). Photographs in lectures: Gestures as meaning-making resources. *Linguistics and Education, 15*, 275–293.

Siegler, R. S. (2006). Microgenetic analyses of learning. In D. Kuhn & R. S. Siegler (Eds.), *Handbook of child psychology* (Vol. 2, pp. 464–510). New York: Wiley.

Siegler, R. S., & Crowley, K. (1991). The microgenetic method: A direct means for studying cognitive development. *American Psychologist, 46*, 606–620.

Tsui, C.-Y., & Treagust, D. F. (2003). Genetics reasoning with multiple external representations. *Research in Science Education, 33*, 111–135.

van der Aalsvoort, G. M., van Geert, P., & Steenbeek, H. W. (2009). Microgenetic methodology: Possibilities with regard to research on learning and instruction. In K. Kumpulainen, C. Hmelo-

Silver, & M. Cesar (Eds.), *Investigating classroom interaction: Methodologies in action* (pp. 203–229). Rotterdam, The Netherlands: Sense.

Venville, G. J. (2008). Effective biology analogies. In A. G. Harrison & R. K. Coll (Eds.), *Using analogies in middle and secondary science classrooms: The FAR guide – An interesting way to teach with analogies* (pp. 113–116). Thousand Oaks, CA: Corwin Press.

Watson, J. D., & Crick, F. H. C. (1953a). Molecular structure of nucleic acids: A structure for deoxyribose nucleic acid. *Nature, 171*(4356), 737–738.

Watson, J. D., & Crick, F. H. C. (1953b). Genetical implications of the structure of deoxyribonucleic acid. *Nature, 171*(4361), 964–967.

Woski, S. A., & Schmidt, F. J. (2002). DNA and RNA: Composition and structure. In T. M. Devlin (Ed.), *Textbook of biochemistry with clinical correlations* (5th ed., pp. 27–92). New York: Wiley-Liss.

Chapter 18
Multiple Representations in Modeling Strategies for the Development of Systems Thinking in Biology Education

Roald Pieter Verhoeff, Kerst Th Boersma, and Arend Jan Waarlo

Introduction

Biologists try to bring order in the endless variety of life's structures and processes. In doing so, they trace the process of evolution from relatively simple life forms preceding eukaryotic cells to the complex multicellular organisms living together in similar complex ecosystems. Today, biological research is often considered to have entered the era of post-genomics research in which the complexity of life is explained via an integrated approach from many disciplines including bioinformatics, evolutionary biology, and genomics. This transdisciplinary approach to the study of the complex physical and chemical organization of life is typical to a *systems thinking approach* as von Bertalanffy (1945, 1950) already articulated in the 1930s[1] with his General Systems Theory. For biological researchers, systems thinking is a basic conceptual framework underlying their daily work on complex and dynamic living systems. In molecular biology, for example, systems biology refers to the integration of experimental and computational approaches to understand and predict complex cellular functions (Alberghina, 2007), and evolutionary biology—traditionally engaged in searching for similarities in anatomy,

[1] Bertalanffy developed his General Systems Theory first via lectures, beginning in the 1930s and later via publications, starting in 1945.

R.P. Verhoeff (✉)
Department of Philosophy and Science Studies, Radboud University Nijmegen, Nijmegen, The Netherlands
e-mail: r.verhoeff@science.ru.nl

K. Th Boersma
Faculty of Science, Utrecht University, Utrecht, The Netherlands
e-mail: k.t.boersma@uu.nl

A.J. Waarlo
Centre for Science and Mathematics Education, Utrecht University, Utrecht, The Netherlands
e-mail: A.J.Waarlo@uu.nl

D.F. Treagust and C.-Y. Tsui (eds.), *Multiple Representations in Biological Education*, Models and Modeling in Science Education 7, DOI 10.1007/978-94-007-4192-8_18,
© Springer Science+Business Media B.V. 2013

embryology, and physiology—has been integrated with comparing proteins and genome sequences between organisms (Moore, 2007).

Following the transformation of studies of virtually all life processes, many educationalists consider systems thinking as a metacognitive skill that enables students to understand and cope with these new scientific advancements that reach our society. Systems thinking is one of the skills required by the Dutch examination syllabus for biology; students should be able to demonstrate an understanding that biological relations are complex by nature and often cannot be explained in a monocausal way. Students should also be able to relate biological phenomena at various levels of organization to one another.

Knippels (2002) proposed a strategy based on systems thinking for genetics education: the "yo-yo strategy." This yo-yo strategy copes with complexity by explicitly distinguishing the levels of biological organization starting from the concrete organism level and by descending and ascending these levels. Explicating the levels makes the transect nature of genetics transparent to students and provides an insight into where hereditary phenomena, processes, and structures occur at the different levels of biological organization.

Accepting that systems thinking ought to be a major component of the upper secondary school, biology curriculum obviously has implications for the content and structure of the entire biology curriculum. At present, several topics in the Dutch biology curriculum—cell biology, behavior, and ecology, for example—are limited to only one level of biological organization; systems thinking requires that topics be defined to cover different levels of biological organization. In this approach, the use of models is essential because in biology structures and processes at different levels of biological organization are often abstracted into models. In particular, at the molecular and cellular levels, models are used to enable aspects of a system—which are either complex or not directly perceivable through the senses—to be rendered more readily visible. Moreover, models are potentially valuable learning and teaching tools for developing a scientific way of thinking (Gilbert, 1993). In other words, a systems thinking approach to biology education should (1) engage students in exploring horizontal and vertical relationships between concepts from the molecular up to and including the societal or population level and (2) challenge them to use visualizations and other models to construct knowledge in a so-called model-based learning trajectory (Clement, 2000). This approach covers four elements of a systems thinking competence as listed in Table 18.1 (Boersma, Waarlo, & Klaassen, 2010; Verhoeff, Waarlo, & Boersma, 2008). As students are actively engaged in a series of modeling activities, our implementation of systems thinking is referred to as a systems modeling approach.

In this chapter, we report a critical appraisal of our systems modeling approach in three parts. First, we lay a theoretical foundation under our modeling strategy by articulating different characteristics of models or representations and the emergent modeling approach which prescribes the sequence in which these models should be placed in a bottom-up educational strategy. Second, we articulate two studies that both designed and evaluated the development of a learning and teaching strategy that engaged students in developing multiple representations of living systems with increasing complexity. The first study we address here focused on the

Table 18.1 Four elements of a systems thinking competence for biology education (Verhoeff et al., 2008)

1. Being able to distinguish between the various levels of organization—that is, cell, organ, and organism—and to match biological concepts with specific levels of biological organization
2. Being able to interrelate concepts at a specific level of organization (horizontal coherence)
3. Being able to link biology concepts from different levels of organization (vertical coherence)
4. Being able to think back and forth between abstract visualizations (models) and real biological phenomena

development of an initial systems model in cell biology education (Verhoeff, 2003; Verhoeff et al., 2008), and the second study employed computer modeling as a tool in the understanding of the dynamics in ecosystems (Westra, 2008; Westra, Boersma, Savelsbergh, & Waarlo, 2008). Finally, we formulate more general recommendations about the use of multiple representations in the development of systems modeling. To this aim, we address in this chapter the following question:

- In what way does the nature and sequence of the multiple systems representations in the two tested modeling strategies contribute to students' learning about complex and biological phenomena?

Systems Modeling Approach

The General Systems Theory (von Bertalanffy, 1945, 1950) emphasizes the hierarchical structure and open nature of biological systems and states that biological systems—that is, cells, organisms, populations, and ecosystems—are complex, but highly organized entities in constant interaction with their environment. In biology, the complexity of the natural world is often reduced to representations that focus on those features that are essential for understanding a certain process or structure. This is particularly the case in the field of cell biology where phenomena at the molecular level are not directly perceivable with the eye. But in ecology, modeling has been considered as an essential approach to understand (pieces of) ecological reality and being able to forecast developments in an ecosystem (Likens, 1985), for example, distributions and abundance of organisms.

Over the last 20 years, a considerable number of theoretical and empirical articles on the use of models and modeling in science education have been published, recognizing the functionality of models in scientific thinking. For example, Gilbert (1993) considered models as integral to thinking and working scientifically because models are science's products, methods, and its major learning and teaching tools. Because of the large variety of models used in science and science education, typologies of models—as presented by Coll and Taylor (2005), Gilbert and Boulter (2000), Harrison and Treagust (2000), and Gilbert and Treagust (2009)—are helpful in characterizing selected models. An important dimension seems to be the distinction between (or the continuum from) idiosyncratic mental models to analogical, scientifically accepted consensus models of Gilbert and Boulter or symbolic models of Harrison and Treagust. A learning pathway should lead students from their idiosyncratic mental models, via intermediate models, toward a theoretical target model (Clement, 2000). During such a trajectory, several authors (e.g., Abell & Roth, 1995; Coll & Taylor, 2005) have argued that it is preferable that students construct and critique their own models, and not to introduce expressed models developed by others, because that would more effectively support their conceptual development.

Both ideas—a learning pathway ranging from mental models to a theoretical model and the construction of the models by the students themselves—come together in what Gravemeijer (1999) defined for mathematics education as *emergent modeling*. In his view, models fulfill a bridging function between the phenomenological appearances of mathematics in reality, on the one hand, and the formal mathematics, on the other. He described (p.160) a pathway in which a *model of* an informal scientific activity (the mental model) emerges and gradually develops into a formal *model for* scientific reasoning. From this perspective, an animal represented by a schematic drawing is considered a *model of* that animal since the animal still can be recognized, for example, by its outline. However, if we represent the same animal by a rectangle connected via a double arrow with the space outside, we apply a *model for* understanding that animal, indicating that the animal can be considered as an open system with an input and output. According to the emergent modeling approach, students should first be engaged in developing *models-of* that refer to a specific situation or phenomenon. Second, *models-for* should be employed to enable students to focus on interpretations and solutions independently of situation-specific images (Gravemeijer, 1999). In practice, this means that students can mentally move from "discussions on general activities and reasoning" (p. 160) to situation-specific problem-solving activities which eventually facilitate formal scientific reasoning no longer dependent on the support by models.

Besides the deliberate use of the yo-yo strategy (Knippels, 2002) in both the general and situation-specific activities, the idea of emergent modeling was recognized afterward in the modeling strategy of Verhoeff (2003) and deliberately applied in the modeling strategy of Westra (2008) (see also Boersma & Waarlo, 2009). A third component became apparent in reflection on both modeling strategies and can be labeled as the employment of multiple representations by Gilbert and Treagust (2009).

A categorization of the different models or representations with increasing abstractness that were employed in the two modeling strategies is presented in Table 18.2. Categories 2–4 represent the biological phenomenon (category 1) and can be classified as *models-of,* while categories 5–8 refer to biological phenomena and can therefore be classified as *models-for*. This categorization will guide our analysis in the next paragraph.

Modeling Strategies of the Two Studies

Research Approach

Both the two modeling studies (Verhoeff, 2003; Westra, 2008) presented here followed the same research design approach. This implies that a design of a learning

Table 18.2 Model categories with increasing abstractness (C1–C8)

Category	Description	Type
C8	Computer model 2 specification of computer model 1 by quantification of relations between variables, presented as graphic output	
C7	Computer model 1 referring to a category of biological phenomena	
C6	Bridging model relating an abstract model to a computer model	Models-for
C5	Abstract model referring to a category of biological phenomena	
C4	Schematized drawing or animation of a biological phenomenon or category of biological phenomena	
C3	Figurative drawing of a biological phenomenon or category of biological phenomena	Models-of
C2	Photo or film of a biological phenomenon	
C1	Biological phenomenon	

and teaching strategy was elaborated in a scenario indicating the desired behavior of students and the teacher and the expected learning outcomes (for a more elaborate description of the design research approach, see Bulte, Westbroek, De Jong, & Pilot, 2006). The scenario was tested in two or three design cycles in school practices of different schools and adapted where the learning and teaching process did not unfold as expected (see also Verhoeff et al., 2008). Both studies focused on preuniversity biology education (students aged 16–18 years old).

In both studies, the systems modeling approach consisted of an integration of the three components introduced in the preceding section—a learning pathway ranging from mental models to a theoretical model, the active engagement of students in the construction of models, and the yo-yo strategy developed by Knippels (2002). The yo-yo strategy aimed at a reduction of the complexity and abstractness of biological subjects by sequencing the biological content according to levels of biological organization. Although the design of the modeling strategies in both studies consisted of the same integrated emergent modeling components, some major differences are worth noting here. In the strategy of Verhoeff (2003), the starting point was that students constructed their own models whenever possible. In the study of Westra (2008)—due to the complexity of computer modeling, and particularly to students' difficulties with establishing the underlying mathematical relationships—students primarily focused on performing previously defined computer modeling tasks.

In the two following paragraphs, both modeling strategies are introduced, each followed by a critical reflection on their successful and unsuccessful characteristics.

The Development of Systems Thinking in Cell Biology Education

The study of Verhoeff started with the premise that purposeful application of systems thinking provided a way to address the acquisition of coherent understanding of cell biology. However, systems thinking was not only considered a tool for

Table 18.3 Sequence of modeling activities to introduce systems thinking in cell biology education

	Modeling activity (name)	Description (students and teachers actions)
1	General orientation on the cell as basic unit of life	Teacher introduces the cell as basic unit of all organisms and discusses with students how cells maintain themselves, leading to an interest in free-living cells
2	Developing a model of free-living cells	Based on the observations of *Paramecia* through a light microscope, students draw free-living cells fulfilling the fundamental life processes. Students question whether their model applies to their body cells as well
3	Developing a general 2-D model of cells	Students compare their drawings of cells with electron micrographs. The teacher introduces the orderly representations of the general and structural characteristics of cells in their textbooks. Students further explore these representations, including the organelles and adjust their initial drawings
4	Building a 3-D model of a plant cell	Using textbook and internet, different student groups construct a 3-D model of a specific organelle and connect it to other organelles in order to be able to place the organelle in a large 3-D model of a plant cell. Students also present their findings to the other students
5	Explication of systems thinking	A computer-aided program guided students in exploring the process of digestion at the level of the organism, organ, and cell. Each level is generalized into a structural unit in continuous exchange with its environment and as functional subsystem of a system at higher level of organization
6	Application of the hierarchical systems model	Students explore and model the process of breast-feeding by interrelating the different levels of organization, guided by the hierarchical systems model, thereby recognizing the benefits of thinking back and forth between the different levels

developing coherent cell biological knowledge but it also constituted a desired learning outcome of the learning and teaching strategy as outlined in Table 18.3.

The first step in the strategy was to acquire a basic notion of the cell and its organization, implicitly developed from a systems perspective. Subsequently, the development of a systems concept was facilitated by a referral to the acquired notion of the cell "as a system" and furthering insights using a systems perspective. According to this approach, a reasonable motive to introduce the systems concept is evoked when students discover that structures and processes at different levels of biological organization can be abstracted into the same model representing a "living system." Finally, the integration of the different levels of organization in a hierarchical systems model would constitute the final step in understanding the cell as a functional unit of the organism. In a developmental research approach, a learning and teaching process was optimized in several research cycles; each cycle focused on testing the strategy in the classroom and on reflection and adjustment of the designed learning and teaching activities in close cooperation with biology teachers. The learning and teaching strategy was optimized in four subsequent case studies with preuniversity students at two different schools (see Verhoeff, 2003; Verhoeff et al., 2008).

The study of Verhoeff resulted in a modeling strategy in which multiple representations of the cell were (a) used and developed with increasing complexity and abstractness and (b) integrated in a general systems model and thus closely resembled the emergent modeling strategy invented by Gravemeijer (1999). In the six modeling activities constituting the strategy, students were continuously engaged in thinking back and forth between visual representations differing in abstractness, including biological objects themselves. The representations that students either constructed themselves or retrieved from their schoolbooks or the Internet are shown in Table 18.4. As Table 18.4 shows, the models in our learning trajectory could be divided into four different categories of abstractness (C2–C5 in Table 18.2), in addition to the biological phenomenon itself (C1).

The first four modeling activities successfully engaged students in exploring and modeling complex interrelations within the cell. Based on observations of a unicellular organism (biological object) and electron micrographs of cells, students constructed schematized drawings representing all "living cells" and eventually an abstracted hierarchical systems model representing all living systems (see category 5 in Table 18.4). In terms of emergent modeling, the first four modeling steps consisted of referential activities, in which models-of cells referred to the activity of exploring cells and the basic life functions they fulfill.

The fifth modeling activity—which explored the functional relation of cells with higher levels of organization—marked a problematic stage. Students experienced an interruption with the earlier modeling activities in which they were actively engaged in developing modeling themselves, resulting in a general 3-D model of cells. Modeling had become a tool for exploring biological (cell) components and their interrelations. Now, they were engaged in a general activity in which models-for systems thinking made possible an interpretation of living systems independently of the imagery they had studied so far. It marked an interruption in three ways: (1) they were not constructing their own models anymore, but followed a modeling computer program; (2) after exploring the cellular level and dealing with cellular representations, this activity entailed abstracting structures and processes at the organ and organism level as well, combined with integrating these abstractions in a hierarchical systems model; and (3) In guiding the students during this exploration, the computer model did only employ representations in the higher categories of abstractness (C3–C5) so that students *lost contact* with concrete representations.

In hindsight and from an emergent modeling perspective, the fifth modeling phase could be characterized as formal systems theoretical reasoning, although it still depends on the support of the general model-for systems thinking as students employ it to explore a new phenomenon. Based on our empirical findings in the classroom, the focus of the improvement of the strategy should lie on engaging students in exploring the functional relationship between cells, organs, and the organism, for example, by exploring the process of endocrine regulation. Students

Table 18.4 Models on different degrees of abstractness used in the learning and teaching strategy on systems thinking in cell biology education (Verhoeff, 2003)

Category	Description	Exemplary model
5	Abstract model referring to a category of biological phenomena	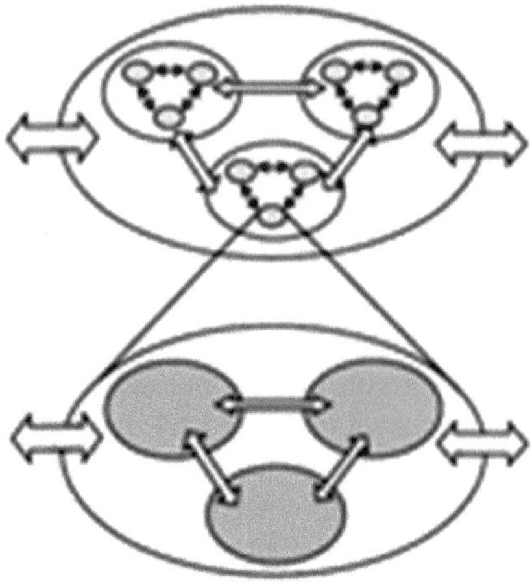
4	Schematized drawing or animation of the biological phenomenon or category of biological phenomena	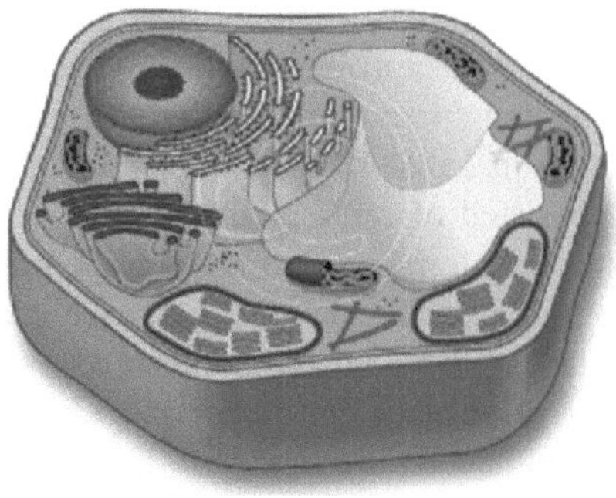
		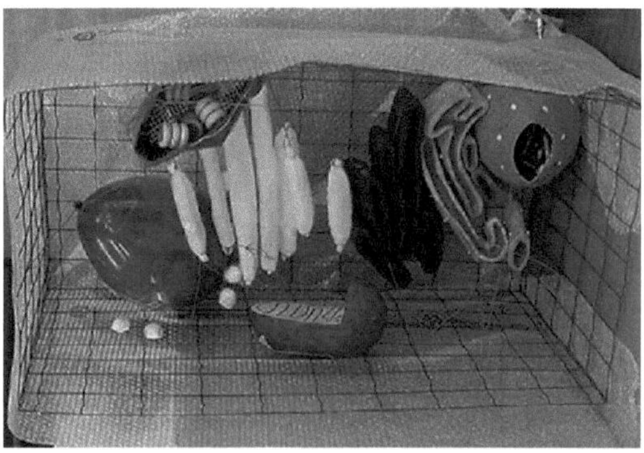
3	Figurative drawing of the biological phenomenon	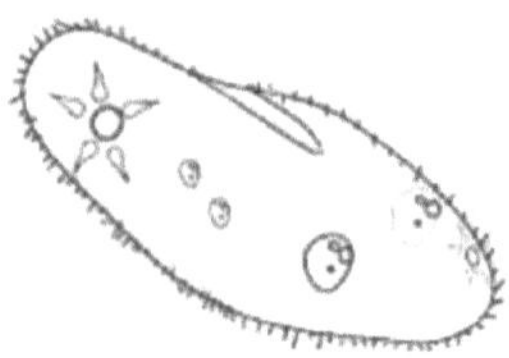 

(continued)

Table 18.4 (continued)

Category	Description	Exemplary model
2	Photo or film of the biological phenomenon	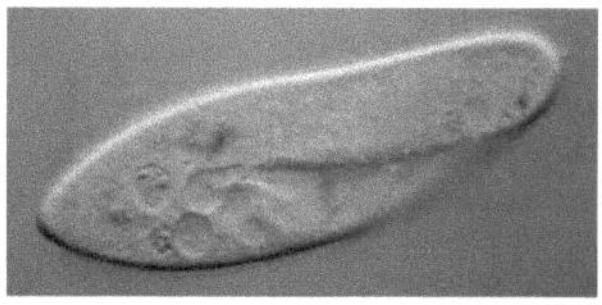 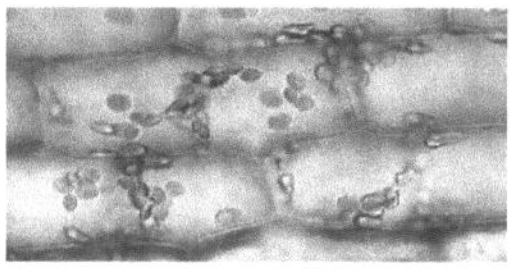
1	Biological phenomenon	

could elaborate their model of the cell as a basic unit of life to a model-for understanding the cell as a functional unit of the organism and thus for addressing the vertical and horizontal coherence in natural phenomena. In this way, students are challenged to first extend their own models to focus on the cell as part of a larger (structural) organization that can fulfill its function for the living organism. Eventually their model could be tested or applied in exploring a new phenomenon like breast-feeding at the cellular level up to the organism level.

Although the hierarchical systems model was introduced too abruptly in our tested strategy, it proved helpful in exploring biological phenomena crossing several levels of organization during the sixth modeling activity. This activity employed representations with high abstractness as well, but it was successful in engaging students in formal systems reasoning. In our case study, the hierarchical systems model was successfully used as a tool to acquire a coherent understanding of the process of breast-feeding—by exploring the horizontal coherence at the cellular, organ, and organism levels—and to interrelate the concepts from different levels of organization (vertical coherence). The six activities that make up the modeling trajectory are depicted in Fig. 18.1. Key element in each activity is about thinking back and forth between abstract visualizations (models) to real biological phenomena (see C1–C5 in Table 18.4).

The Strategy on the Understanding of Dynamic Behavior

The study of Westra (2008) focused on the development of a learning and teaching strategy about ecosystem behavior using modeling and systems thinking in authentic practices. The intention was to use computer modeling for clarifying the dynamics at three levels of biological organization: organism, population, and

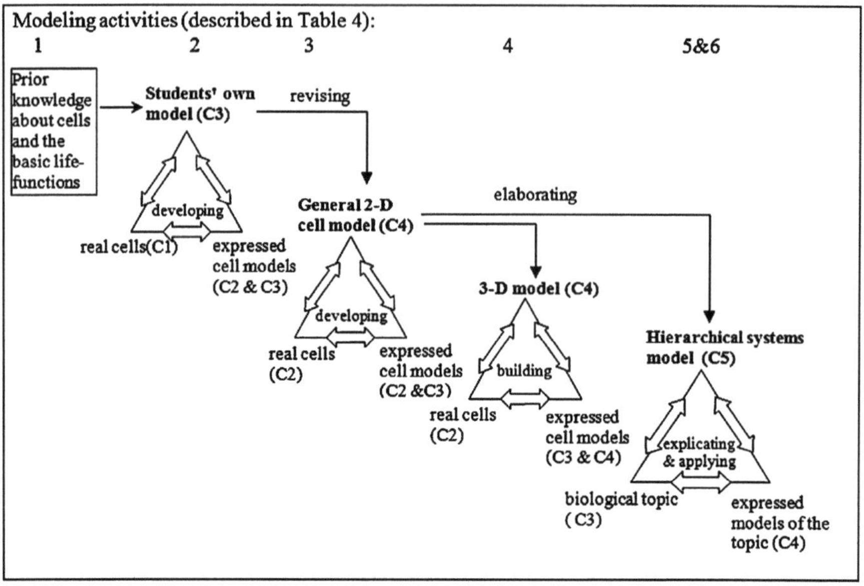

Fig. 18.1 Six modeling activities (C1–C5) from Table 18.3 in the trajectory from prior knowledge toward the hierarchical systems model via intermediate models differing in abstractness

ecosystem. Consequently, computer modeling and systems thinking were considered as prerequisites for the development of ecological concepts in the study.

The following sequence of authentic practices was built into the design of the learning and teaching strategy: (1) researchers working at the Netherlands Institute for Ecology working on mussel cultures, (2) ecologists working on managing rabbits in the dunes, and (3) ecologists working on the reduction of overcrowding of African elephants. The first two practices were used for the acquisition and extension of a dynamic concept of ecosystem, whereas the last one was used for testing students' conceptual understanding. Starting from this sequence of authentic practices, the yo-yo strategy (Knippels, 2002) was used to structure the sequence of modeling activities from the level of the organism to the level of the ecosystem. The strategy of emergent modeling (Gravemeijer, 1999) helped in shaping the learning and teaching processes from students' understanding of concrete organisms to the symbolic language of the selected modeling tool (Powersim) (Table 18.5).[2] An overview of the resulting sequence of modeling activities is presented in Table 18.5, whereas a number of models of different abstractness used in the modeling activities are shown in Table 18.6.

[2] The modeling tool which we refer to here is a software program called *Powersim Constructor Lite* and has been developed by Powersim Software AS (www.powersim.com). In educational settings, the program can be used, free of charge, on a noncommercial basis as in the study of Westra (2008).

Table 18.5 Modeling activities in the learning and teaching strategy on the understanding of dynamic ecosystem behavior (Westra, 2008)

	Modeling activity	Description
1	Exploration of the mussel culture in the Easter Scheldt	Teacher introduces the mussel culture in the Easter Scheldt and asks which factors may influence the optimization of a mussel culture. Students mention external factors influencing the growth of a mussel
2	Feeding and excretion of the mussel	The teacher shows an animation of the feeding of the mussel. The students describe the filter feeding, going from the inhalant region, via the gills to the mouth, and the discharge of water, by comparing a dead mussel with a schematic drawing and finally determine the dry weight of a mussel
3	Computer modeling 1: the growth of a mussel	The students compare some models of a mussel with a Powersim model of the mussel's filtering and dissimilation, with a quantitative change in dry weight, and build the computer models in dyads, making use of the instructions in their work book
4	Computer modeling 2: the growth of a mussel population	The teacher introduces a mussel population. Students identify in dyads the factors that should be included in the model and build a Powersim model of populations with different densities by extension of the model of the growth of a single mussel
5	Computer modeling 3: the growth of a mussel population in an ecosystem	The teacher introduces the concept of ecosystem and shows a Flash animation showing what happens with the plankton near the mussel. The students identify other species and build a Powersim model of an ecosystem by extension of the model of the mussel population, with birds as predators
6	Computer modeling 4: changes in the density of a rabbit population in the dune ecosystem	The students compare the complexity and dynamics of the Easter Scheldt mussel cultures with the dune ecosystem and have to determine factors that may raise the density of the rabbit population. Next, students explore two ready-made models, a simple Lotka-Volterra model for interspecific competition and a more complex model with a varying carrying capacity (Westra, 2008, p. 153)

A specific complication in the sequence of modeling activities is that the abstract systems model of a mussel (category 5, see Table 18.2) is not matching with the symbols of the graphic modeling tool of Powersim (category 7). Consequently a so-called bridging model (category 6) was introduced, which demonstrates both the characteristics of the systems model and the symbols used in the modeling tool.

Table 18.6 Models on different degrees of abstractness used in the learning and teaching strategy on ecosystem behavior (Westra, 2008)

Category	*Description*	*Exemplary model*
8	Computer model 2, specification of computer model 1 by quantification of relations between variables, represented as graphic output	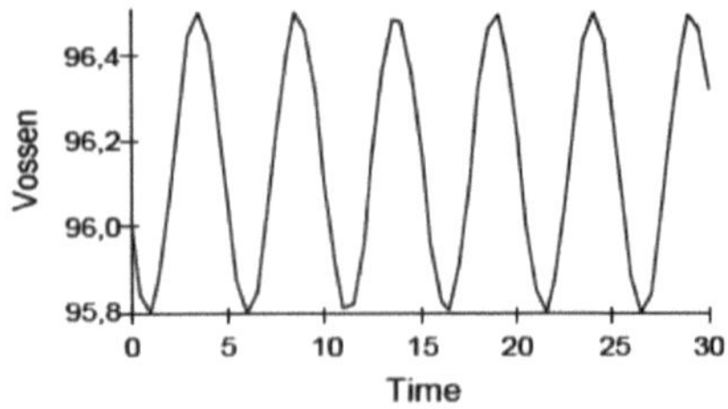
7	Computer model 1 referring to a category of biological phenomena	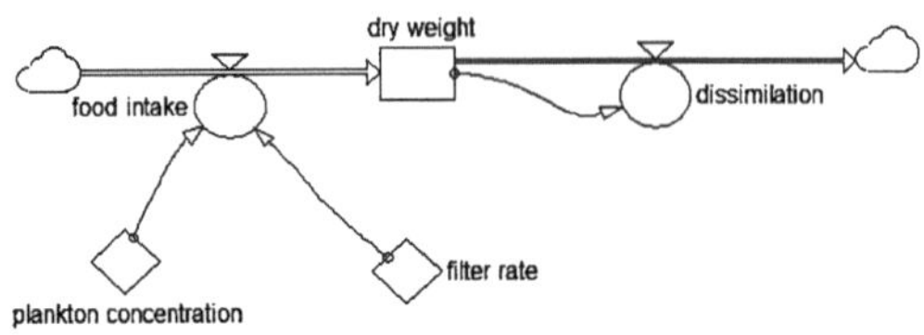
6	Bridging model relating the abstract model to the computer model	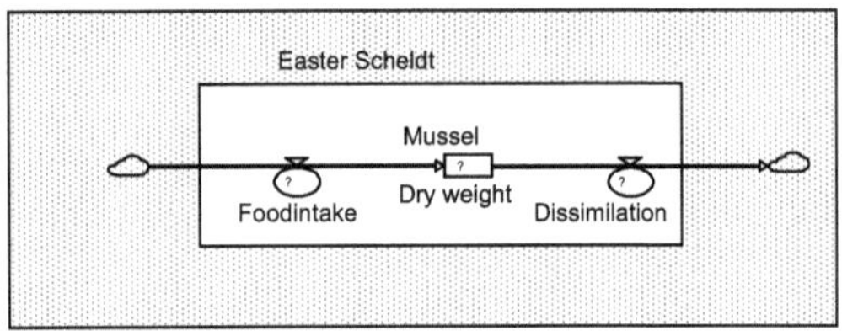
5	Abstract model referring to a category of biological phenomena	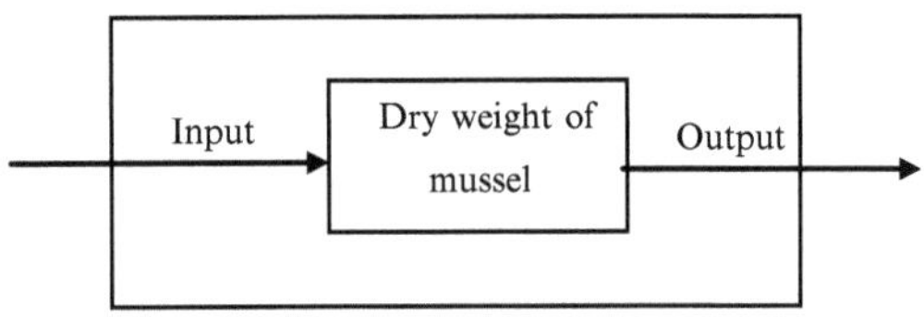
4	Schematized drawing or animation of a biological phenomenon or category of biological phenomena	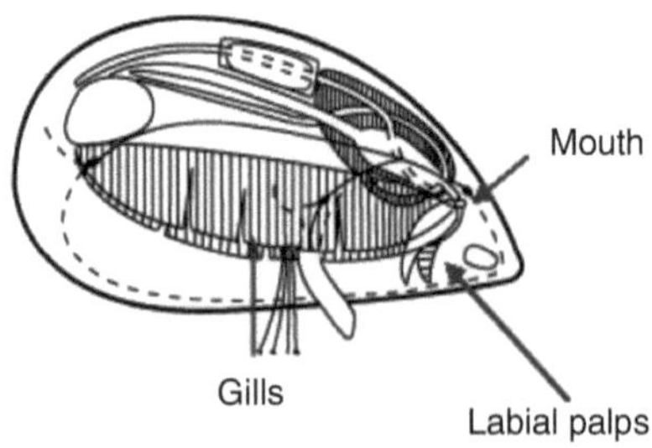
3	Figurative drawing of a biological phenomenon	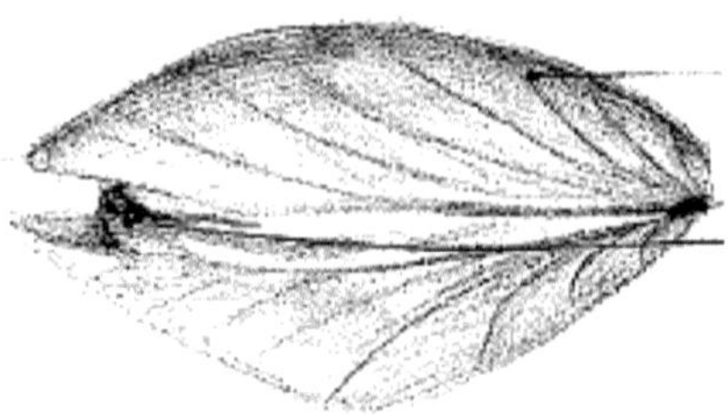

(continued)

Table 18.6 (continued)

Category	*Description*	*Exemplary model*
2	Photo or film of a biological phenomenon	
1	Biological phenomenon	

Table 18.7 Modeling structure of the learning and teaching strategy on ecosystem behavior

Authentic practice	Modeling activity (see Table 18.5)	Levels of biological organization	Biological objects and categories of models
(1) Ecological research on mussels	1	Organism	5. Abstract model
	2	Organ	1. Biological object
		Organism	4. Schematized drawing
	3	Organism	2/3. Photo with arrows representing input and output
			5. Abstract model
			6. Bridging model
			7. Computer model 1
			8. Computer model 2
	4	Organism	5. Abstract model
		Population	7. Computer model 1
			8. Computer model 2
	5	Organism	4. Schematized drawing
		Population	5. Abstract model
		Ecosystem	7. Computer model 1
			8. Computer model 2
(2) Nature management on rabbits	6	Population	7. Computer model 1
		Ecosystem	8. Computer model 2

Westra's study succeeded in designing a sequence of learning and teaching activities based on the three strategies mentioned above. The resulting modeling structure is indicated in Table 18.7.

As shown in Table 18.7, the sequence of modeling activities 3–5 constituted a complete succession of levels of biological organization from the level of the organism to the level of the ecosystem, as prescribed by the yo-yo strategy. Furthermore, as shown in Table 18.7 in modeling activity 3, models of all categories of abstractness were used, from a photo to computer models, including a bridging model. In the modeling activities 4–6, a bridging model was not used anymore, suggesting that it was not required to explain the relation between the computer model and the abstract systems model. When looking at the multiple representations in the strategy, it can be noticed that apart from the first activity, all modeling activities employed multiple representations. In modeling activities 2, 3, and 5, "models-of" biological objects are used together with "models-for," whereas in modeling activities 4 and 6, only "models-for" are used.

When the strategy was tested in the classroom practice, several shortcomings were noticed during the post-instructional analysis. Both the researcher and the participating teachers put much emphasis on students' computer modeling competences and much effort in finding a workable balance between engaging students in developing their own models and the complex endeavor of mathematizing the ecological models' underlying causal relations. By this one-sided focus, too little attention was given to the development of students' conceptualization of ecosystems. That was demonstrated already in the first lesson during which the teacher introduced—without discussion and without explicitly relating these concepts to students' prior knowledge of ecosystems—the concepts *dynamic*s, *complexity*, and *level of biological organization.* Although a discussion on these concepts was planned in the reflection on modeling activity 5, they were not put into practice before their transfer to the second practice on dune management.

Consequently, the aim that students would develop a dynamic conception of ecosystems by means of computer modeling was not attained, and it was concluded that the development of the concepts (ecosystems, dynamics, and complexity) during the lessons was problematic (Westra, 2008). Consequently, the students did not reach an extended view on the complexity and dynamics of ecosystems.

Considering the fact that not all modeling activities were performed as planned, it seemed probable that the outcomes would have been more satisfying if the scenario could have been followed completely. On the other hand, it seemed probable that the problems were partly due to mistakes in the scenario as well. A first issue in this study that needed improvement was that an adequate conceptual development would not only have required an explicit, stepwise development of the ecological concepts but also an attempt to relate the results from the modeling activities to empirical data.

That brings us to a second issue for improvement of the strategy. When working with the computer model in our study, many students had problems in recognizing the meaning of the natural phenomena being investigated. For example, many students did not have any idea of the real value of the dry weight of a mussel and were not alarmed by values such as 1 kg. Even more alarming was that it was concluded that many students lost contact with natural phenomena during modeling. Westra suggested that this might have been caused by students not really understanding the abstract nature of the model. Furthermore, many students had difficulties with the symbolic language of the modeling tool and with formalizing relations between components.

Westra's study seemed to indicate that computer models like the model of Powersim entail many difficulties equally for students and teachers. Consequently, in planning a learning and teaching trajectory on the development of a dynamic conception of ecosystem by means of computer modeling, it should be recommended to separate initially students' conceptual development of ecosystems from the development of their systems modeling competence. In such a learning and teaching trajectory, a computer tool should be introduced when students' conceptual development gets stuck by difficulties in representing the outcomes of complex interactions in ecosystems.

An important factor in the relation between models of populations and ecosystems and natural phenomena—which should be addressed in the first part of an adapted learning and teaching trajectory—is that in many cases populations and ecosystems have no clear systems boundaries. Westra (2008) reported that students had difficulties with the idea of systems boundaries and in particular with the systems boundaries of populations. This observation indicates that more emphasis is required on the development of systems models of populations and ecosystems and their matching with reality.

Discussion

Looking back at the strategies of the two studies that have been tested in classroom practice, we can articulate some recommendations for the design of and research on modeling strategies in biology, and in particular on the role of multiple representations in the development of systems thinking.

In both studies, models and modeling activities were used to develop students systems thinking and related systems concepts. In the design of the learning and teaching trajectories of both studies, two strategies developed prior to the studies for defining and sequencing modeling activities were implemented:

- The yo-yo strategy (Knippels, 2002) to define a sequence of learning and teaching activities from the level of the organism to the level of the ecosystem
- The strategy of emergent modeling (Gravemeijer, 1999) to structure the sequence from *models-of* with increasing abstractness to *models-for*

Both studies demonstrated that an integrated modeling approach, based on the above two strategies, needs to carefully consider the employment of multiple representations to facilitate both students' conceptual understanding of the topic at hand and understanding the way these different representations are instrumental in acquiring this understanding. Eventually the aim was to promote students learning at a metacognitive level, that is, students should be enabled to engage in the scientific practice of using models as tools for observation, exploration, synthesis, and, to a lesser extent, prediction of the behavior of biological systems. In other words, multiple representations were employed to guide students learning about biological phenomena in combination with acquiring an understanding of nature of science as an enterprise that is largely concerned with extending and refining (systems) models (Gilbert, Boulter, & Rutherford, 1998).

In the study of Verhoeff (2003), the active engagement of students in constructing and revising cellular models was a successful element in students' conceptual learning. Key to this success was the recurring process of comparing more abstract or general models with familiar representations of biological phenomena. Although students could successfully apply the hierarchical systems model to a previously unexplored phenomenon like breast-feeding, the missing link in the emergent modeling strategy (Gravemeijer, 1999) proved to be the step in which systems thinking was formalized into formal scientific reasoning of using models-for without the support of rather concrete models-of. This raises questions about the desired metacognitive systems thinking competence of students.

In the study of Westra (2008), the lack of support by models-of was even more apparent. Students were so involved in computer modeling that many of them lost contact with the natural phenomena they were actually modeling. This "loss of contact" was certainly intensified by the difficulties they had with the symbols and representations of the modeling tool. The shortcomings in both studies were already noticed during the studies. However, the analytical focus in this chapter on the use of multiple presentations in an emergent modeling approach has provided us more insight into the *dos* and *don'ts* of a modeling trajectory that fosters students' conceptual learning and their modeling competence, both of which are essential to scientific practice.

Four *design criteria* proved to be essential are as follows:

Stepwise and explicit navigation between different levels of biological organization, taking the level of the organism as a starting point. Without sufficient elaboration of a certain level of organization (e.g., the population level in Westra's study), conceptual difficulties will arise at other organizational levels (i.e., understanding of the concept population is a prerequisite for understanding the dynamics within an ecosystem).

Stepwise change from rather concrete models to models with higher abstractness. If students are not enabled to relate an abstract model to empirical phenomena or representations of these phenomena (as in Westra's study), students might fail to understand the usefulness of the abstract model (metacognition) and/or lose contact with empirical phenomena. The careful employment of multiple representations of different degrees of abstractness facilitates such a gradual development.

Stepwise development of models with greater generality. If in the development of a general (systems) model the generality of the models is not gradually increasing (as in Verhoeff's study), students will not be convinced of the need of using a general model (metacognition).

Consideration of the perceptibility of certain systems characteristics—such as the systems boundary (in Westra's study) or the vertical coherence between systems at different organizational levels (in Verhoeff's study). If no explicit attention is given to the matching of systems characteristics to empirical phenomena (as in Westra's study and during Verhoeff's fifth modeling activity), students might experience difficulties in acquiring coherent understanding of these phenomena.

It should be noticed that these four design criteria were not the input in the development of the modeling strategies of both studies, but the output of the analysis presented in this chapter. It is expected that these four design criteria will be helpful in solving the experienced difficulties with the uninterrupted learning trajectory from the perspective of the students. In the development of systems thinking in cell biology education, the interruption was apparent after the development of the 3-D model-of cell and before the introduction of the hierarchical systems model-for systems thinking. The learning trajectory focusing on the development of a dynamic conception of ecosystems in Westra's study was not sufficiently articulated, and the students' difficulties with computer modeling were underestimated. To address these shortcomings will certainly result in trajectories that would take more—or maybe even much more time—for attaining the desired learning outcomes.

References

Abell, S. K., & Roth, M. (1995). Reflections on a fifth-grade life science lesson: Making sense of children's understanding of scientific models. *International Journal of Science Education, 17*, 59–74.

Alberghina, L. (2007). Systems biology: Definitions and perspectives. In R. van Driel (Ed.), *Systems biology; A grand challenge for Europe* (pp. 8–9). Strasbourg, France: European Science Foundation.

Boersma, K. T., & Waarlo, A. J. (2009). On the theoretical input and output of 'design research' in biology education. In M. Hamann, A. J. Waarlo, & K. T. Boersma (Eds.), *The nature of research in biological education: Old and new perspectives on theoretical and methodological issues* (pp. 463–479). Utrecht, The Netherlands: CD-Bèta Press.

Boersma, K., Waarlo, A. J., & Klaassen, K. (2010, July). *The feasibility of systems thinking in biology education*. Paper presented at the 8th ERIDOB conference, Braga.

Bulte, A. M. W., Westbroek, H. B., De Jong, O., & Pilot, A. (2006). A research approach to designing chemistry education using authentic practices as contexts. *International Journal of Science Education, 28*(9), 1063–1086.

Clement, J. (2000). Model based learning as a key research area for science education. *International Journal of Science Education, 22*(9), 1041–1053.

Coll, R. K., & Taylor, I. (2005). The role of models and analogies in science education. *International Journal of Science Education, 27*, 183–198.

Gilbert, J. (1993). *Models and modeling in science education*. Hatfield, UK: Association of Science Education.

Gilbert, J., & Boulter, C. (Eds.). (2000). *Developing models in science education*. Dordrecht, The Netherlands: Kluwer.

Gilbert, J. K., Boulter, C., & Rutherford, M. (1998). Models in explanations, part 1: horses for courses? *International Journal of Science Education, 22*(9), 891–894.

Gilbert, J. K., & Treagust, D. (2009). Towards a coherent model for macro, submicro and symbolic representations in chemical education. In J. K. Gilbert & D. Treagust (Eds.), *Multiple representations in chemical education* (pp. 333–350). Dordrecht, The Netherlands: Springer.

Gravemeijer, K. P. E. (1999). How emergent models may foster the constitution of formal mathematics. *Mathematical Thinking and Learning, 1*(2), 155–177.

Harrison, A. G., & Treagust, D. F. (2000). A typology of school science models. *International Journal of Science Education, 22*, 1011–1026.

Knippels, M. C. P. J. (2002). *Coping with the abstract and complex nature of genetics in biology education*. Utrecht, The Netherlands: CD-Bèta Press.

Likens, G. E. (1985). *An ecosystem approach to aquatic ecology: Mirror lake and its environment*. New York: Springer.

Moore A. (2007). *New biology for new curricula*. Observations from the 6th international workshop on science education 17–19 May 2007. Heidelberg, Germany: European Molecular Biology Organization.

Verhoeff, R. P. (2003). *Towards systems thinking in cell biology education*. Utrecht, The Netherlands: CD-Bèta Press.

Verhoeff, R., Waarlo, A. J., & Boersma, K. T. (2008). Systems modelling and the development of coherent cell biological knowledge. *International Journal of Science Education, 30*(4), 543–568.

von Bertalanffy, L. (1945). Zu einer allgemeinen Systemlehre [Toward a general systems theory]. *Blätter für deutsche Philosophie, 3/4*.

von Bertalanffy, L. (1950). The theory of open systems in physics and biology. *Science, 3*, 23–29.

Westra, R. H. V. (2008). *Learning and teaching ecosystem behaviour in secondary education*. Utrecht, The Netherlands: FIsme Press.

Westra, R. H. V., Boersma, K. T., Savelsbergh, E. R., & Waarlo, A. J. (2008). Towards understanding ecosystem behaviour through systems thinking and modeling. In M. Hamann, M. Reiss, C. Boulter, & S. D. Tunnicliffe (Eds.), *Biology in context: Learning and teaching for the twenty-first century* (pp. 205–216). London: Institute of Education, University of London.

Chapter 19
Conclusion: Contributions of Multiple Representations to Biological Education

David F. Treagust and Chi-Yan Tsui

This Volume and Biology Education in the Twenty-First Century

Our book project began in 2009 with the intent to bring together international biology educators and biology education researchers who are involved in improving biological education from the perspective of multiple representations. It was also our goal that this volume would be able to address how biological education could meet the challenges of the twenty-first century, in which the breakthroughs in biological research would necessitate the integration of research and education with global economics and human social structures (Kress & Barrett, 2001).

Over the first decade of the twenty-first century, there have been numerous reports calling for reforms of science and biology education in high schools and universities. For example, Labov, Reid, and Yamamoto (2010) argued, based on the US National Science Council's (2009) report, that there is a need to rethink and restructure high school and undergraduate biology education, making it more relevant and accessible to more, if not all, students. In a similar manner, there have been calls for reforms in the science curriculum in many other countries, particularly in Australia (Tytler & Prain, 2010), the UK (e.g., Reiss, Millar, & Osborne, 1999), and Germany (e.g., Fischer, Kauertz, & Neumann, 2008). In these reforms, biology takes a central role because of the rapid development and advances in the biological sciences since the Human Genome Project for which the twenty-first century is often known as the century of biology (Carey, 1998; Kress & Barrett, 2001).

It was with this background that we proposed to international scholars three research questions for writing their chapters (see Box 19.1) to which their chapters

Unless stated otherwise, the term *multiple representations* in this chapter refers to multiple external representations (MERs) used by Ainsworth (1999).

D.F. Treagust (✉) • C.-Y. Tsui
Science and Mathematics Education Centre, Curtin University, Perth, WA 6845, Australia
e-mail: D.Treagust@curtin.edu.au; C.Tsui@curtin.edu.au

D.F. Treagust and C.-Y. Tsui (eds.), *Multiple Representations in Biological Education*, Models and Modeling in Science Education 7, DOI 10.1007/978-94-007-4192-8_19,

© Springer Science+Business Media B.V. 2013

Box 19.1 Research Questions Suggested by the Editors to the Chapter Authors in 2010

1. In what ways does your research involve the use of multiple representations in biology teaching and learning?
2. Do you have any particular emphasis on one or more of the following multiple external representations in your research in biological education and why:
   - Using analogies, metaphors, visualizations, language, and others
   - At the macro, micro/submicro and/or symbolic levels
   - Along hierarchically organized levels from molecules to the biosphere
3. What pedagogical functions of multiple external representations in biology does your research show that can enhance teaching and learning biological concepts (i.e., in helping students construct their mental models or internal representations of such concepts)?

in this volume have responded in various ways. This volume is unique for its rich collection of empirical studies and theoretical expositions on the utility and effectiveness of using multiple representations in biological education that fill a gap in the literature in science education. Some of the themes of the 17 chapters (Chaps. 2 to 18) are common; yet they differ in both the content areas and contexts within which learning and teaching take place in different languages in more than ten countries.

Seeking a Unifying Theoretical Model for Teaching and Learning with MERs in Biological Education

In the Introduction chapter, we commenced with Ainsworth's (1999, 2006) functional taxonomy of multiple external representations (MERs) and our view that learning with multiple representations involves three dimensions: modes of representations, levels of representations, and domain knowledge of biology. We then proposed our theoretical cube model for examining and interpreting the themes and theoretical positions of the chapter authors. We contended that the different MERs used by chapter authors can be accommodated within the three dimensions of our theoretical cube model. Furthermore, we believed that our three-dimensional cube model can be used to examine and interpret the chapters in terms of translation between the external representations of biological knowledge in various ways for achieving one or more of the complementing, constraining, and constructing functions of MERs for learners (Ainsworth, 1999). We also believed—compared to other theoretical frameworks—that the functional taxonomy of MERs can provide a more useful unifying framework to explore how MERs of biological phenomena can enable learners to construct their understanding in terms of reasoning and internal representations or mental models (Gentner & Stevens, 1983).

In this chapter, we present an overall synthesis of the themes of the chapters before we conclude how MERs can contribute to biological education in the twenty-first century. The themes are generally in line with the view that learning with external representations (a cognitive science perspective) and constructivist learning (a science education perspective) have a primary commonality in terms of agency (McKendree, Small, Stenning, & Conlon, 2002) or a sense of empowerment that is an important part of scientific literacy (Anderson, 2007). Useful representations always have embedded information that requires learners to engage in deep thinking about the represented knowledge, often in collaboration with others, for deeper understanding. For example, "reasoning with an abstract representation of a situation can be more effective than reasoning with a concrete situation alone...a good representation system captures exactly the features of a problem that are important rather than representing everything" (McKendree et al., 2002, p. 60). This view supports our rationale for bringing together the perspectives from cognitive science and science education in this volume.

Enhancing Learning with MERs

The utility of Ainsworth's MER functions is explicitly referred to or is illustrated by seven chapter authors who explain the benefits and costs of learning with MERs in different ways in terms of visualization of textbook diagrams (Eilam), phylogenetic tree thinking (Halverson and Friedrichsen), learning genetics reasoning (Tsui and Treagust), comprehension of biotechnological tools (Yarden and Yarden), deconstructing and decoding textbooks complex process diagrams (Griffard), using analogy and gesture for understanding DNA double helix (Srivastava and Ramadas), and learning through translations across representations (Schönborn and Bögeholz). Most other chapters use similar ideas for discussing how MERs can support learning in other content areas of biology. We discuss the common themes in the sections that follow.

Visual Representations and MER Functions

Visualizations or visual representations are highlighted by all chapter authors as the most common recurring theme, although different visual modes of representation were used in their research on learning and teaching involving MERs: drawings, pictures, photographs, diagrams, images, videogames, animations, simulations, and symbolism/symbols (see Table 19.1). In most of these studies, visual representations are often deployed simultaneously or concurrently with the verbal representations (auditory, textual, sentential, discursive, etc.) to maximize the utility and effectiveness in achieving one or more of the pedagogical functions of MERs, particularly the complementary functions.

Table 19.1 Visual modes of representations discussed in the chapters of this volume

	Visual modes used/mentioned								
Chapter authors	Visualizations	Pictures/ photos	Images	Drawings	Diagrams	Symbolism/ symbols	Animation/ simulations	Videogames	Major findings
Anderson et al.	√		√		√	√[a]	√		Symbolism is pedagogically useful for teaching of the submicroscopic world of biology
Roth and Pozzer-Ardenghi		√[a]	√	√					Reading pictures is a social practice, and learning is rooted in social interactions
Eilam	√	√	√	√	√[a]				Static visual representations need appropriate design and implementation for useful learning
Liu and Hmelo-Silver		√		√	√				Conceptual representations in hypermedia afford/constrain how learners set goals of/monitor/ evaluate their learning
Yarden and Yarden	√[a]	√	√			√	√		Animations are useful for student understanding of biotechnological method PRC with teachers' support
Schönborn and Bögeholz	√[a]	√		√		√			Translation across MERs is essential for constructing biological knowledge and useful for teaching
Horwitz							√	√[a]	Videogame activities help students hone their reasoning skills in learning evolution by natural selection
Clément and Castéra		√	√[a]						Textbook analyses show that determinism in representing human genetics is related to values in sociocultural contexts

Griffard	√	√	√		√[a]	√	√	Premedical students need representational competence to decode complex process diagrams for meaningful learning
Halverson and Friedrichsen	√				√	√		Representational competence for tree thinking in learning evolution can inform curricula and instruction design
Schwartz and Brown		√	√			√		MERs can scaffold student thinking across photosynthesis and plant cellular respiration and their interconnections
Wong et al.		√			√	√	√	The case study of scientific research during the SARS crisis revealed new NOS features for biology teacher education
Buckley and Quellmalz		√	√	√	√	√	√[a]	Computer-based simulations are useful for teaching and assessing learning of human body systems/genetics/ecosystems
Tsui and Treagust	√	√	√	√	√	√		Computer-based MERs help students' understanding of concepts and reasoning in genetics but not for all students
Niebert et al.		√		√	√	√		Representations in the perceptible mesocosm foster student understanding of abstract phenomena of cell division and climate change

(continued)

Table 19.1 (continued)

	Visual modes used/mentioned								
Chapter authors	Visualizations	Pictures/ photos	Images	Drawings	Diagrams	Symbolism/ symbols	Animation/ simulations	Videogames	Major findings
Srivastava and Ramadas	√	√			√[a]		√		Gesture and analogy are useful in enhancing undergraduates' mental visualization of the 3-D double helix of DNA
Verhoeff et al.	√	√	√	√[a]	√	√	√		Modeling studies show that the use of MERs helps secondary students develop systems thinking in cell biology/ecology

[a]Major focus of the chapter

Visual representations are not new in science education—they have been used in science textbooks for hundreds of years. For example, *Orbis Sensualium Pictus*—one of the commonly recognized first modern science textbooks for children published in Germany in 1658—included extensive scientific illustrations (Buxton & Provenzo, 2011). Visual representations are now considered to be very important in learning and teaching with models and modeling in science education (e.g., Gilbert, Reiner, & Nakhleh, 2008). Further, learning how to visually represent ideas is important for students in learning science as illustrated by Ainsworth, Prain, and Tytler (2011) who asserted that drawing as an activity plays an important role in students learning science—to engage in science learning, to learn to represent science, to reason in science, to use it as strategy for learning science, and to communicate ideas in science. In a similar way, the chapter authors illustrate various learning outcomes in which visualizations in MERs can support student understanding of biology (see Table 19.1).

From multi-representational perspectives, we consider visualizations as a major group of modes of representation—one of the three dimensions in our theoretical cube model—which can demonstrate how biological knowledge across the unifying themes of living systems is represented at different levels (see Fig. 1.2 in the Chap. 1). The more specific functions of using visual modes of representations in supporting learning in the various chapters are discussed in the following sections.

Fostering Conceptual Understanding of Content Knowledge of Biology

Many of the chapter authors have shown that using a variety of representations can help students construct a deeper conceptual understanding of biology. These MERs include hypermedia-based conceptual representations that foster learners' co-construction of biological knowledge (Liu and Hmelo-Silver); complex process diagrams for premedical students' understanding of the complex concepts of molecular biology (Griffard), dynamically linked MERs in *BioLogica* that help students develop reasoning in genetics (Buckley and Quellmalz; Tsui and Treagust), animations that promote student comprehension of biotechnological methods (Yarden and Yarden), and interactive computer videogames that enable 4th grade students to develop deep understanding of the concepts underlying the theory of evolution by natural selection (Horwitz).

Several chapter authors also have illustrated how the use of MERs in university biology teaching and research can enable a better understanding of biology in a variety of domains: for example, biotechnological methods (Yarden and Yarden), molecular biology (Griffard; Halverson and Friedrichsen), photosynthesis and plant cellular respiration (Schwartz and Brown), evolutionary biology (Halverson and Friedrichsen), genetics (Buckley and Quellmalz; Tsui and Treagust), and human body systems (Liu and Hmelo-Silver; Buckley and Quellmalz). In addition, multiple representations are illustrated in experts' views of the knowledge structure of biology and teachers' professional development by various chapter authors (e.g., Roth and Pozzer-Ardenghi; Yarden and Yarden; Wong et al.; and Srivastava and Ramadas). As discussed in Griffard's chapter, some latest biology textbooks use MERs in several ways to enhance learning and teaching: for example, multilevel perspectives to show macro and micro views of complex biological structures, process figures to illustrate complex processes in series of small steps, and color consistency to organize and clarify complex concepts.

Constructing Deeper Understanding in Terms of Scientific Reasoning

In terms of the third pedagogical function of MERs for constructing deeper understanding, seven chapters have included reasoning skills as the major outcome in various content domains and at different levels of education.

To teach elementary students to develop scientific reasoning skills for understanding evolution by natural selection, Horwitz's computer videogames *Evolution Readiness* provide motivating interactive learning environments for young learners based on previous research studies that have pointed to possible affordances for learning (see a review of videogames in Owston, 2012). In secondary schools, Tsui and Treagust's case studies investigated the development of students' six types of genetics reasoning, whereas Buckley and Quellmalz's large-scale studies explored model-based reasoning while learning with computer-based simulations. In the domains of cytology and ecology, Verhoeff et al. focus on secondary students' reasoning in systems thinking as the learning outcome. For learning at the university level, Halverson and Friedrichsen's study investigated how students learned about evolution using phylogenetic tree thinking. Wong et al. studied scientists' reasoning in searching for the causative agent of the SARS disease and used the case study of the authentic scientific research for professional development of preservice and in-service biology/science teacher education to promote deeper understanding of nature of science.

Developing Representational Competence and Other Skills

Another recurring theme is about the competence and skills for learning biology using MERs. In particular, three chapters describe representational competence—for learning biotechnological methods (Yarden and Yarden), for deconstructing and decoding complex process diagrams (Griffard), and for comprehending and constructing phylogenetic trees (Halverson and Friedrichsen). Three other chapters focus on competence for translating across representations of biological structures and functions (Schönborn and Bögeholz; Schwartz and Brown; and Srivastava and Ramadas). Several chapters focus on the skills of reading and interpreting visualizations: static visualization skills for reading textbook diagrams (Eilam), dynamic visualization skills for simulation-based representations (Yarden and Yarden; Buckley and Quellmalz), and reading pictures and other inscriptions from Vygotsky's sociocultural perspectives (Roth and Pozzer-Ardenghi). Verhoeff et al.'s, Srivastava and Ramadas's, and Buckley and Quellmalz's chapters are common in their focus on modeling skills. Both chapters by Anderson et al. and Verhoeff et al. focus on systems thinking skills which we discuss in more detail later in this chapter.

Development of representational competence stands out among these chapters as the most important outcome in learning with MERs. Representational competence, as Halverson and Friedrichen's chapter points out, is domain-specific and can have as many as seven levels. For example, in their chapter, the representational competence is about reading and building phylogenetic trees in a novice-expert continuum in terms of seven levels—*no use*, *superficial use*, *simplified use*, *symbolic use*, *conceptual use*, *scientific use*, and *expert use*. Accordingly, evolutionary biologists' representational competence is at the expert level, enabling them to quickly interpret and deeply understand the phylogenetic trees and use multiple representations to solve phylogenetic problems, explain evolutionary phenomena, and make predictions.

Enhancing the Quality of Teaching: Achieving Pedagogical Functions of MERs

There are several groups of MERs for biological knowledge suggested by some chapter authors that appear to be increasingly important for biology teaching and biology teacher education in the twenty-first century. We believe that these warrant further discussion in synthesizing the themes of the chapters and in drawing conclusions for this volume.

Anthropocentric or Human-Centered Representations to Constrain Interpretations of Biological Phenomena

In terms of the second pedagogical function of MERs for constraining interpretation or misinterpretation of a more abstract representation using a less abstract one, we identified in the Introduction (Chap. 1) two similar themes common to a number of chapters—mesocosmic and anthropocentric representations. We now subsume both into one single theme—anthropocentric or human-centered representations.

Niebert et al. argue, from the perspective of learning through source-to-target mapping, that the perceptible mesocosm should lie in common source domains of biology. This is because mapping from these less abstract source domains in mesocosm (e.g., schemata based on bodily experience) to the more abstract target domains in microcosm (e.g., cell division) or in macrocosm (e.g., climate change) is easier for students to understand biological knowledge. In other words, in Niebert et al.'s example, a representation at the meso level (e.g., breaking a bar of chocolate) serves to constrain the interpretations of the abstract representation of biological phenomena (e.g., cell division). Similarly, several other chapter authors argue that learners always find representations closely related to humans or anthropocentric representations useful for understanding complex and abstract biological knowledge: *self as referent* (Schwartz and Brown), *bodily experience* (Niebert et al.), and *gesture*s or *body positions* (Roth and Pozzer-Ardenghi; Srivastava and Ramadas). More recent studies also include the use of haptic representations in scaffolding learning of molecular biology (e.g., Bivall, Ainsworth, & Tibell, 2011).

Despite the usefulness of anthropocentric representations in biology instruction, some critics call on educators to be cautious about anthropocentric thinking, particularly in environmental education where anthropocentrism has recently been a focus of philosophical discussion that this human-centered thinking might not help students develop the right relationship with nature (e.g., Carvalho, Tracana, Skujiene, & Turcinaviciene, 2011). Unfortunately this is all too common. As Bonnett (2007) notes, nature is "seen essentially as a resource, an object to be intellectually possessed and physically manipulated and exploited in whatever ways are perceived to suit (someone's version of) human needs and wants" (p. 710). Therefore, such human-centered thinking might justify the exploitation of nature by and for humankind, as well as possibly mask the social and political dimensions behind the biology-based societal problems (e.g., Bell & Russell, 2000). It follows that the possible bias in using anthropocentric representations should not be overlooked by biology teachers and biology teacher educators.

Systems Representations for the Interconnectedness of the Curriculum

Multiple external representations (MERs) are relevant to improving school biology in that this notion can be used to address the perennial critique of the deficit in the interconnectedness of knowledge in school biology curricula (Buckley and Quellmalz) and shortfalls in the systemic transfer of knowledge across multiple levels of biological organization (e.g., Schönborn & Bögeholz, 2009; Schönborn & Bögeholz's chapter). Some other chapter authors also take a systems view of the interconnectedness that focuses on one of the unifying themes in living systems: evolution of organisms from simple to complex forms (Halverson and Friedrichsen), information transfer from DNA to subcellular organelles through a hierarchically organized biological structures to the whole organisms (Buckley and Quellmalz; Tsui and Treagust), and energy transfer from the sun to producers, consumers, and decomposers through the hierarchically organized ecosystems (Schwartz and Brown).

Furthermore, the notion of using MERs for learning is also in keeping with *systems biology* (Vidal, 2009)—the latest development of biological science—that aims at identifying the systems level understanding of life phenomena in the post-genomic age. Addressing this issue is the study reported by Verhoeff et al. on the importance of models and modeling activities to develop students' systems thinking and related systems concepts in secondary schools. Indeed, for over a decade, MERs have been used with increasingly powerful information and communications technology (ICT) (see examples from various disciplines in van Someren, Reimann, Boshuizen, & de Jong, 1998) that is now ubiquitously available in many schools and homes for learning. Indeed, ICT has revolutionized the way people learn and how they communicate their ideas through electronic discourses and resources.

Philosophical, Cultural, Social, and Political Impacts on Representations of Biological Education and Nature of Science

Three chapters illustrate philosophical, cultural, social, and political impacts on representations of biology and biological education. Roth and Pozzer-Ardenghi discuss reading pictures as a social practice from anthropological and social-psychological perspectives. Clément and Castéra examine genetic determinism in textbooks from 16 countries. Wong et al. portray the social and political factors in Hong Kong scientists' research on Severe Acute Respiratory Syndrome (SARS) virus that threatened the world as a dangerous pandemic in 2003; their case study was subsequently used in biology and science teachers' professional development programs for understanding nature of science.

From sociocultural perspectives rooted in anthropology and social psychology, Roth and Pozzer-Ardenghi discuss reading photographic pictures as a social practice and learning from pictures as social interactions. Their research in this area over 15 years has been conducted in North America, Brazil, and Korea. For example, their high school textbook analysis indicated that pictures or photographs are a useful resource in forming a link between scientific inscriptions such as a table, graph, or formula, and students' everyday experience but that additional scaffolding is needed for more effective learning. Their chapter also explores pictures in university lectures and how scientists read photographic images. Given that sociocultural perspectives (e.g., Vygotsky, 1968, 1978) have been increasingly popular as a theoretical framework in science education research and practice (e.g., Lemke, 2001; Tsaparlis & Papaphotis, 2009), Roth and Pozzer-Ardenghi's framing of reading pictures as learning through social interactions has important implications for all levels of biological education.

Clément and Castéra examine the representations of genetic diseases in French biology textbooks and report their content analysis on how textbooks across 16 European and other countries depicted twins and metaphorized genetics. Their chapter has identified on a macro level, how genetic determinism is used to represent genetics in textbooks from these countries with different languages, ideologies, cultures, and religions. One of the interesting findings was that biology textbooks in several East European countries are still influenced by the political ideologies of the former USSR, for example, the pseudoscientific ideas of Lysenkoism. This cross-country textbook analysis provides rare and valuable insights into biological education in the non-English-speaking world.

Wong et al. portray Hong Kong scientists' crucial success in identifying a coronavirus as the agent for causing SARS among other key episodes during the SARS outbreak in 2003. SARS was a previously unknown but highly contagious and deadly disease that first appeared at the end of 2002 in southern China. In this very urgent hunt for the causative agent, the scientists used models and modeling across multiple facets and perspectives—from rumors to in-depth studies, from puzzling observations to administrative decisions to quarantine all affected individuals, and from research evidence to political decisions to ban the sale of wild animals for food. Yet there was an untold political decision that had constrained scientific research and delayed the prevention of SARS from spreading across the world. Chinese officials initially covered up the truths about SARS for political reasons by censoring reports in the media about this mysterious disease to avoid public fear and instability during the leadership change in the ruling Communist Party. SARS cases continued to increase for months in early 2003 and spread to Beijing (Abraham, 2004; Loh, 2004). On April 8, *Time* magazine reported online what a Chinese army doctor in Beijing revealed that there were many more SARS cases than the official figures and the situation was very serious. Thereafter, China belatedly took immediate and drastic actions to stop the SARS contagion from becoming a deadly global pandemic (Jakes, 2003; Lemonick & Park, 2003). The lack of free flow of information alongside the bureaucratic red tape is obviously counterproductive to scientific research, and lessons must be learned from the SARS crisis (e.g., Ding & Wang, 2003). Unfortunately, similar tragic happenings continue to occur in China. For example, the delay in investigating the scandal of melamine[1]-contaminated milk products—just before the Beijing Olympic Games in August 2008—resulted in the death of several babies and illness of many children who developed kidney stones (cf. Spencer, 2008). Scientists appear to be helpless and powerless in the face of political impact on research and free flow of information. This is important for a deeper understanding of nature of science.

These three chapters remind biology teachers and biology teacher educators that the external representations of phenomena in biological research, and education may be compromised by philosophical, cultural, social, and political factors that are

[1] Melamine is a nitrogen-rich, toxic industrial material (*1, 3, 5-triazine-2, 4, 6-triamine*) illegally added to milk products in China to increase their apparent protein content.

often overlooked. Consequently, we believe that representations of biological knowledge must be interpreted within a broader context related to the surrounding cultures and the societal or political factors in order to construct a deeper understanding of nature of science.

Teaching Biology in the Non-English-Speaking World

Many chapter authors in this volume have languages other than English as their first language. The authors themselves and the participants in their studies are well represented in terms of the *three concentric circles* in Kachru's (as citied in Martin & Siry, 2011) *model of world Englishes*. In particular, the cross-country analysis of textbooks from 16 countries by Clément and Castéra—comparing the representation of genetics in terms of genetic determinism—is notable and is otherwise unknown to the community of science educators in the English-speaking world.

This volume reminds readers that many of today's students are learning biology in languages other than English and the authors' studies also involved the use of other languages for learning and teaching biology—for example, German, French, Indian (Marathi and Hindi), Hebrew, Dutch, Arabic, Portuguese, Chinese (Cantonese), and Korean. Whereas many English language learners (ELLs) in US schools are learning biology in English (their L2) (MacSwan & Rolstad, 2005), secondary students in Germany and other countries within the European Union are also increasingly using English (their L2) for learning content subjects (Wannagat, 2007). This trend of using English for science education is on the increase in some Chinese universities (Tong & Shi, 2012). For English being used for learning and teaching biology, there is ample research evidence that the bilingual approach is useful to better support ELLs to learn the content knowledge in their L2 (e.g., Kroll & Hermans, 2011).

Tsui and Treagust touch on the possible benefits of bilingual representations for ELLs in Hong Kong when these students learned genetics in English (their L2). When learning English (L2)-taught content subjects such as science and biology, ELLs could capitalize on the rich resources of their prior knowledge of biology in their first language (L1), which is often overlooked (e.g., MacSwan & Rolstad, 2005; Tong & Shi, 2012). From a psycholinguistic perspective, for ELLs whose L2 is not proficient enough, learning biology in L2 depends on effective mediation of their L1 to conceptually process their L2 learning (Kroll & Hermans, 2011). As such, bilingual representations of biological knowledge can be useful to serve one or more pedagogical functions of MERs. Professional development of biology teachers for developing their proficiency in both students' L1 and L2 is also important for effective teaching using the bilingual approach (Wannagat, 2007). This area warrants further research in science and biological education because English is, and is expected to be, the lingua franca of science in the twenty-first century and beyond.

Multiple Methods of Assessment to Inform Teaching with MERs

The chapter authors in this volume have illustrated how learning and teaching with MERs in biology need to be critically examined and assessed, particularly at the university level, so that instructors and professors, as well as undergraduate and graduate students including student teachers, can more effectively understand and use multiple representations in their teaching. Examples of assessment include the use of hypermedia for assessing learning about human body systems (Liu and Hmelo-Silver), online reflective journal entries for assessing learning of evolutionary tree thinking (Halverson and Friedrichsen), clinical interviews and paper-and-pencil tests to assess learning about complex processes diagrams of molecular biology in university textbooks (Griffard), and analysis of pictures in textbooks and gestures in lectures (Roth & Pozzer-Ardenghi). In Srivastava and Ramadas's chapter, they report the use of in-depth microgenetic method using interview-cum-teaching and observations to assess undergraduates' mental visualization of 3-D double helical structure of DNA. Observations of secondary students' and teachers' modeling actions were used for assessing systems thinking (Verhoeff et al.). Tsui and Treagust report the use of a two-tier diagnostic test for evaluating secondary student understanding of genetics reasoning. Two-tier tests (Treagust, 1988) have been developed and used in evaluation of several biology domains such as osmosis and diffusion (Odom & Barrow, 1995) and genetics (Tsui & Treagust, 2010); however, no two-tier diagnostic tests are yet available to specifically assess learning with MERs in biology—this can be an area for further research.

To assess learning in computer-based learning environments, online assessments of outcomes are usually used. As illustrated in this volume, to evaluate student understanding of genetics from the multi-representational learning environment *BioLogica*, online pretests and posttests were used to evaluate student understanding in terms of six types of genetics reasoning (Tsui and Treagust). Similarly, built-in online assessment was used for assessing undergraduates' learning of cognitive and metacognitive skills for co-constructing their knowledge about human body systems (Liu and Hmelo-Silver). Computer data logging that can track student learning also was used for evaluating different outcomes of student learning from interactive computer programs on human body systems, genetics, evolution, and ecology (Buckley and Quellmalz; Horwitz; Tsui and Treagust). For noncomputer learning environments, clinical interviews and paper-and-pencil tests remain the common reliable and valid approaches to assess student learning about biological processes (Eilam; Griffard) and evolutionary tree thinking (Halverson and Friedrichsen).

As already discussed in the Introductio (Chap. 1), evaluation of multimedia learning environments requires appropriate methods for specific research questions within particular learning contexts. Besides conventional experimental research designs, the more useful methods appear to be computer modeling, case studies, ethnographic studies, and microgenetic studies (Ainsworth, 2008). Our review indicates that the variety of methodologies reported in the chapters for assessing student learning and evaluating student understanding have rightly pointed in this direction. These should inform biology teachers and biology teacher educators on how MERs can be effectively used to support learning.

Contributions to Biological Education in the Twenty-First Century

From the preceding review and synthesis of the themes arising from the chapters in this volume, we have discussed a number of issues of learning and teaching with MERs, as well as methodologies for assessment. These are relevant to the future directions for biological education.

In the committee-authored report of the US National Research Council (2009) about the new biology in the twenty-first century, the committee identified four major areas of societal challenges—food, environment, energy, and health—as directions for biological research which would involve integration of scientific information, theory, and technology about complex problems, deeper understanding of biological systems, and biology-based solutions to societal problems, as well as feedback and benefits to contributing disciplines and to education (Labov et al., 2010) (see Fig. 19.1).

We believe that this goal is also part of the challenge for teachers and students at all levels to, respectively, teach and learn biology with multiple representations. As illustrated by the chapters in this volume, visualization skills, reasoning skills, tree building skills, representational competence, and systems thinking skills all appear to be useful, and even crucial, for learning biology in the twenty-first century from the wide variety of multiple representations in biology textbooks, online resources, and school lessons or university lectures. It is equally important to educate new biologists for solving the world's biology-based societal problems as well as to educate all students with diverse learning needs for promoting scientific literacy in modern societies.

As noted by Labov et al. (2010), the new biology in the twenty-first century involves complex interdisciplinary problems that will require biologists to incorporate "emerging theory, new technologies, fundamental findings from basic research in the life sciences" and to integrate into biology "physical sciences, mathematics, and engineering [that] could enable biology to contribute to rapid progress in practical problem-solving" (p. 11). New biologists also need to have "deep knowledge in one discipline and a 'working fluency' in several" (p. 13).

Finale

While gratefully acknowledging the excellent contributions of the chapter authors from around the world to *Multiple Representations in Biological Education*, we must say that it has been a great privilege for us to edit their chapters and that our many e-mail communications and interactions are very useful. We are grateful to

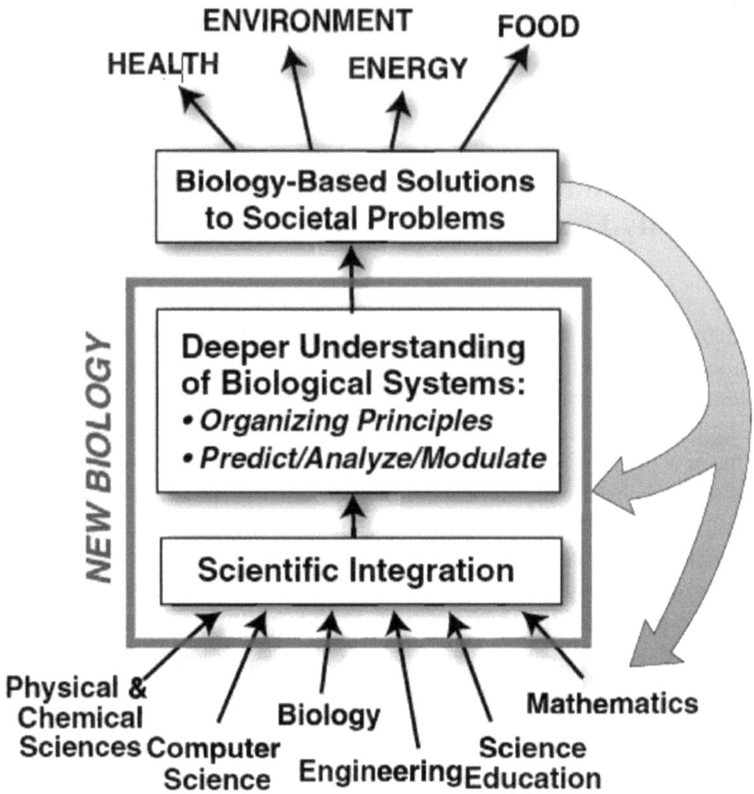

Fig. 19.1 New biology for the twenty-first century (National Research Council, 2009). Reprinted with permission

John Gilbert, Bernadette Ohmer, Shaaron Ainsworth, Kathleen Fisher, Anat Yarden, and Kristy Halverson who have provided us valuable advice and help in one way or another in completing this volume.

We look forward to a revitalized biological education with which people can create a better world—"a more peaceful and prosperous world, where their children can live healthy, happy lives. . ." (Ferris, 2010, p. 290)—where research in biology and biological education can contribute to solving the major challenges of humankind, such as food, environment, energy, and health (National Research Council, 2009). We also envision a more scientifically literate citizenry, a more connected international community of biology educators, a more ecologically balanced global environment, and a more socially just and democratic global community (cf. Rindermann, 2008).

We hope this volume will be a timely reference for biology education researchers, biology teachers, and biology teacher educators, as well as postgraduates of science education around the world. We also hope that this collection of research reports and theoretical expositions in the area of multiple representations can encourage more studies in this direction so that biology educators can better harness the resources in the repertoire of multiple external representations (MERs) for improving biological education. We believe that *Multiple Representations in Biological Education* can make a small contribution in this direction.

References

Abraham, T. (2004). *Twenty-first century plague: The story of SARS*. Hong Kong: Hong Kong University Press.

Ainsworth, S. (1999). The functions of multiple representations. *Computers in Education, 33*(2/3), 131–152.

Ainsworth, S. (2006). DeFT: A conceptual framework for considering learning with multiple representations. *Learning and Instruction, 15*(3), 183–198.

Ainsworth, S. (2008). How should we evaluate multimedia learning environments? In J.-F. Rouet, R. Lowe, & W. Schnotz (Eds.), *Understanding multimedia documents* (pp. 249–265). Dordrecht, The Netherlands: Springer.

Ainsworth, S., Prain, V., & Tytler, R. (2011). Drawing to learn in science. *Science, 333*, 1096–1097.

Anderson, C. W. (2007). Perspective in science learning. In S. K. Abell & N. G. Lederman (Eds.), *Handbook of research on science education* (pp. 3–56). Mahwah, NJ: Lawrence Erlbaum.

Bell, A. C., & Russell, C. L. (2000). Beyond human, beyond worlds: Anthropocentrism, critical pedagogy, and the poststructuralist turn. *Canadian Journal of Education, 25*(3), 188–203.

Bivall, P., Ainsworth, S., & Tibell, L. A. (2011). Do haptic representations help complex molecular learning? *Science Education, 95*(4), 700–719.

Bonnett, M. (2007). Environmental education and the issue of nature. *Journal of Curriculum Studies, 39*(6), 707–721.

Buxton, C. A., & Provenzo, E. F. (2011). *Teaching science in elementary & middle school*. Thousand Oaks, CA: Sage.

Carey, J. (1998). We are now starting the century of biology. *Business Week*. Retrieved from http://www.businessweek.com/1998/35/b3593020.htm

Carvalho, G. S., Tracana, R. B., Skujiene, G., & Turcinaviciene, J. (2011). Trends in environmental education images of textbooks from western and eastern European countries and non-European countries. *International Journal of Science Education, 33*(18), 2587–2610.

Ding, K., & Wang, X. (2003). "Feidian" de jingshi yu woguo de jiaoyu [The warning and enlightenment SARS gave us and the education in China]. *Journal of Nantong Institute of Technology (Social Sciences), 19*(3), 1–4.

Ferris, T. (2010). *The science of liberty: Democracy, reason and the laws of nature*. New York: HarperCollins.

Fischer, H. E., Kauertz, A., & Neumann, K. (2008). Standards of science education. In S. Mikelskis-Seifert, U. Ringelband, & M. Brückmann (Eds.), *Four decades of research in science education: From curriculum development to quality improvement* (pp. 24–42). Münster, Germany: Waxmann.

Gentner, D., & Stevens, A. L. (Eds.). (1983). *Mental models*. Hillsdale, NJ: Lawrence Erlbaum.

Gilbert, J. K., Reiner, M., & Nakhleh, M. (Eds.). (2008). *Visualization: Theory and practice in science education*. New York/London: Springer.

Jakes, S. (2003). Beijing's SARS attack. Retrieved from TIME World website: http://www.time.com/time/world/article/0,8599,441615,00.html

Kress, W. J., & Barrett, G. W. (Eds.). (2001). *A new century of biology*. Washington, DC: Smithsonian Institution Press, in association with the American Institute of Biological Sciences.

Kroll, J. F., & Hermans, D. (2011). Psycholinguistic perspectives on language processing in bilinguals. In M. S. Schmid & W. Lowie (Eds.), *Modeling bilingualism: From structures to chaos* (pp. 15–36). Amsterdam, The Netherlands/Philadephia, PA: John Benjamins.

Labov, J. B., Reid, A. H., & Yamamoto, K. R. (2010). Integrated biology and undergraduate science education: A new biology for the twenty-first century. *CBE-Life Science Education, 9*, 10–16.

Lemke, J. L. (2001). Articulating communities: Sociocultural perspectives on science education. *Journal of Research in Science Teaching, 38*(3), 296–316.

Lemonick, M. D., & Park, A. (2003, May 5). The truth about SARS. *Time*, 28–37.

Loh, C. (Ed.). (2004). *At the epicentre: Hong Kong and the SARS outbreak*. Hong Kong, China: University of Hong Kong Press.

MacSwan, J., & Rolstad, K. (2005). Modularity and the facilitation effect: Psychological mechanisms of transfer in bilingual students. *Hispanic Journal of Behavoiral Sciences, 27* (2), 224–243. doi:10.1177/0739986305275173.

Martin, S. N., & Siry, C. (2011). Networks of practice in science education research: A global context. *Journal of Research in Science Teaching, 48*(6), 592–623.

McKendree, J., Small, C., Stenning, K., & Conlon, T. (2002). The role of representations in teaching and learning critical thinking. *Educational Review, 54*(1), 57–67.

National Research Council. (2009). *A new biology for the 21st century*. Washington, DC: National Academic Press.

Odom, A. L., & Barrow, L. H. (1995). Development and application of a two-tier diagnostic test measuring college biology students' understanding of diffusion and osmosis after a course of instruction. *Journal of Research in Science Teaching, 32*(1), 45–61.

Owston, R. D. (2012). Computer games and the quest to find their affordances for learning. *Educational Researcher, 41*(3), 105–106.

Reiss, M. J., Millar, R., & Osborne, J. (1999). Beyond 2000: Science/biology education for the future. *Journal of Biological Education, 33*, 68–70.

Rindermann, H. (2008). Relevance of education and intelligence for the political development of nations: Democracy, rule of law and political liberty. *Intelligence, 36*(4), 306–322.

Schönborn, K. J., & Bögeholz, S. (2009). Knowledge transfer in biology and translation across external representations: Experts' views and challenges for learning. *International Journal of Science and Mathematics Education, 7*, 931–955.

Spencer, R. (2008). China accused over contaminated baby milk. *The Telegraph*. Retrieved from http://www.telegraph.co.uk/news/worldnews/asia/china/2963808/China-accused-over-contaminated- baby-milk.html

Tong, F., & Shi, Q. (2012). Chinese–English bilingual education in China: A case study of college science majors. *International Journal of Bilingual Education and Bilingualism, 15*(2), 165–182.

Treagust, D. F. (1988). Development and use of diagnostic tests to evaluate students' misconception in science. *International Journal of Science Education, 10*(2), 159–169.

Tsaparlis, G., & Papaphotis, G. (2009). High school students' conceptual difficulties and attempts at conceptual change: The case of basic quantum chemical concepts. *International Journal of Science Education, 31*(7), 895–930.

Tsui, C.-Y., & Treagust, D. F. (2010). Evaluating secondary students' scientific reasoning in genetics using a two-tier diagnostic instrument. *International Journal of Science Education, 32* (8), 1073–1098.

Tytler, R., & Prain, V. (2010). A framework for re-thinking learning in science from recent cognitive science perspectives. *International Journal of Science Education, 32*(15), 2055–2078.

van Someren, M. W., Reimann, P., Boshuizen, H. P. A., & de Jong, T. (Eds.). (1998). *Learning with multiple representations*. London: Pergamon.

Vidal, M. (2009). A unifying view of 21st century systems biology. *FEBS Letters, 583*(2009), 3891–3894.

Vygotsky, L. S. (1968). *Thought and language*. Cambridge, MA: The MIT Press.

Vygotsky, L. S. (1978). *Mind in society: The development of higher psychological processes*. Cambridge, MA: Harvard University Press.

Wannagat, U. (2007). Learning through L2: Content and Language Integrated Learning (CLIL) and English as Medium of Instruction (EMI). *International Journal of Bilingual Education and Bilingualism, 10*(5), 663–682. doi:10.2167/beb465.0.

About the Authors

Trevor R. Anderson is Associate Professor in the Divisions of Chemistry Education and Biochemistry, Department of Chemistry, Purdue University, USA. He is a member of the IUBMB Educational Committee and the editorial board of *Biochemistry & Molecular Biology Education (BAMBEd)*. In 2007 he was the recipient of the Monash University, Biomedical Sciences "Scholar in Residence in Education" award. His Visualization in Biochemistry Education (VIBE) research group focuses on four interrelated areas: (1) visual literacy and the development and assessment of visual skills, (2) science inquiry and reasoning about experiments, (3) concept inventories and reasoning with core concepts and representations, and (4) curriculum and faculty development. Central to his faculty development activities has been his workshops on scholarship, his design of curriculum change models, his development of the CARD e-resource (http://www.card.unp.ac.za), and his authorship of the Bridging-the-Gap series in *BAMBEd* aimed at encouraging the application of educational research to teaching practice. E-mail: ander333@purdue.edu

Kerst Th Boersma is Emeritus Professor of Biology Education at Utrecht University in the Netherlands and a member of the Royal Netherlands Academy of Arts and Sciences. After a PhD in palaeontology he was educational publisher and teacher trainer of secondary biology teachers for 10 years and worked for 13 years at the National Institute for Curriculum Development, as head of the departments for science education and research and development. In 1997 he was appointed as Professor in Biology Education at Utrecht University where he was managing director of the Freudenthal Institute for Science and Mathematics Education from 2006–2008. His research interests are conceptual development, context-based biology education, and systems thinking. From 2005–2010 he was chairing the National Biology Curriculum Innovation Board, which developed examination programs for upper secondary context-based biology education. He retired in 2008 but is still involved in the guidance of PhD studies. E-mail: k.t.boersma@uu.nl

D.F. Treagust and C.-Y. Tsui (eds.), *Multiple Representations in Biological Education*, Models and Modeling in Science Education 7, DOI 10.1007/978-94-007-4192-8,
© Springer Science+Business Media B.V. 2013

Susanne Bögeholz is Professor of Biology Education at Georg-August-Universität Göttingen (Germany). She is a member of the executive board of "ZeUS" Göttingen, a center for empirical education research, and speaker of a German Science Foundation-funded PhD Research Training Program. Susanne has taught Biology, Pedagogics, and German as a foreign language in Germany, France, and Hungary before joining the Leibniz-Institute for Science Education (Kiel, Germany). She received a PhD in biology education with a thesis on the influence of nature experience and environmental knowledge on pro-environmental action. Later she worked on mainstreaming Education for Sustainable Development (ESD). Her research interests are in socio-scientific decision-making skills, curricular and extra-curricular ESD with a focus on biodiversity conservation. Susanne uses conceptual, qualitative, and quantitative social science methods including the probabilistic modeling of student cognitive skills. In 2011, Susanne received the "Erhard Friedrich-Preis Naturwissenschaften," a national award for outstanding contributions to science education. E-mail: sboegeh@gwdg.de

Mary H. Brown is Professor at Lansing Community College, Lansing, Michigan, USA, and is the coordinator for all science education courses, including nature study. She currently specializes in integrated science for education majors. Mary teaches biology methods to pre-service teachers for Central Michigan University, and online environmental science for in-service teachers at Western Michigan University. She completed her undergraduate and graduate degrees at Central Michigan University, completed postgraduate studies at Rutgers State University, and received her Ph.D. in Science Education from Western Michigan University. For more than 30 years she has taught in all science disciplines. Mary has been active in state science organizations, has presented at state and national conferences, and is the recipient of numerous citations for merit and excellence. She is a past President of Michigan Community College Biologists and a founding member of Michigan Association of Community College Teacher Education Programs. E-mail: brownm@lcc.edu

Barbara C. Buckley is a Senior Research Associate in the Science, Technology, Engineering, & Mathematics (STEM) program of WestEd (USA). Prior to joining WestEd, Dr. Buckley was research fellow with the MISTRE group at the University of Reading (UK) and a research scientist at the Concord Consortium (USA), playing a central role in the *Modeling Across the Curriculum* project. At TERC, she contributed to the *GLOBE* and *Global Lab* projects. She is Principal Investigator of the NSF-funded *SimScientists Human Body Systems* project that will design, develop and test simulation-based instructional modules for high school science classrooms and co-Principal Investigator of the NSF-funded *Calipers II* project that is developing simulation-based assessments in biological and physical science for middle-school. Dr. Buckley holds an A.M. and a Ph.D. in Education from Stanford University, an M.S. in Biology from Wright State University, and a B.S. in Medical Technology from Michigan State University. E-mail: bbuckle@wested.org

Jérémy Castéra obtained his PhD in 2010 from the University of Lyon 1 in France. His thesis work in didactics of biology was based on the analysis of biology textbooks and teachers' conceptions about human genetics with their comparison between France and 18 other countries. He was also involved in the BIOHEAD-Citizen project (2004–2008) where he collaborated with the researchers of 18 other countries. Today, Jérémy has obtained a 3-year postdoctoral fellowship, supported by the Estonian Research Mobility Scheme in conjunction with the FP7 Marie Curie COFUND "People." Since October 2010, he has worked in the Science Education Centre of the University of Tartu in Estonia, with new research interests in the field of educational technologies. E-mail: jcastera@ut.ee

Maurice W.M. Cheng (B.Pharm., P.C.Ed., M.Ed., Ph.D.) is Assistant Professor in the Division of Science, Mathematics and Computing Education, Faculty of Education, The University of Hong Kong, the Hong Kong SAR, China. His major research interests are teaching and learning of science and chemistry, with a particular focus on the roles and uses of diagrams, mental visual representations, scientific models and modeling. Grounded on his research, he completed two projects on science teachers' professional development. The teaching materials developed and videos of classroom teaching can be found at and learningscience.edu.hku.hk and http://web.edu.hku.hk/knowledge/projects/science/qef_2010/d6/main.html. At the moment, he is developing an iOS Application for the learning of chemical bonding at school level. His teaching areas include science and chemistry education, and learning psychology. He has taught in secondary school, and serves the official curriculum and assessment committees (science and chemistry subjects) of Hong Kong. E-mail: mwcheng@hku.hk

Pierre Clément is Honorary Professor at the University Lyon 1 in France, where he did most of his research and teaching. He started in animal biology, analyzing ultrastructures, life cycles and behaviors among natural clones of rotifers. He worked then in didactics of biology. From 1990 to 1996, he was President of AEDB (Association Européenne de Didactique de la Biologie) and was then in the staff of ERIDOB (European Research in Didactics of Biology) and co-President of ARDIST (Association pour la Recherche en Didactique des Sciences et des Techniques). He was co-chair of the ESERA 2011 Conference in Lyon (*Science Learning and Citizenship*). In the board of IOSTE (International Organization for Science and Technology Education), he represents Mediterranean countries. He was coordinating the FP6 European project BIOHEAD-Citizen (*Biology, Health and Environmental Education for Better Citizenship*) involving 19 countries. He is now enlarging this research to other countries for international comparisons. E-mail: Pierre.Clement@univ-lyon1.fr, clement.grave@free.fr

Lynn du Plessis has taught Biology (Life Science) and General (Natural) Science in secondary schools for over 33 years. She was Head of the Biology (Life Science) Department at a prestigious independent school for 20 years. In her role as Co-ordinator of the GET Phase (Grades 8 and 9), she guided teachers in the introduction of the new curriculum. She also edited a series of school textbooks at the GET Level. From her experiences in both the teaching and publishing arenas, she became aware of the need for guidelines for presenting unambiguous diagrams and also of difficulties experienced by students when interpreting diagrams. She was awarded her Ph.D. in Science Education at the University of KwaZulu-Natal in 2007 for her research into students' difficulties with the interpretation of arrow symbolism in diagrams. She presented papers on this topic at several local and international conferences. E-mail: aubreydup@mweb.co.za

Billie Eilam is Professor of Education in the University of Haifa in Israel. Her research focuses on cognitive aspects of visualization in teaching, learning, and curriculum, particularly in the science domain (biology), and on applying theories to authentic learning situations. She co-authored several books in Hebrew on the subjects of ecology and inquiry skills, publishes widely in leading refereed journals and presents regularly at international conferences. Billie's authored book: *The Dual Function of Visual Literacy in the Teaching Profession: Implications for Teacher Education* is soon to be published by Cambridge University Press. Throughout her career, Billie has been involved in junior and high school biology teaching (as teacher and mentor), in educational projects and in teachers' professional development. Billie was past Head of the Department of Teacher Education, currently heads the curriculum track, and served on several national committees in education. She won several grants for developing curriculum materials and conducting research on learning. E-mail: beilam@edu.haifa.ac.il.

Patricia Friedrichsen is Associate Professor at the University of Missouri in the United States. She is jointly appointed in the Division of Biological Sciences and the Department of Learning, Teaching & Curriculum. She earned a Masters degree in Curriculum and Instruction from the University of Nebraska and a Ph.D. in Science Education from Penn State University, both in the USA. She taught biology in secondary schools for 13 years, and currently teaches an undergraduate large-enrollment (400+ students) introductory biology course. Her research interests include pedagogical content knowledge development of secondary science teachers and evolution education. She received the Association for Science Teacher Education (ASTE) Innovation in Teaching Science Teachers Award in 2003. She presents her work at national and international conferences, and has published in *Science Education* and *Journal of Research in Science Teaching*. Additionally, she is a frequent contributor of practitioner articles in *American Biology Teacher*. E-mail: FriedrichsenP@missouri.edu

Phyllis B. Griffard is Senior Lecturer in Biology at the Weill Cornell Medical College in Qatar. She teaches introductory biology in the pre-medical program in a branch campus of Cornell University. As a doctoral candidate in biochemistry at Purdue University she earned a PhD in Science Education with Jim Wandersee at Louisiana State University. As Director of the Scholars Academy at the University of Houston-Downtown, she spearheaded student mentoring and research participation initiatives to improve retention of underrepresented minorities in science careers. A biologist and educator, she has presented her work at science education conferences and a Gordon Research Conference and published on the topics of minority retention, graphic representation, conceptual development and nature of science. She has procured funding for these and other undergraduate research projects. Her teaching gives significant attention to concept maps and complex process diagrams, graphic representations that are of particular value in biology. E-mail: phyllisgriffard@gmail.com

Harald Gropengiesser is Professor of Biology Education and managing director of the Institute for Science Education at Leibniz Universität Hannover, Germany. He is a civil engineer and taught surveying for 2 years at Moshi Technical School, Tanzania. After further studies he taught biology and chemistry for 10 years at a Gymnasium (grammar school) in Bremen, Germany. For 4 years Harald was the editor of the German Biologists Association journal. He was awarded his graduation to Dr. rer. nat. in biology education on Educational Reconstruction of Seeing from the Carl von Ossietzky University of Oldenburg, Germany. At the Goethe University of Frankfurt-am-Main, as a Professorship for 2 years he was in charge of biology education. His Habilitation dissertation on How to Understand Learners' Thinking was accepted at the Carl von Ossietzky University of Oldenburg. Harald is a co-editor of the leading German university textbook for biology education. E-mail: gropengiesser@idn.uni-hannover.de

Abindra S. Gupthar is Professor of Biochemistry at the University of KwaZulu-Natal in South Africa. He received his undergraduate training in Biochemistry at the University of Durban-Westville, Durban, South Africa, and postgraduate biotechnology education at the University of the Witwatersrand, Johannesburg, specializing in yeast hybridization studies at Ph.D. level. His published work includes genetic manipulation of D-xylose fermenting yeasts, growth kinetics and killer yeast properties. Abin has taught a wide range of topics in molecular biology for 24 years and has developed a keen interest in biochemical education. In 2008, he obtained a second doctorate from the newly formed University of KwaZulu-Natal for his work on student difficulties with the visual and symbolic language used in molecular biology. He continues to investigate the nature of learning difficulties in molecular biology in an attempt to develop remedial strategies. E-mail: gupthara@ukzn.ac.za

Kristy L. Halverson is Assistant Professor in the Department of Biological Sciences at the University of Southern Mississippi in the United States. She earned a Masters degree in Ecology and Evolutionary Biology from Iowa State University and a Ph.D. in Curriculum and Instruction: Science Education and a Minor Certificate in College Teaching from the University of Missouri, both in the USA. She currently teaches introductory and upper-level biology as well as courses in human learning and qualitative research. Her research interests include investigating how college students learn biology using visualizations, such as phylogenetic trees, and exploring effective science communication to improve science literacy among the general public and promote citizen science. Her past research has been presented at national and international conferences for science and science education and published in *Science Education*, *International Journal of Science Education*, *Evolution: Education and Outreach*, and the *American Biology Teacher*. E-mail: Kristy.Halverson@usm.edu

Cindy E. Hmelo-Silver is Professor of Educational Psychology at Rutgers, the State University of New Jersey in the United States where she coordinates the Learning Sciences PhD specialization. She received her PhD from Vanderbilt University. Her current research interests focus on how people learn about complex phenomena and how technology can help support that learning. As part of this work, she studies problem-based learning and collaborative knowledge construction. Her current work focuses on helping middle school students learn about aquatic ecosystems in a technology-rich learning environment. Professor Hmelo-Silver has published numerous journal articles, book chapters, and edited several books in these areas and is co-editor of *Journal of the Learning Sciences*. E-mail: cindy.hmelo-silver@gse.rutgers.edu

Paul Horwitz is a theoretical physicist with broad interests in the application of technology to science and mathematics education. He directs the Modeling Center at the Concord Consortium (USA) and is also the CEO of Educational Network Services (ENS), a for-profit company he founded in 2003 to commercialize educational technology resulting from research. His own research has used computer-based interactive models to create challenging learning games that pose problems and then monitor and react to students' actions. He was the Principal Investigator on the ThinkerTools Project, which pioneered the use of such games for teaching Newtonian mechanics. RelLab, a simulated "Relativity Laboratory" that he designed, won two EDUCOM Higher Education Software Awards. He directed the design and implementation of GenScope as well as its successor program BioLogica—multi-level models of genetic processes ranging in scale from DNA to populations. Under his direction, ENS is currently working with Teachers College Columbia University to develop MathemAntics™, a sequence of games intended to teach mathematics concepts to children from pre-school to third grade. E-mail: phorwitz@concord.org

Tracy L. Hull obtained her Masters in Biochemistry Education from the School of Biochemistry, Genetics and Microbiology at the former University of Natal (now KwaZulu-Natal) in Pietermaritzburg, South Africa. Her studies were done within the former Science Education Research Group (SERG) involving a project entitled, "Students' Use of Diagrams for the Visualization of Biochemical Processes." Subsequently she left academia to pursue a business career as a specialist sales representative in the pharmaceutical industry and her personal goal of completing ironman events.

Lei Liu is a Research Scientist in the Cognitive and Learning Sciences group at Educational Testing Service (ETS) in the USA. Before joining ETS, she was a post-doctoral fellow and research director at the University of Pennsylvania leading a large-scale NSF grant on secondary science curriculum and assessment reform. She earned her PhD degree from Rutgers University in 2008 and her dissertation won the best Doctoral Research Award from the National Association of Research in Science Teaching and the Outstanding Dissertation Award from the Graduate School of Education at Rutgers University. Her research has drawn heavily on cognitive and constructivist learning theories, and focused on investigating how different learning factors may affect cognitive and metacognitive learning processes and science learning progressions in culturally diverse contexts. Her current research focuses on designing innovative assessments, learning environments, and professional development to help learning and teaching sciences. E-mail: lliu001@ets.org

Kai Niebert is a Science Educator at the Institute of Science Education at the Leibniz Universität Hannover (Germany). He is involved in post-doctoral work at Curtin University's Science and Mathematics Education Centre (Perth, Australia), Penn State University (Pennsylvania, USA) and the Leibniz Universität Hannover. Kai earned his doctorate degree with research on students' and scientists' conceptions of climate change. For his research he received the Public Communication of Science Award of the Helmholtz Foundation and the Young Researcher Award of the German Association for Biology Education. Kai's research focuses on the role of metaphors, analogies, and experience in understanding science. He has developed learning environments and communication strategies based on the analysis of students' and lay peoples' source of understanding science. Kai is vice-president of the German Friends of Nature (NFD) and member of the Executive Committee of the Germany League for Nature (DNR). E-mail: niebert@idn.uni-hannover.de

Lilian Pozzer-Ardenghi is a McGill University Science Education scholar whose research interests include inscriptions and how these are used in both printed school materials and classroom lectures. On this topic, she has contributed scientific articles and book chapters, and co-authored the book *Critical Graphicacy, Understanding Visual Representations in Practices in School Science*, with W.-M. Roth and J.-Y. Han. Her research also focuses on gestures and body positioning and orientation, and how these are integrated with other semiotic resources, such as speech and inscriptions, in science lectures. Her latest book, *Staging and Performing Scientific Concepts: Lecturing is Thinking with Hands, Eyes, Body, and Signs*, compiles articles dealing specifically with this topic. More recently, her work has included socio-cultural issues, particularly the concepts of identity and agency, in the investigations of classroom communication and interaction and science teaching and learning. E-mail: lilian.ardenghi@mail.mcgill.ca

Edys S. Quellmalz serves as the Director of Technology Enhanced Assessments and Learning Systems in the Science, Technology, Engineering, and Mathematics Program at WestEd. She is Co-Principal Investigator of Calipers II: Using Simulations to Assess Complex Science Learning, a five-year assessment project funded by the Discovery Research K–12 program at the National Science Foundation (NSF). She also is leading the design of embedded assessments for the NSF-funded Rutgers Bioinformatics project. Both of these projects will produce technology-based assessments of challenging science knowledge and inquiry skills. Prior to joining WestEd, Quellmalz served as Director of Assessment Research and Design in the Center for Technology in Learning (CTL) at SRI International and has over 35 years experience consulting on numerous state, national, international, student, and teacher assessment programs. Dr Quellmalz graduated from UCLA, where she received a BA in psychology, and an MA and PhD in educational psychology. Email: equellm@wested.org

Jayashree Ramadas is Professor of Science Education and Centre Director of the Homi Bhabha Centre for Science Education, Tata Institute of Fundamental Research, Mumbai, India. She obtained an M.Sc. in Physics from the Indian Institute of Technology, Kanpur, 1976, and a Ph.D. in Science Education from the Homi Bhabha Centre for Science Education, 1981. Professor Ramadas's research has been in cognitive studies of science learning and development and evaluation of school science curricula. She led the research and fieldwork-based Homi Bhabha Curriculum for Primary Science, titled "Small Science," which is published in four languages. For this work she was awarded the Third World Academy of Sciences Regional Prize 2011. Professor Ramadas was a member of the National Curriculum Framework 2005 Science Focus Group and the Inter-Academy Panel's International Oversight Committee for Inquiry-Based Science Education. She is a member of the IUPAP International Commission on Physics Education. Her current research is concerned with visuo-spatial thinking in science (including the use of diagrams and gestures). She guest-edited a 2009 special issue of the *International Journal of Science Education* on this emerging research area. E-mail: jr@hbcse.tifr.res.in

Tanja Riemeier is Junior Professor of Biology Education at the Institute for Science Education at the Leibniz Universität Hannover (Germany). Before completing her dissertation on how to learn cell biology, she was a teacher for biology and chemistry for several years. Tanja's research focuses on students' learning processes while teaching biological aspects. Furthermore, she is interested in students' processes of argumentation and reasoning. E-mail: riemeier@idn.uni-hannover.de

Wolff-Michael Roth is Research Professor at Griffith University in Australia and a Fellow of the American Association for Advancement of Science and the American Educational Research Association. Having received a Masters in physics, he taught mathematics, science, physics and computer science at the middle and high school levels during the 1980s and early 1990s. He researches mathematics and science learning across the life span, in formal educational, informal, and workplace settings. His recent publications include *Passibility: At the Limits of the Constructivist Metaphor* (Springer, 2011) and *Geometry as Objective Science in Elementary School Classrooms: Mathematics in the Flesh* (Routledge, 2011). He is the 2009 recipient of the Distinguished Contributions to Science Education through Research Award from the National Association for Research in Science Teaching and the 2006 Whitworth Award for Education Research from the Canadian Education Association. In 2011, he received an honorary doctorate from the University of Ioannina, Greece. E-mail: w.roth@griffith.edu.au

Konrad Schönborn obtained a PhD in biochemistry education in 2005 from the University of KwaZulu-Natal (South Africa), and concurrently taught physics, chemistry, and biology at a secondary school. From 2006 to 2008, he received postdoctoral fellowships from the University of Göttingen (Germany) and Linköping University (Sweden), respectively. Konrad is currently a Senior Lecturer in Visual Learning and Communication at Linköping University, and Coordinator of the Swedish National Graduate School in Science, Mathematics, and Technology Education Research (FontD). His research explores learners' interactions with multiple external representations, such as the influence of multisensory virtual models on visualizing, processing, and constructing conceptual understanding about molecular and cellular phenomena communicated at different levels of biological and chemical organisation. The research is conducted within the Division of Media and Information Technology, which is affiliated to Visualization Center C in Norrköping, a cutting-edge center that develops, investigates, and disseminates various visualization forms across scientific and public domains. E-mail: konsc@itn.liu.se

Reneé Schwartz is Associate Professor of Biological Sciences at Western Michigan University, USA, with a joint appointment in the Mallinson Institute for Science Education. She earned a Bachelor's degree in Genetic Biology and a Masters in Molecular Biology, before earning a PhD in Science Education from Oregon State University, USA. Her research focuses on the teaching and learning of nature of science and scientific inquiry. She combines her science background and her teaching/learning interests by studying learners' developing epistemological views of science through engaging in authentic and scientific inquiry practices and reflective opportunities. Reneé develops contextualized nature of science and inquiry instruction in science content courses for future teachers and future STEM faculty. She has funded projects from the National Science Foundation and the Howard Hughes Medical Institute. Renee is a long-time member of the National Association for Research in Science Teaching, serving on the Board of Directors in 2009–2012. E-mail: r.schwartz@wmich.edu

Anveshna Srivastava is a Research Scholar at Homi Bhabha Centre for Science Education, Tata Institute of Fundamental Research, Mumbai, India. Her current research involves the use of microgenetic research design to capture the trajectory of learning in undergraduate students while they engage in a clinical-interview-cum-teaching session to understand the structure and function of DNA molecule. Her research interests include the use of gesture and mental imagery for understanding of complex molecular biology concepts. She has presented part of her work at the 2011 Gordon Research Conference at Rhode Island, USA. She has a flair for teaching and interacting with young children and conducts classes for undergraduate students on their demand at weekends. She is a member of the Commonwealth Association of Science, Technology and Mathematics Educators (CASTME)—Asia, Cognitive Science Society and Society of Biology. She obtained a B.Sc. degree in Botany (Hons.) and an M.Sc. degree in Biochemistry. E-mail: anveshna@hbcse.tifr.res.in or anveshna.sriv@gmail.com

David F. Treagust is Professor of Science Education at Curtin University in Australia and a Fellow of the Society of Biology and the American Educational Research Association. He taught secondary science for 10 years in schools in England and Australia before working at universities in the USA and Australia. His research interests include students' ideas about science, their conceptual change learning, and how these ideas can contribute to curriculum design, assessment, and teachers' classroom practice. With John Gilbert, he is the co-editor of *Multiple Representations in Chemical Education*. David was President of the National Association for Research in Science Teaching (NARST) (1999–2001), Managing Director of the Australasia Science Education Research Association (2003–2010), the 2005 recipient of the NARST Outstanding Paper Award for a research report on students' learning of biology with multiple representations, and the 2006 recipient of NARST's Distinguished Contributions to Science Education through Research Award. E-mail: D.Treagust@curtin.edu.au

Chi-Yan Tsui received his BSc (Hons) in biology and chemistry and Postgrad DipEd from the Chinese University of Hong Kong. His other postgraduate qualifications in education were from the UK and New Zealand, and he was awarded his PhD in science education from Curtin University in Australia. Chi-Yan taught secondary biology and science for 21 years and was head of the Biology Department in a Hong Kong school. He worked in science teacher education and professional development in Australian universities. With extensive classroom research experience, he has presented and published his work at national and international conferences and in journals. He was the 2005 recipient of the NARST Outstanding Paper Award for his research on learning genetics with multiple representations. As University Associate of Curtin University, he is an online science educator based in Hong Kong and Auckland (New Zealand) and experienced editor for Curtin University, international science educators, and science and mathematics education journals. E-mail: C.Tsui@curtin.edu.au

Roald Pieter Verhoeff is Assistant Professor in Science Communication at the Department of Philosophy and Science Studies of the science faculty at Radboud University Nijmegen, the Netherlands. He studied Biology at Utrecht University, the Netherlands, and in 2003 he finished his PhD project there with his thesis: *Towards systems thinking in cell biology education*. After his PhD he developed practical expertise as a genetic counselor. As postdoctoral researcher, he worked on a project DNA-labs for Citizenship from 2006 to 2009 at the Freudenthal Institute for Science and Mathematics Education at Utrecht University. In this project he studied learning and communication processes in cancer genomics practices to form a basis for designing new learning methods for secondary education. Since November 2009, he has been teaching science students in science communication and coordinating the Masters degree track at Radboud University. He is also project manager of the Nijmegen Science Hub, which acts as a bridge between academic research and primary education. E-mail: r.verhoeff@science.ru.nl

Arend Jan Waarlo is Senior Lecturer in Biology Education at Utrecht University in the Netherlands and holds a special chair of the Dutch Genetic Alliance regarding genetics and health communication. His PhD thesis was on biology teaching and health education. In 1999 and 2000, he was also employed at the University for Humanistics in Utrecht, the Netherlands. He educated biology teachers, life-stance and moral education teachers, professionals in health and environmental education, and was engaged in the Utrecht master's degree programme Science Education and Communication. He participated in various curriculum development and research projects. From 2000 to 2004 he was a member of the Academic Committee of the European Researchers in Didactics of Biology (ERIDOB). His current research interest is in acquiring genomics-related opinion-forming competence. Since 2008 he has also been Principal Investigator in Genomics Education and Communication at the Centre for Society and the Life Sciences in the Netherlands. E-mail: A.J.Waarlo@uu.nl

Alice Siu Ling Wong received her BSc degree in Physics with First Class Honors from the University of Hong Kong, the Hong Kong SAR, China and received her DPhil degree in Physics from the University of Oxford (UK). She then worked as a Croucher Foundation Postdoctoral Fellow in the Clarendon Laboratory and a college Lecturer at Exeter College, Oxford (UK). She is currently Assistant Dean and Associate Professor in the Faculty of Education at the University of Hong Kong. Her key research interests include nature of science, socioscientific issues and teacher professional development. She is also the TIMSS National Research Coordinator (science component) for Hong Kong. She serves on the editorial boards of *Science & Education* and *Canadian Journal of Science Education*. She is a founding member of the East-Asian Association for Science Education. She is also the leading author of *New Senior Secondary Physics at Work*, the most popular physics textbook in Hong Kong. E-mail: aslwong@hku.hk

Anat Yarden is Associate Professor of Science Education and Head of the Biology Group at the Department of Science Teaching at the Weizmann Institute of Science in Israel. She holds a PhD in Molecular Biology (Weizmann Institute of Science), and postdoctoral training in Genetics (Stanford University). The primary theme in all of her activities has been the attempt to adapt practices employed by scientists, to the processes by which students and teachers accumulate and advance their knowledge within the discipline of biology. Toward this end, her research group adapted the use of primary literature as well as authentic bioinformatics tools and laboratory experiments for the teaching and learning of biology in high schools. In addition to publishing numerous research papers, Anat has co-authored several textbooks, including a textbook on cutting edge molecular biology for 12th grade students. Anat headed the Academic Committee of the European Researchers in Didactics of Biology (ERIDOB) organization (2008–2010). E-mail: anat.yarden@weizmann.ac.il

Hagit Yarden is a Postdoctoral Research Fellow in the Department of Education in Technology and Science, Technion Institute of Technology, Israel. She received her undergraduate biology degree from the Faculty of Biology, Technion and her graduate qualifications in science education from the Faculty of Education, Tel Aviv University and the Feinberg Graduate School of the Weizmann Institute of Science, Israel. Her Ph.D. Thesis is concerned with learning and teaching biotechnological methods using interactive animations. Hagit taught biology majors for the matriculation examination in high school for 6 years and in the last 3 years she had designed and taught novel courses at the Rothschild–Weizmann program for Excellence in Science Teaching at the Feinberg Graduate School of the Weizmann Institute of Science. With broad research experience in educational technologies and educational reforms in higher education she has presented and published her work at national and international conferences and in journals. E-mail: yardenh@technion.ac.il

Valerie W.Y. Yip is a Post-doctoral Research Fellow at the Faculty of Education, The University of Hong Kong, the Hong Kong SAR, China. She had taught biology and environmental studies for 8 years in a Hong Kong school before she studied her PhD at University of Cambridge (UK). She has worked for teacher professional development in science education. Her research interests include students' ideas about science, language issues of science literacy, teacher learning and liberal studies implemented in the local context. E-mail: valyip@hku.hk

Index

A
Abd-El-Khalick, F., 225
Abrams, E., 208
Ainsworth, S., 272, 288, 313, 350, 351, 355, 364
Anderson, C.W., 357
Anderson, T.R., 21–24, 29
Animation
  biotechnology education (*See* Biotechnology education and animations)
  computer-based simulations, 263
  food web, 257
  freeze frame, 258
  online resources, 248
Anthropocentrism, 14
Anzelmo, J., 208
Apple Classroom of Tomorrow (ACOT), 250
Aquarium, 63
Atkins, M., 96
Atkins, P., 239
Ausubel, D.P., 95, 103
Azevedo, R., 78

B
Bahar, M., 311
Beckwith, J., 151
Beilock, S.L., 314
Bell, A.C., 219
Bell, R., 226
Biological knowledge
  bilingual representations, 361
  biological concepts, 112–113
  biological fundamentals, 112–113
  biological terms, 112
  Delphi approach, 114–115
  ecology domain, 117–118
  evolution domain, 118–119
  external representations, 114
  formal and informal, 57
  genetics domain, 118
  horizontal translation processes, 119–121
  HTM, 10
  mesocosmic representations, 13
  students' effective translation, 123–124
  translation processes and communication competencies, 113
  underlying biological principles, 112–113
  vertical translation processes, 119–123
BioLogica projects
  mesocosmic representations, 13
  model-based genetics learning
    chromosomes, 254
    GenScope project, 253
    hypermodel, 254
    methods, 255–256
    multilevel simulation, transmission genetics, 254
    representations, 254
    scaffolding, 254–255
  Monohybrid activity, 284–287
  and multiple representations
    activity scripts, 273
    external representations, 272
    hypermodel, 273
    internal representations, 272
    learning goals, 273–275
    log files, 276
    MERs, 272, 273
    model-based learning, 276

D.F. Treagust and C.-Y. Tsui (eds.), *Multiple Representations in Biological Education*, Models and Modeling in Science Education 7, DOI 10.1007/978-94-007-4192-8,
© Springer Science+Business Media B.V. 2013

Biology education
 CRM model (*see* Conceptual-reasoning-mode (CRM) model)
 MERs (*see* Multiple external representations (MERs))
 pictures
  air bubble, 49
  communication, 39
  double cones and rods, 49
  high school textbooks, 42–43
  inscriptions, 40–41
  lectures, 43–46
  perception, 50
  print and online media, 42
  single cone, 49
  social interactions, 50–51
  social origin, 46–47
  social-psychological approach, 40
  transcript fragment, 47–48
  transcription convention, 51
  T's assessment, 48–49
  uses, 40
 theoretical model teaching and learning, 350–351
 in twenty-first century, 349–350, 363, 364
Biotechnology education and animations
 cognitive theory, multimedia learning
  active-processing assumption, 95
  dynamic visuals, 96
  information delivery theory, 95
  limited-capacity assumption, 95
  prior knowledge/spatial ability, 96
  static and dynamic visualizations, 95
  static display, 96
  verbal and pictorial information, 94
  visual and verbal representations, 94
  working memory, 96
 continuous version, 97
 pedagogical characteristics
  constructivist teaching, 103–104
  hands-on point of view, 102
  students' attention, 103
  teacher, role of, 101–102
 sequential version, 97
 spatial contiguity principle, 97
 students' comprehension
  DNA restriction enzyme digestion, 98–101
  PCR method, 97–98
Blari, L., 226
Blissett, G., 96
Bodner, G., 226
Boersma, K.T., 126
Bögeholz, S., 111, 114
Bonnett, A., 358
Boulter, C.J., 250, 333
Bourreée, F., 116
Bowen, G.M., 41, 44, 46
Brown, M.H., 209, 211
Brumby, M., 187
Buckley, B.C., 250, 356, 357
Burkett, C., 75

C

Calipers projects, simulation-based assessments
 Calipers II project
  content and inquiry targets, 258
  curriculum-embedded formative assessment, 256
  data inspector, 258
  Ecosystem suite, 256, 257
  freeze frame, animation, 258
  learning management system, 259
  model-based learning and evidence-centered design, 256, 257
  population simulation, 258, 259
  summative benchmark assessments, 259
 Calipers I project, 256
 methods, 259–260
Capra, F., 204
Carvalho, G.S., 158
Castéra, J., 359–361
Catley, K.M., 65
Cavallo, A., 197
Cell biology education
 hierarchical systems model, 336, 337
 learning and teaching strategy, 337–339
 modeling activities sequence, 336
 six modeling activities, 339, 340
 Verhoeff study, 334, 337
 vertical and horizontal coherence, 339
Cell division
 experiences with representations, 308
 false reasoning, 294
 learning episode, 293–294
 schemata
  experiential basis, 296, 297
  reexperience and reflection, 297
  shaping understanding, 295–296
 striving, effective representation, 294–295
 uncovering hidden hand, 305–307
Character viewpoint gestures, 314
Chargaff, 327
Clement, J.J., 263

Clément, P., 148, 151, 158, 359–361
Climate change
  conceptions, 298, 299, 308
  schemata
    beginning interview, teaching experiment, 303–305
    container-flow schema, 300–301
    container schema, 298, 300
    learning environments design, 301–302
    modeling carbon cycle, 303–304
    modeling conceptions, 302–303
    modeling story, 305
    reconstructing conceptions, 304
    source-path-goal schema, 300
  uncovering hidden hand, 305–307
Clothespin model, 316, 323
Cognitive skills. *See* Hypermedia
Cognitive theory, multimedia learning
  active-processing assumption, 95
  dynamic visuals, 96
  information delivery theory, 95
  limited-capacity assumption, 95
  prior knowledge/spatial ability, 96
  static and dynamic visualizations, 95
  static display, 96
  verbal and pictorial information, 94
  visual and verbal representations, 94
  working memory, 96
Collins, A., 90
Collins, S., 115
Coll, R.K., 333
Color blindness, 59
Complex biology learning
  BioLogica projects (*see* BioLogica projects, Model-based genetics learning)
  Calipers projects (*see* Calipers projects, simulation-based assessments)
  classroom realities, 262
  genetic variation, 261
  hypermedia, 261
  policy level, 262
  Science for Living projects, circulatory system (*see* Science for Living (SFL) projects)
  SimScientists program, 261
  standards and assessments, 262
  student interactions with representations, 263–264
  student learning, 262–263
Complex process diagrams
  biology students
    domain-specific representation strategies, 179
    icons and signs, 179
    in-depth clinical interviews, 176
    qualitative analysis, 176–177
    seed germination process, 178
    semi-structured interview, 177–178
    viral replication, 173, 177
  conceptual knowledge, 176
  deconstruction
    microRNAs diagram, 172–174
    nitrogen cycle diagram, 170, 172–173
    pictures, arrows and text, 171
  eye-tracking tools, 176
  graphic skills, 176
  limitations, 180
  MERs, 167–169
  pedagogical recommendations
    content knowledge, 181
    learning goal, 180
    model complete decoding, 180
    production process, 181–182
  representational competence, 179
  semiotics, 169–170
  textbook diagrams, 174–175
Concept Inventory for Evolution Readiness (CIER), 140–143
Conceptual-reasoning-mode (CRM) model
  classification, 21, 22
  external representations
    analytical tool, 23–27
    guidelines, 23–24
    sound assessment tasks, 23
  factors, 20, 21
  remediation strategy, 33–36
  student reasoning difficulties
    cardiac cycle, 28–29
    DNA-strand symbolism and function, 29–30
    immunoglobulin G, 30–31
    metabolic pathway, 31–33
  Venn diagram, 20, 21
Conceptual representations. *See* Hypermedia
Cook, M.P., 176, 186
Crawford, B., 226
Crick, F.H.C., 311
Cromley, J.G., 78, 176
Crowley, K., 315
Cube model
  HTD, 11
  HTM, 9–10
  limitations, 12–13
  VTL, 10–11
Cummins, C., 208
Cyclical representations, 65

D
3-D bell jar model, 60
Deconstruction
  microRNAs diagram, 172–174
  nitrogen cycle diagram, 170, 172–173
  pictures, arrows and text, 171
Diagrams
  complex process diagrams (*see* Complex process diagrams)
  noncladogenic, 65
  tree, 64, 67
  two-step diagram, 70
  Venn diagram, 20, 21
Didactic Transposition Delay (DTD), 155
DNA structure, mental visualization
  analogy, 327
  background preparation, 318–319
  data analysis
    sessions video recordings, 319
    understanding helical structure, 323–324
    understanding ladder structure, 320–323
    "+ve" transitions context, 325–326
  demographic information, 315, 316
  double-helical structure, 312
  3-D structure, 326
  interview-cum-teaching sequence, 315
  microgenetic research design, 315
  molecular biology, 311
  multiple representations
    DNA backbone and nitrogenous base pairs, 316–318
    role in learning, 313–314
  nucleoside introduction, 319
  palm gesture (*see* Palm gesture)
  screening test, 315
  structural-functional linkages, 311
  students' reasoning processes, 315
  textbook representations, 312–313
2-D pictures, 66
3-D structure
  DNA structure, mental visualization, 326
  scales, 66–67
Dual coding theory, 4
Duarte, S., 148
Duit, R., 35
Duschl, R., 115
Dynamic ecosystem behavior
  bridging model, 344
  computer modeling, 339–340
  learning and teaching strategy, 340–343
  natural phenomena, 344, 345
  populations, 345
  post-instructional analysis, 344

E
Ecology
  horizontal translation, 119–120
  vertical translation, 121
Effective learning environments representations
  afford experiences, 308
  denote conceptions, 308
  depict schemata, 309
Eilam, B., 64, 65
Evolution
  animal learning activities
    ecosystems, experiments, 138
    natural selection, 138
    predators and prey, 138
    variations and adaptations, 137
    virtual ecosystem, 137
  CIER, 140–143
  domain, 118
    horizontal translation, 120–121
    vertical translation, 122
  Evolution Readiness project, 142–143
  negative feedback loop, 130
  off-line activities and teacher support, 138–139
  plant learning activities
    environmental changes, 135–136
    environmental variable, 136
    mystery plant adaptation, 134–135
    transfer exercise and test, 136
    virtual field, 133
    virtual greenhouse, 133–134
  plant model
    high-light environment, 131
    inheritance, variation, and fitness, 130–131
    population, 130
  readiness materials, 139–140
  trial curriculum, 139
External representations (ERs)
  biological knowledge, 114
  BioLogica projects, 272
  conceptual-reasoning-mode model
    analytical tool, 23–27
    guidelines, 23–24
    sound assessment tasks, 23
  SFL projects, circulatory system, 253
  visualization, 55

F

Ferguson, W., 90
Fisher, K., 364
Food web, 257
Forissier, T., 148, 151, 158
Fraga, M.F., 156
Friedrichsen, P., 356, 357
Function-centered hypermedia
  coding and analysis procedures, 81–82
  dyads, 85, 86
  functional episodes, 86–87
  knowledge construction and metacognition, 86–89
  opening screen, 79
  qualitative results, 84
  quantitative results, 83–84
  understanding structures and behaviors, 82

G

Genetics
  after-school program, 289
  bilingual learners, 288–289
  code-mixing, classroom, 288
  data collection and analysis
    Australian study, 277–278
    Hong Kong study, 278
  domain, 117–119
    horizontal translation, 119
    vertical translation processes, 121
  gene conceptions and genetics reasoning
    Hong Kong *vs*. Perth students reasoning, 280–283
    identification, 278–279
    multidimensional conceptual change framework, 270
    ontological conceptual change, 270
    two-tier test, 270
    types of reasoning, 271–272
  genetics literacy, 270
  human genetics (*see* Human genetics)
  multilevel thinking, 269–270
  multiple representations and *BioLogica*
    activity scripts, 273
    external representations, 272
    hypermodel, 273
    internal representations, 272
    learning goals, 273–275
    log files, 276
    MERs, 272, 273
    model-based learning, 276
  research approach, 276
  school context, 276–277
  students' log file analysis
    Mei-yee's log file, 286–287
    MERs, 283
    Monohybrid activity, 284–287
    specifications, 282
    Verbatim transcription, audio recordings, 285
  visual-graphical representations, 287
  writing with confidence, 279–280
Gentner, D., 314
Gertzog, W.A., 270
Gilbert, J.K., 198, 229, 238, 239, 333, 364
Goel, A.K., 77
Goldin-Meadow, S., 314, 318, 327
Gould, S.J., 151
Gravemeijer, K.P.E., 334, 337
Greene, J.A., 78
Griffard, P.B., 356
Gupthar, A.S., 29

H

Hackling, M.W., 272
Haddadi, N., 148
Halverson, K.L., 197, 356, 357, 364
Hansell, M.H., 311
Harrison, A.G., 333
Haslam, F., 219
Hegarty, M., 56
Hermans, D., 289
Hewson, P.W., 270
Hickey, D.T., 270, 271
Hmelo-Silver, C.E., 77
Hodson, D., 227
Hoffmann, L., 115
Horizontal translation
  ecology domain, 119–120
  evolution domain, 120–121
  genetics domain, 120
Horizontal translation across modes of representations (HTM), 9–10
Horizontal translation across the domain knowledge of biology (HTD), 11
Horwitz, P., 254, 256, 276, 356
Hull, T.L., 31
Human-centered representations, 14
Human genetics
  genetic determinism
    additive epresentation, 149
    Darwinian selection, 150–151
    environmental influence, 148
    epigenetics, 149–150
    innatist ideas, 148
    interactive epresentation, 149
    phenotype and genotype, 147
    social challenges, 151
    stochastic process, 151

Human genetics (*cont.*)
systemic approach, 151
genetic diseases, French textbooks
Didactic Transposition Delay, 155
environmental influence, 153–154
monogenic diseases, 152–153
occurrences, 153
human twins image, 156
metaphor genetic program
Cyprus, 158
Estonia, Poland, and Hungary, 159
Finnish textbooks, 157
France, 158
in Germany, 158
Italian textbooks, 158–159
Lithuanian textbooks, 159
Maltese textbooks, 157
Moroccan, Lebanese, and Senegalese textbooks, 158
Portuguese textbooks, 157
Romanian textbook, 159
Tunisian textbooks, 158
Hundhausen, C.D., 75
Hypermedia
coding and analysis procedures, 81–82
function-centered hypermedia
coding and analysis procedures, 81–82
dyads, 85, 86
functional episodes, 86–87
knowledge construction and metacognition, 86–89
opening screen, 79
qualitative results, 84
quantitative results, 83–84
understanding structures and behaviors, 82
nonlinear information, 75
representational tools, 76
structure-behavior-function, 77
structure-centered hypermedia
coding and analysis procedures, 81–82
dyads, 85, 86
functional episodes, 86–87
knowledge construction and metacognition, 86–89
opening screen, 79
qualitative results, 84
quantitative results, 83–84

I
In-the-plane-of-the-backbone gesture, 321–323
Ioannides, C., 307

J
Jacquard, A., 151
Johnstone, A.H., 311

K
Kachru, 361
Kahn, A., 151
Kamiejski, R., 148
Kamin, L.J., 148
Keller, E.F., 148
Kinchin, I.M., 113
Kindfield, A.C.H., 270, 271
Knippels, M.C.P.J., 332, 335
Kozma, R.B., 113, 179, 186, 189, 197, 198
Kroll, J.F., 289
Kuechle, J., 113
Kwan, J., 227

L
Labov, J.B., 349, 363
Lawrence, A.J., 272
Learning management system (LMS), 259
Lederman, N.G., 225, 226
Lewontin, R.C., 148, 151
Liu, L., 77, 228, 229
Lord, T.R., 188

M
Malacinski, G.M., 94
Marbach-Ad, G., 105, 311
Marino, S., 188
Mathai, S., 314
Mayer, R.E., 264
Mental images, 61
Merriam, S.B., 276
Mesocosm, 13, 306–307
Metacognitive skills. *See* Hypermedia
Metaphorical gestures, 314
Michel, P., 116
Millar, R., 115
Moore, R., 187
Moos, D.C., 78
Morrison, J.B., 95
Multimedia learning
active-processing assumption, 95
dynamic visuals, 96
information delivery theory, 95
limited-capacity assumption, 95
prior knowledge/spatial ability, 96
static and dynamic visualizations, 95

static display, 96
verbal and pictorial information, 94
visual and verbal representations, 94
working memory, 96
Multiple external representations (MERs)
abstraction, 5
anthropocentric/human-centered representations, 14
assessment methods, 362–363
biological knowledge (*See* Biological knowledge)
biology, domain knowledge of, 8
cognitive load theory, 7
cognitive tasks, 6
complex process diagrams, 167–169
cube model
HTD, 11
HTM, 9–10
limitations, 12–13
VTL, 10–11
design parameters, 6
DNA structure, 313–314
dual coding theory, 4
functions of, 6
graphs, 5
higher-order learning, 4
learners' interpretation, 5
learning enhancement
conceptual understanding, content knowledge, 355–356
deeper understanding, scientific reasoning, 356
representational competence and other skills, 356–357
visual representations, 351–355
levels of representations, 7–8
mesocosmic representations, 13
modes of representations, 7
multiple representations and *BioLogica,* 272, 273
quality of teaching
anthropocentric/ human-centered representations, 357–358
curriculum interconnectedness, 358–359
non-English-speaking world, 361
philosophical, cultural, social, and political impacts, 359–361
students' log file analysis, 283
systems representations, 14–15
tables, 5
theoretical model, 9
in twenty-first century, 15–16

N
Nature of science (NOS), 225–227
Niebert, K., 357, 358
Noncladogenic diagrams, 65
Novick, L.R., 65

O
Observer viewpoint gestures, 314
Ohmer, B., 364
O'Mahony, P., 158
Osborne, J., 115

P
Padalkar, S., 314
Pallant, A., 105
Palm gesture
DNA backbone representation, 316, 317
3-D structure visualization, 326
helical structure, 323–324
instructional and diagnostic tool, 326–327
with M4 model, 320
nitrogenous base, 316, 317
nucleoside, 319
observer viewpoint gesture, 327
with palm and straightened fingers, 317, 318
understanding ladder structure, 320–323
"+ve" transition, 325
Patrick, M.D., 197
Photosynthesis and cellular respiration
Ann case representations
interconnections, 212–213
misconceptions, 215
multiple ecological levels, 213–214
nested systems, 214–215
ATP and NADH, 219
interconnections
biological nested systems, 207
chemical bond energy, 204–205
gross anatomical structures, 206
interrelated system components, 205
multiple ecological levels, 205, 206
radiant energy, 204
intuitive conceptions, 207–208
Jay case representations
interconnections, 215–216
misconceptions, 217
multiple ecological levels, 216–217
nested systems, 217
preservice teachers, 208
clarifying interviews, 209, 212

Photosynthesis and cellular respiration (*cont.*)
cognitive interviews, 209
data collection, 209
explaining set, 209–212
Scaffolding connections
guides and signposts, 221–222
self as first referent intuitive conception, 220
systems, 204
Phylogenetic tree representations
components
apomorphies and common ancestry, 190
main branch, 190, 191
number and location of nodes, 190, 191
proximity of organisms, 190, 191
representations styles and types, 190
tree readers, 189
content knowledge mastery, 186
data analysis, 189
data collection, 188–189
interpretations, 189
representational competence framework
conceptual use, 194
expert use, 196
levels, 196
milestones, 190–191
no use, 192
scientific use, 194–196
simplified use, 193
superficial use, 192, 193
symbolic use, 193–194
research design, 188
scientific literacy, 185
spatial reasoning, 186
visual representations, 185, 186
Piaget, J., 46
Pictures
air bubble, 49
communication, 39
double cones and rods, 49
high school textbooks, 42–43
inscriptions, 40–41
lectures, 43–46
perception, 50
print and online media, 42
single cone, 49
social interactions, 50–51
social origin, 46–47
social-psychological approach, 40
transcript fragment, 47–48
transcription convention, 51
T's assessment, 48–49
uses, 40
Posner, G.J., 270
Pozzer-Ardenghi, L., 326, 359
Prain, V., 355
Price, F., 105
Pyramidal representation, 67

Q
Quellmalz, E.S., 256

R
Ramadas, J., 314, 357, 362
Ratcliffe, M., 115
Rea-Ramirez, M.A., 263
Reid, A.H., 349
Reiser, B.J., 80
Rose, S., 148, 151
Rotbain, Y., 105
Roth, V.M., 41, 44, 46
Roth, W.-M., 326, 359
Russell, J., 113, 179, 186, 189, 197, 198

S
Salmi, L.R., 116
Samarapungavan, A., 226
SARS. *See* Severe acute respiratory syndrome (SARS)
Scaffolding connections
guides and signposts
language, 221
tracing, organizing, and mapping, 222
self as first referent intuitive conception, 220
Scales
3-D structure, 66
size, 63
temporal, 64
Schäfer, M.S., 158
Schemata
mesocosm, 306
teaching cell division
experiential basis, 296, 297
reexperience and reflection, 297
shaping understanding, 295–296
understanding climate change
beginning interview, teaching experiment, 303–305
container-flow schema, 300–301
container schema, 298, 300
learning environments design, 301–302
modeling carbon cycle, 303–304
modeling conceptions, 302–303
modeling story, 305

reconstructing conceptions, 304
source-path-goal schema, 300
Schönborn, K.J., 21–24, 111, 114
Schwartz, R.S., 209, 211, 226
Science for Living (SFL) projects
laboratory simulations, 250
Life Lab, 250
main interface screen, 249
methods
ACOT, 250
consensus models, 253
data analysis, 251
data collection, 250
diverse data sources, 250–251
expressed models, 253
external representations, 253
Joanne's learning, 252
mental models, 252
model-based learning, 252, 253
preconceptions test, 251
semi-structured interviews, 251
natural phenomena, 249
NoteBook, 250
Pump Lab, 250
representations, 248–249
two screen views, 249, 250
Scientific models
curriculum goals, Hong Kong Science Education, 225–227
in curriculum guide
content analysis, 237
exemplar phenomena, 239
physical artifacts, 238
scientific enterprise processes/ outcomes, 239–240
virtual artifacts, 238–239
SARS (*see* Severe acute respiratory syndrome (SARS))
Self-generated model, 62–63
Séralini, G.E., 151
Severe acute respiratory syndrome (SARS)
hunt, causative agent, 230–231
natural host search, coronaviruses
bats, 232
civet cats, 231–232
genetic analysis, the viral samples, 231
logical reasoning, 232
phylogenetic tree, 233
viral strains, 233
roles and functions, 236–237
series of events, 227
tragic outbreak at Amoy Gardens
building structure, 235
contextual and environmental conditions, 236
cross-disciplinary investigation, 236
epidemiological and environmental investigations, 234
new construction model, 235
peculiar infection pattern and fast transmission rate, 234, 235
series of rare events combination, 236
transmission mode identification
chains of transmission, 229
micro level, 229, 230
new infectious disease, 228
preventive measures, 230
respiratory droplets, 229
Shade, C.K., 65
Shapiro, A., 78
Shaw, J.E., 153
Sicree, R.A., 153
Siegler, R.S., 315
SimScientists program, 261
Size scales, 64
Soderberg, P., 102
Southerland, S., 208, 220
Spatial reasoning, 186
Srivastava, A., 357, 362
Stavy, R., 105, 311
Strike, K.A., 270
Structure-behavior-function (SBF), 77
Structure-centered hypermedia
coding and analysis procedures, 81–82
dyads, 85, 86
functional episodes, 86–87
knowledge construction and metacognition, 86–89
opening screen, 79
qualitative results, 84
quantitative results, 83–84
Superficial interpretation, 67–68
Suthers, D.D., 75
Systems thinking
cell biology education
hierarchical systems model, 336, 337
learning and teaching strategy, 337–339
modeling activities sequence, 336
six modeling activities, 339, 340
Verhoeff study, 334–335, 337
vertical and horizontal coherence, 339
design criteria, 346–347
3-D model-of cell, 347
Dutch biology curriculum, 332
dynamic ecosystem behavior
bridging model, 344
computer modeling, 339–340
learning and teaching strategy, 340–343
natural phenomena, 344, 345

Systems thinking (*cont.*)
populations, 345
post-instructional analysis, 344
elements, 332, 333
genetics education, 332
integrated modeling approach, 345
loss of contact, 346
metacognitive level, 345
model-based learning trajectory, 332
multiple external representations, 14–15
research approach, 334–335
systems modeling approach
categories with abstractness, 334, 335
consensus models, 333
emergent modeling, 334
symbolic models, 333
typologies, 333
yo-yo strategy, 345

T
Taylor, I., 333
Temporal scales, 64
Textbook diagrams
complex process diagrams, 174–175
double helix and ladder structure, 313
immunoglobulin G, 30
intracellular calcium homeostasis, 166
microRNAs representation, 172
Three-dimensional (3-D) model
of human anatomy, 59–60
respiratory system function, 60–61
Tinker, R.F., 105, 276
Treagust, D.F., 113, 176, 219, 229, 270, 313, 333, 356, 361, 362
Treagust, D.P., 35
Tree diagram, 64, 67
Tree thinking. *See* Phylogenetic tree representations
Tsui, C.-Y., 113, 176, 313, 356, 361, 362
Tversky, B., 95
Two-step diagram, 70
Tytler, R., 355

V
Venn diagram, 20, 21
Venville, G.J., 270, 271, 278–280, 316
Verhoeff, R.P., 126, 334, 335, 337, 346, 356, 357, 359
Vertical translation
ecology domain, 121
evolution domain, 122
genetics domain, 121
Vertical translation across levels of representations (VTL), 10–11
Vidal, M., 261
Virtual ecosystem, 137
Virtual field, 133
Virtual greenhouse, 133–134
Visualization
biology domain, 56–57
decorations
diet and cholesterol, 58
opening page, book chapters, 58–59
3-D model
of human anatomy, 59–60
respiratory system function, 60–61
external representations, 55
internal representation, 56
learning difficulties, 56
live ecosystem model, 63
scales
3-D structure, 66
size, 64
temporal, 64
self-generated model, 62–63
superficial interpretation, 67–69
teachers' representations, classroom board, 68–70
Vollmer, G., 307
von Bertalanffy, L., 331
von Glasersfeld, E., 46
Vosniadou, S., 307
Vygotskian, 270
Vygotsky, L.S., 42, 357

W
Waarlo, A.J., 126
Watson, J.D., 311, 316
Westby, E., 226
Westra, R.H.V., 126, 334, 335, 339, 340, 343, 345, 346
Winters, F.I., 78
Wong, S.L., 227, 356, 359, 360
Wood-Robinson, C., 219
Working memory, 96

Y
Yamamoto, K.R., 349
Yarden, A., 364
Yousuf, M.I., 116
Yung, B.H.W., 227

Z
Zazzo, R., 156
Zell, P.W., 94
Zimmet, P.Z., 153

MIX
Papier aus verantwortungsvollen Quellen
Paper from responsible sources
FSC® C105338

If you have any concerns about our products,
you can contact us on
ProductSafety@springernature.com

In case Publisher is established outside the EU,
the EU authorized representative is:
Springer Nature Customer Service Center GmbH
Europaplatz 3, 69115 Heidelberg, Germany

Printed by Libri Plureos GmbH
in Hamburg, Germany